Sequence Stratigraphy Applications to Shelf Sandstone Reservoirs

---

Outcrop to Subsurface Examples

AAPG Field Conference, September 21-28, 1991

By
J.C. Van Wagoner
C.R. Jones
D.R. Taylor
D. Nummedal
D.C. Jennette
G.W. Riley

Published by
The American Association of Petroleum Geologists
Tulsa, Oklahoma, U.S.A.

Published January, 1992

ISBN: 0-89181-815-4

Association Editor: Susan Longacre
Science Director: Gary D. Howell
Publications Manager: Cathleen P. Williams
Special Projects Editor: Anne H. Thomas

Cover design: Elizabeth A. Mitchell

Manuscript organization: Clive Jones and Caroline Peacock, Exxon Production Research

ACKNOWLEDGMENTS

We wish to express appreciation to the following companies for their generous financial support to non-industry participants in this Field Conference:

AMOCO Production Company
P.O. Box 800
Denver, Colorado 80201

BP Exploration
301 St. Vincent Street
Glasgow, Scotland
G2 5DD

Exxon Production Research Company
P.O. Box 2189
Houston, Texas 77252-2189

Marathon Oil Company
P.O. Box 269
Littleton, Colorado 80160-0269

We would also like to thank the following individuals who have contributed to and enhanced this publication:

D. E. Owen; Lamar University, Beaumont, TX
Y. Y. Chen, R. Lander, M. B. Farley; Exxon Production Research, Houston, TX
R. Cole, A. Trevena; Unocal Research, Brea, CA

Cover photo: View of the Campanian stratigraphy looking northeast across an unnamed canyon just west of Blaze Canyon, Book Cliffs, Grand County, Utah (about 3 miles northwest of the town of Thompson). Cliff in the foreground comprises the Desert Member of the Blackhawk Formation, unconformably overlain by the Castlegate Sandstone. The shale slope that lies beyond the trees above the Castlegate is the Buck Tongue of the Mancos Shale. These shales are overlain by cliffs of the Sego Sandstone, on the left side of the photo. In the distance, the Sego Sandstone is at the base of the cliffs, and is overlain by the Neslen and Farrer formations. A description of the facies and sequence stratigraphy of the rocks in this photo are presented in Figure 3-11, DAY THREE of this guide book. Photo taken by A. R. Sprague, Exxon Production Research, Houston, TX.

CONTENTS

DAY FOUR: Nonmarine sequence stratigraphy and facies architecture of the downdip Castlegate Sandstone in the Book Cliffs of western Colorado and eastern Utah.

DAYS THREE AND FOUR, ACCOMPANYING OVERVIEW PAPER:

Sequence stratigraphy and facies architecture of the Desert Member of the Blackhawk Formation and the Castlegate Formation in the Book Cliffs of western Colorado and eastern Utah.

by John C. Van Wagoner

Travel Day

DAY FIVE: Travel from Grand Junction, Colorado to Farmington, New Mexico, including geologic overview stops in the San Juan mountains.

DAY FIVE, ACCOMPANYING OVERVIEW PAPER:

Large-scale sequence stratigraphy of the Phanerozoic of the San Juan Mountain region, Colorado.

by Donald E. Owen

New Mexico Segment

DAY SIX: Facies architecture and sequence stratigraphy of the Tocito, Torrivio and Gallup sandstones, Late Turonian - Coniacian, San Juan Basin, northwestern New Mexico.

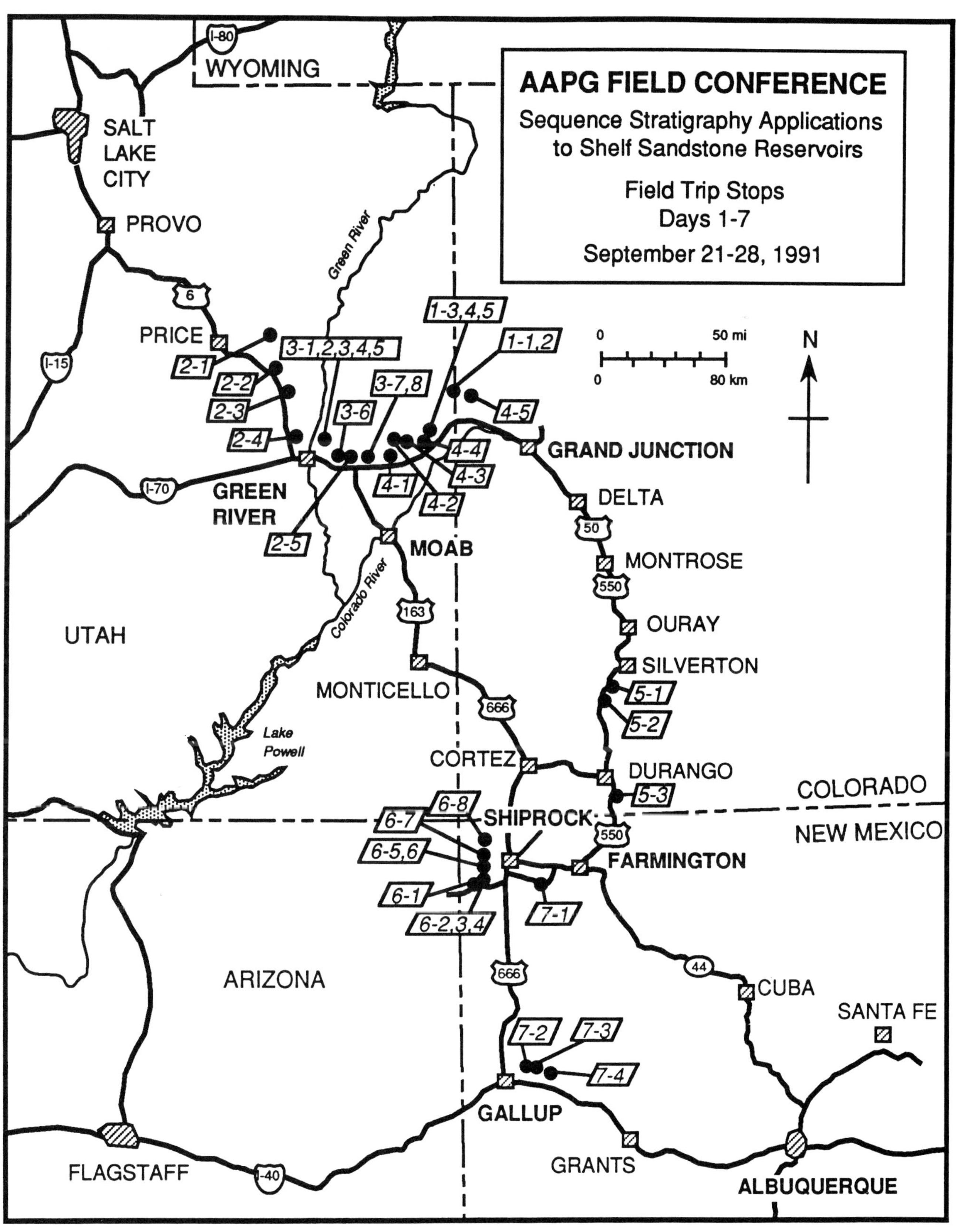
AAPG FIELD CONFERENCE
Sequence Stratigraphy Applications to Shelf Sandstone Reservoirs
Field Trip Stops
Days 1-7
September 21-28, 1991
0 50 mi
0 80 km
N
WYOMING
I-80
SALT LAKE CITY
PROVO
6
PRICE
I-15
Green River
1-3,4,5
1-1,2
3-1,2,3,4,5
2-1
2-2
2-3
2-4
3-7,8
3-6
4-5
4-4
4-3
4-1
4-2
GRAND JUNCTION
I-70
GREEN RIVER
2-5
MOAB
DELTA
50
MONTROSE
550
Colorado River
163
OURAY
UTAH
SILVERTON
MONTICELLO
666
5-1
5-2
Lake Powell
CORTEZ
DURANGO
5-3
COLORADO
6-8
6-7
SHIPROCK
550
NEW MEXICO
6-5,6
FARMINGTON
6-1
7-1
6-2,3,4
ARIZONA
666
44
CUBA
SANTA FE
7-2
7-3
7-4
GALLUP
GRANTS
FLAGSTAFF
I-40
ALBUQUERQUE

ITINERARY

AAPG Field Conference: Sequence-Stratigraphy Applications to Shelf-Sandstone Reservoirs: Outcrop to subsurface examples.

September 21 - 28, 1991: Colorado, Utah and New Mexico

The field conference begins in Grand Junction, Colorado on the afternoon of Saturday 21st September and ends in Albuquerque, New Mexico on the following Saturday, 28th September, late in the afternoon. The accompanying map shows the location of the field stops during the next seven days.

Saturday, September 21: Arrive Grand Junction, Colorado.
Meet at:

Grand Junction Hilton
743 Horizon Drive
Grand Junction, CO 81506
Ph: (303) 241-8888

There will be a field conference overview between 6:00 - 8:00 p.m. in a conference room at the Hilton. It is important that all participants attend this meeting.

Sunday, September 22: Day One: Grand Junction to Green River, Utah.

- Objective: Investigate high-frequency sequence stratigraphy and facies architecture of the Sego Sandstone Member, Price River Formation, Campanian, in the Book Cliffs of western Colorado and eastern Utah .
- Leave Grand Junction Hilton parking lot promptly at 7:30 a.m. Spend evening at:
  River Terrace Motel
  880 East Main
  Green River, UT 84525
  Ph: (801) 564-3401

Monday, September 23: Day Two: Central Utah.

- Objective: Shallow marine facies architecture and parasequence stacking patterns within the Kenilworth Member, Blackhawk Formation, Campanian, in the Book Cliffs of eastern and central Utah.
- Evening in Green River at the Green River Terrace Motel (see Day One).

Tuesday, September 24: Day Three: Green River, Utah to Moab, Utah.

- Objective: Nonmarine sequence stratigraphy and facies architecture of the updip Desert and Castlegate sandstones, Campanian, in the Book Cliffs of western Colorado and eastern Utah
- Spend evening at:
  Super 8 Motel
  889 North Main St.
  Moab, UT 84532
  Ph: (801) 259-8868

Wednesday, September 25: Day Four - Moab to Grand Junction, Colorado

- Objective: Investigate nonmarine sequence stratigraphy and facies architecture of the downdip Castlegate Sandstone Formation, Campanian, in the Book Cliffs of western Colorado and eastern Utah.

- Evening at:

  Grand Junction Hilton
  743 Horizon Drive
  Grand Junction, CO 81506
  Ph: (303) 241-8888

Thursday, September 26: Day Five - Grand Junction to Farmington, New Mexico.

- Objective: Travel day; drive from Grand Junction, Colorado to Farmington, New Mexico through the San Juan mountains with three geologic overview stops en route. Afternoon in Farmington; poster session and overview of New Mexico portion of field conference.

- Bus to Farmington, northwestern New Mexico. Leave Grand Junction at 8:00am from the Hilton parking lot. Meet at conference room in the Holiday Inn at 2:30 p.m. for overview of remainder of trip. Poster session from 3:30 - 7:30 p.m. Evening will be spent at:

  Holiday Inn
  600 East Broadway
  Farmington, NM 87401
  Ph: (505) 327-9811

Friday, September 27: Day Six - Four Corners Platform area, northwestern New Mexico

- Objective: Facies architecture and sequence stratigraphy of the Tocito, Torrivio and Gallup sandstones, Late Turonian - Coniacian, San Juan Basin, New Mexico..

- Evening at Holiday Inn, Farmington (see Day Five)

Saturday, September 28: Day Seven - Farmington to Albuquerque via Gallup, New Mexico

- Objective: Examine the facies and sequence stratigraphy of an exhumed Tocito sandstone body near Farmington, and the sequence stratigraphy of the Gallup Sandstone in the type area, near Gallup.

- Drive to Albuquerque where field conference ends. Spend evening close to historic downtown Plaza at:

  Sheraton Old Town
  800 Rio Grande Blvd.
  Albuquerque, NM 87104
  Ph: (505) 843-6300

PREFACE

This field guide documents the sequence stratigraphy and lithofacies of some of the best exposed Late Cretaceous strata of the Western Interior Seaway in North America. The field guide is organized into two segments. The first, covers the initial four days, examining the sequence stratigraphy of spectacular Campanian outcrops, which form part of the Book Cliffs in western Colorado and eastern Utah. The guide continues in northwestern New Mexico, with the second segment, composed of the final two days of the trip, investigating Turonian and Coniacian strata in the western outcrop belt of the San Juan Basin. A detailed road log, description and sequence-stratigraphic interpretation of the rocks is provided at each field-stop locality. After each day's road log, an accompanying paper is presented, designed to provide a sequence-stratigraphic overview of the rocks examined during the day, drawing on additional outcrop and subsurface observations.

The Book Cliffs and the San Juan Basin provide ideal locations for detailed outcrop studies of sequence stratigraphy, because lithofacies packages (laminae, beds, bedsets) and their important, time correlative bounding surfaces (parasequence and sequence boundaries) can be continuously traced for many miles across the outcrop belt. In this manner, it is possible to construct a precise chronostratigraphic framework, in which to predict the occurrence of reservoir and seal lithofacies, because the sediments deposited responded to relative changes of sea level. Ultimately, using and successfully applying the concepts of sequence stratigraphy can help in the exploration and exploitation of our valuable energy resources.

John C. Van Wagoner

Clive R. Jones

DAY ONE

Grand Junction to Green River, Utah

AAPG FIELD CONFERENCE: SEQUENCE-STRATIGRAPHY APPLICATIONS TO SHELF-SANDSTONE RESERVOIRS: OUTCROP TO SUBSURFACE EXAMPLES

SEPTEMBER 21 - SEPTEMBER 28, 1991

ROAD LOG, DAY ONE: HIGH-FREQUENCY SEQUENCE STRATIGRAPHY AND FACIES ARCHITECTURE OF THE SEGO SANDSTONE IN THE BOOK CLIFFS OF WESTERN COLORADO AND EASTERN UTAH

by

John C. Van Wagoner
Exxon Production Research Company
Houston, Texas

OBJECTIVES: The Lower Sego provides an opportunity to study well-exposed, high-frequency sequences and their systems tracts. Criteria for identification of sequence boundaries will be presented. Sequences and their boundaries will be contrasted with parasequences and their bounding surfaces. The Upper and Lower Sego contain well-exposed tidal deposits within the lowstand systems tracts of high-frequency sequences. These tidal deposits and their relationship to incised valleys and systems tracts will be examined. The incised valleys interpreted to form during relative falls in sea level will be contrasted with distributary channels related to autocyclic mechanisms.

0.0 Leave the parking lot of the Grand Junction Hilton, Grand Junction, Colorado. Turn left onto Horizon Drive. Pass under the I-70 bridge. Turn left onto the entrance ramp for I-70 west.

0.2 Enter I-70 heading west toward the Colorado-Utah State line. For the next 20 miles the Interstate will parallel the Colorado River flowing along the west side of the Grand Valley. The Interstate is built on the gray Cretaceous Mancos Shale. To the west of the Colorado River are the red cliffs of the Colorado National Monument. The Monument is operated by the National Parks Service. These cliffs are the eastern edge of the Uncompahgre Uplift. As you drive north along the Interstate, the steeply dipping eastern limb of the Uncompahgre is clearly visible. This tight monoclinal fold is the result of horizontal compressional tectonics associated with Laramide deformation (Heyman, 1983). The red rocks in the Monument include, from stratigraphically oldest to youngest: the Chinle Formation forming the lower, less resistant slopes, the Wingate Formation forming the massive cliffs up to 400 feet thick, the Kayenta Formation overlain by an unconformity along which the Navajo and Curtis Formations are missing, the Summerville and Entrada Formations, and the Morrison Formation consisting of fluvial sandstones and associated mudstones, within which some of the earliest dinosaur bones in North America were discovered in the late 19th century. The Jurassic Morrison Formation is unconformably overlain by the brown, Cretaceous, coal-bearing Dakota Sandstone. The Dakota caps many of the high mesas within the Monument and forms well-exposed dip slopes along the Interstate in the vicinity of the exit to Mack, Co..

On the east side of the Interstate, the Mancos mudstones lie directly on the Dakota. The thick section of Mancos mudstone is well exposed along the lower half of the Book Cliffs forming the eastern edge of the Grand Valley. The Book Cliffs are spectacularly exposed parallel to I-70 and about 10 miles to the east. The strata in the Book Cliffs along the Grand Valley are all Late Cretaceous, primarily Campanian and Maestrichtian. In the Grand Valley, the sandstones forming the upper half of the Book Cliffs are named from oldest to youngest the Corcoran, Cozzette, and Rollins (Young, 1955). These are all mainly marine strata with minor intervening coal-bearing coastal-plain rocks. A thick coal seam directly overlies the Rollins. Above this coal, the remaining Cretaceous rocks are coal-bearing fluvial and marginal-marine deposits of the Neslen and Farrer Formations.

During the early 20th century the Book Cliffs produced significant quantities of coal. Production peaked near 13,000,000 short tons per year during the period from 1915 to 1920 (Erdmann, 1934). There are four important coal seams along the Book Cliffs in western Colorado: the Anchor, Palisade, Cameo, and Carbonera seams.

The name "Book Cliffs" has been used in print since at least 1854 when it was used in the book published by Captain Gunnison describing his exploration of this area. In 1861 it was used on War Department maps of this area (Erdmann, 1934). The name is said to have been derived from the regular way in which beds of alternating shale and sandstone are piled upon one another like books lying on a table. Gannett (1877) and Campbell (1922) state that the name was given because of the fancied resemblance of the sandstone cap and curved shale slope below to the edge of a bound book (Erdmann, 1934).

14.3 Cross the Colorado River.
16.5 Pass Exit 15 for Loma and Rangley.
20.4 Exit I-70 at Exit 11 to Mack.
20.6 Turn right at bottom of exit ramp.
21.2 Cross the Denver and Rio Grande Railroad tracks.
21.3 Turn left on Highway 6 toward Mack.
22.1 Cross Mack Wash.
23.8 Turn right on 8 road toward Baxter Pass.
26.6 Turn left on S road. The paved road bends sharply left.
29.7 The paved road bends sharply right. Continue on the gravel road toward the north. Cross a cattle guard onto BLM land. You are driving on the Mancos Shale. This road provides access for service vehicles to inspect the many wells and pipelines of a gas field in this part of the Book Cliffs producing from the Dakota. Drive carefully on these gravel roads.
32.1 Cross a bridge over an arroyo.
32.6 Cross another small bridge.
32.7 Turn left at the fork in the road. Drive into Prairie Canyon. Gray Mancos Shale forms the low hills and ridges on either side of the road. Some of the ridges are capped with Quaternary Alluvium. The Museum of Western Colorado has excavated marine dinosaur skeletons from the Mancos in this area.
35.0 The low hills to the east and west are capped with Quaternary Alluvium. The high cliffs straight ahead are made up of the the Buck Tongue of the Mancos Shale at the base, overlain by the Lower Sego, Anchor Tongue of the Mancos Shale, and the Upper Sego, capped by fluvial and marginal-marine strata of the Neslen Formation.
36.1 Pass a corral on the left.
36.6 The low hills of Mancos below the cliff are capped with the Castlegate Sandstone. In this canyon, the Castlegate is composed of distal lower-shoreface, hummocky strata arranged in seven parasequences. A sequence boundary rests on top of this distal parasequence set. To the east of the road, the Castlegate is capped with a thin, red oolite bed up to two feet thick. This bed thickens toward the east and is best exposed in the next canyon to the east, West Salt Creek Canyon.

37.9 STOP ONE. Pull over to the side of the road for an overview of the stratigraphy of the distal Castlegate, Buck Tongue, and Sego Sandstone. The stratigraphy at the mouth of Prairie Canyon is summarized in Figure 1-1. Also refer to the photo of the cliff face along the west side of Prairie Canyon in Figure 1-2a. At this stop, the Castlegate consists of thin, distal, very fine-grained, hummocky-bedded sandstones and interbedded mudstones. These shelf parasequences are the basinward equivalents of thick fluvial and shoreline deposits that we will examine on Day 3, STOPS TWO, THREE, FOUR. On top of the Castlegate in this area is a 1- to 3-foot thick bed of medium- to coarse-grained, carbonaceous sandstone containing well preserved, large plant fragments, abundant spores and pollen, and large, fragile plant cuticles. Marine dinoflagellates have also been recovered from this interval. This assemblage suggests a quiet, brackish-water environment resting directly on top of the open-marine strata of the distal Castlegate.

The Buck Tongue is a deeper water mudstone resting sharply on the top of the Castlegate marking a regionally extensive parasequence set boundary. Several backstepping parasequences occur in the lower portion of the Buck Tongue at the outcrop, but they are best observed on well logs in the subsurface. Several distal parasequences stacked in a progradational pattern occur at the top of the Buck Tongue.

The base of the thick sandstone at the top of the Buck Tongue along the cliffs to the north is the base of the Lower Sego Sandstone (Fig. 1-2a). This base is a regionally extensive sequence boundary marked by truncation and a basinward shift in facies. In this area, the Lower Sego contains 5 high-frequency sequences. Each sequence consists of a lowstand incised valley filled with tidal deposits and a transgressive to highstand systems tract composed of thin, shelfal parasequences.

The Anchor Tongue is primarily mudstone in this area with thin, very fine-grained hummocky beds. Based on correlation from other areas within the Book Cliffs, there is a sequence boundary within the Anchor.

The lower part of the Upper Sego consists of sharp-based, very fine-grained hummocky-bedded sandstones up to 6 feet thick. These beds rest sharply on the Anchor, but no evidence of a sequence boundary exists between the hummocks and the underlying thin sandstones and mudstones of the Anchor. A significant sequence boundary separates the hummocky beds below from incised valley-fill tidal deposits above. In many places this sequence boundary truncates the underlying hummocks. To the west of Prairie Canyon the Upper Sego contains two sequences. The lowstand portion of the lower sequence contains fluvial strata overlain by a flooding surface. Lower-shoreface hummocky strata rest on this parasequence boundary. The lowstand portion of the upper sequence contains tidal strata.

A major sequence boundary rests directly on top of the Upper Sego in Prairie Canyon and separates this unit from the coal-bearing fluvial and marginal-marine rocks of the Neslen above. This boundary can be followed throughout the Book Cliffs and locally removes the entire Upper Sego by erosion, juxtaposing coals on the Anchor Shale.

At the end of the overview, return to the vehicles and continue driving north into Prairie Canyon.

39.2 Cross a cattleguard. Fifty feet ahead, turn left on a dirt road into Jim Canyon, a branch off of Prairie Canyon.

39.7 Pass a green gas scrubber facility. The cylindrical towers are filled with a dessicant to remove water from the produced gas before it is introduced into the pipeline. Drive straight across the pad and continue west on the dirt road.

40.2 Drive through a Z-shaped turn with the Lower Sego to the right side of your vehicle. Drive out of the turn and park away from the cliff walls on the dirt road.

STOP TWO. Walk down the dirt road, retracing the route you drove in until you reach the base of the Sego and the top of the Buck Tongue. This contact between a medium-grained sandstone and the open-marine thin sandstones and mudstones of the Buck Tongue is a sequence boundary. The cross bedding within the Lower Sego at this stop is interpreted to have formed in a tidal environment. Cross-bed sets and the criteria for a tidal interpretation are reviewed in Figure 1-2b, c and in Figure 1-4. Figure 1-3 is a measured section of the strata exposed at this stop. Facies interpretations and the sequence stratigraphy of these rocks are presented in this figure. Using this measured section, study the stratification in this outcrop ending at the top of the Lower Sego above the vehicles.

Four sequences are developed in the Lower Sego in Jim Canyon. Each sequence exhibits the same general associations of facies summarized by the following description: the sequence boundary is marked by a basinward shift in facies and truncation. Medium-grained, high-energy subtidal sandstones, commonly with associated siderite rip-up clasts lying on the sequence boundary, rest erosionally on either shelf mudstones and thin, wave-rippled sandstones (see Fig. 1-5a) or on lower-shoreface hummocky beds (see Fig. 1-4b). The tidal strata are arranged in upward-thickening parasequences (Fig. 1-5b) interpreted as tidal bars within a tide-dominated delta forming the lowstand systems tract. Regional correlations within the Lower Sego indicate that the tidal bars fill incised valleys. These correlations also demonstrate that the surfaces interpreted as sequence boundaries have a regional extent, ruling out an interpretation of tidal-inlet fill for the tidal deposits. Regional correlations of the sequences and sequence boundaries within the Lower Sego, Anchor, and Upper Sego are illustrated in two cross sections (Figs. 3, 4) in the accompanying overview paper on the Sego entitled *High-Frequency Sequence Stratigraphy and Facies Architecture of the Sego Sandstone in the Book Cliffs of Western Colorado and Eastern Utah.* In the remainder of the road log this paper will be referred to as "the Sego paper". The distribution of the Lower Sego sequence boundaries and associated facies in Prairie Canyon are illustrated in Figure 1-6.

The tidal deposits of the lowstand systems tract are sharply overlain by a significant flooding surface marking a change from tide-dominated to wave-dominated depositional conditions. Hummocky-bedded sandstones (34 feet in Fig. 1-3) or open-marine mudstones and thin hummocky-bedded sandstones (130 feet in Fig. 1-3) rest on the flooding surface. These surfaces are interpreted as parasequence set boundaries and separate the lowstand systems tract below from the transgressive or highstand systems tracts above. Commonly, but not always, the initial parasequence on the major flooding surface backsteps. Lowstand systems tracts are not always overlain by major flooding surfaces. The lowstand in sequence 4 (Fig. 1-6) is overlain by a minor flooding event which superimposes slightly deeper water tidal deposits on shallower water tidal strata. In this case the minor flooding event is coincident with a sequence boundary, sequence-boundary 5, which can only be identified in Jim Canyon by correlation from another area where sequence-boundary 5 is more obviously expressed (see Figs. 3 and 4 in the Sego paper).

After studying the outcrop and discussing the observations and interpretations, return to the vehicles and continue heading west on the dirt road.

40.7 Turn the vehicles around at the production pad for Graham Resources San Arroyo Unit 30, Sec.21 T16S R26E. Return to Prairie Canyon on the same road that you drove in on.
42.2 Rejoin Prairie Canyon road. Turn right and drive south toward Mack.
48.6 Cross the cattle guard and turn right on the main gravel road to Mack.
51.7 Rejoin the paved road.
54.7 Turn right onto 8 road.
57.5 Turn left onto Highway 6 and proceed toward Mack.
60.1 Turn right toward I-70.
60.8 Turn right and enter entrance ramp for I-70 west. Merge onto I-70 and drive west toward the Colorado-Utah border.
62.0 Cross over the Denver and Rio Grande Railroad on a highway bridge. The fluvial Dakota Sandstone outcrops on either side of the road.
64.0 For the next 50 miles you will be crossing the northern edge of the Uncompahgre Uplift. Salt-cored anticlines and associated synclines are found within this uplift. Northwest-trending anticlines and synclines of probable Laramide age (approximately 70 ma.) intersect the Book Cliffs in this area.
67.0 The Morrison Formation of Jurassic crops out on either side of the Interstate.
69.8 Passing Exit 2 for Rabbit Valley. Some of the earliest dinosaur skeletons found in North America were found in this valley. The La Sal Mountains can be seen in the distance to the southwest. The La Sals are a Tertiary laccolith.
71.7 Leave Colorado and enter Utah. A broad vista of the Book Cliffs can be seen to the north and northwest across the Mancos Shale.
76.3 Passing Exit 225 to Westwater.
77.7 The La Sals can be seen again to the southwest.
81.5 Exit I-70 at Exit 220. At the top of the exit ramp, turn right toward the Book Cliffs.
82.0 Turn right at the stop sign. Follow the BLM sign to Hay/East/Middle Canyon. You are now heading east on the old paved road paralleling I-70.
83.0 Cross the old highway bridge over Westwater wash.
83.3 Begin driving on the gravel road which now bends to the north and parallels the wash.
87.0 Turn right at the fork in the road and the road up the hill out of the Westwater Wash valley. At the top of the hill you will be driving across a broad, flat Mancos Shale surface.
91.6 Pass a gas scrubber on the right.
91.7 Turn left at the T junction and drive north toward the Book Cliffs. Follow the sign to East/Hay/Middle Canyons. The road is still crossing the Mancos.
95.6 The Castlegate Sandstone crops out ahead as the sandstone cliffs overlying the Mancos Shale. In this area, the Castlegate consists of a thick succession of 1-to 2-foot thick hummocky-bedded sandstones and thin interbedded sandstones arranged in 6 or 7 parasequences, each 10-to 15-feet thick. The top of the Castlegate in this area is a low-relief erosional surface and is interpreted to represent a sequence boundary. Relative to the Castlegate in Prairie Canyon, this Castlegate outcrop represents a more proximal position within the parasequence set.
95.9 Pass the Westwater gas compressor plant on the left.
96.2 Follow the dirt road around the bend to the left. Do not drive straight through the fence line. The pasture on the other side of the fence is private property.
96.4 Cross Westwater wash.
97.0 Turn left off of the main dirt road onto a jeep trail just before the main dirt road bends right and climbs a hill onto the top of the Castlegate. Proceed straight along the jeep trail. The cliff on the north side of the jeep trail is the Lower Sego Sandstone.
97.4 Stop to study the Lower Sego.

STOP THREE. Facing north, you will see a natural amphitheater in the cliff face. A shale slope on the west side of the amphitheater allows you to walk up to the base of the Lower Sego. Using the measured sections and other figures provided in this guidebook, study the facies and stratigraphic relationships in the Lower Sego.

The sequence stratigraphy of this stop has been correlated to Prairie Canyon. Study the two cross sections, Figures 3 and 4 in the accompanying Sego paper. At this stop, sequence-boundaries 2, 5, and 6 are well exposed; sequences 3 and 4 have downlapped and are represented here as surfaces coincident with the parasequence set boundary on top of the lowstand of sequence 2. This area is an interfluve for sequence 1. Figure 1-7a, b shows truncation and onlap associated with sequence-boundary 6. Shallow subtidal and intertidal rocks arranged in upward-shoaling tidal bars fill an incised valley . The sequence boundary at the base of the valley incises into thin, hummocky-bedded sandstones and interbedded mudstones interpreted to have been deposited in a distal lower-shoreface environment (Fig. 1-7b). This juxtaposition of very shallow-marine strata directly on top of deeper marine strata is a basinward shift in facies. Along this outcrop, 50 feet of incision occurs at the base of sequence-boundary 6. The tidal deposits in sequence 6 represent the lowstand systems tract. These deposits are overlain by shelf mudstones of the Anchor Tongue (Figs. 3 and 4, accompanying Sego paper). As in Prairie Canyon, this juxtaposition of facies represents a parasequence set boundary and defines the top of the lowstand. The flooding surface at the top of the lowstand of sequence 6 is remarkably planar over a wide area of outcrop with no detectable lag or "transgressive" deposits.

Sequence-boundary 5 is marked by a 5-foot thick upper fine-grained sandstone resting sharply on marine mudstone. The sandstone contains abundant siderite rip-up clasts, *Ophiomorpha* burrows and remnants of 2- and 3-dimensional cross-beds with clay drapes. To the west, this sandstone bed thickens to 18 feet. Sequence-boundary 2 occurs at the base of the reddish upper fine-grained sandstone cropping out at the base of the amphitheater at this stop. Figure 1-7c illustrates this unit with the number 2. Examination of Figure 4 from the Sego paper shows that the red sandstone is the distal end of a lowstand tidal-bar complex. The incised valley edge for this complex is spectacularly exposed at the Saddleback East location on the cross section (Fig. 4 and 10 in the Sego paper). As the lowstand tidal-bar complexes in the Lower Sego are traced basinward, they change facies into bioturbated, upper fine-grained red to reddish gray sandstones, commonly containing abundant red siderite or hematitic rip-up clasts and nodules up to 3 inches in diameter. These red sandstones gradually become shalier in a basinward direction until they become indistinguishable from the Mancos Shale, except for a horizon of sideritic or limonitic nodules marking the correlative conformity of the sequence boundary. Significantly, the lowstand tidal deposits do not change facies into hummocky strata in a seaward direction. The hummocky strata are always spatially and temporally separated from the tidal deposits by a parasequence set boundary.

At the end of the stop, turn the vehicles around and drive east along the jeep trail.

97.8 Shift into 4-wheel drive, turn sharply right and ascend a hill to climb up on top of the Castlegate following a rutted jeep trail.

97.9 Stop at the top of the hill for **OPTIONAL STOP FOUR.** Walking to the eastern edge of the Castlegate cliff, you can see three sequence boundaries in the Lower Sego across the valley. The strata along this cliff are illustrated in Figure 1-8a. The truncation at the base of the youngest boundary, sequence-boundary 6, observed at the previous stop, is well developed and cuts down from the mouth of the canyon toward the north or up the canyon. This truncation can be easily seen if you follow the thickness change of the well-developed black shale bed, cropping out in the center of the cliff (number 3 on Fig. 1-8a), from the mouth of the canyon, up the canyon towards the north. Notice that the shale bed gets thinner and finally disappears to the north due to truncation by the sequence boundary at the base of the overlying sandstone. At this position, the thick sandstone bed above the shale rests directly on top of the sandstone bed below the shale. Paleogeographic data such as channel orientations and paleocurrent directions indicate that in the Lower Sego, north is up-depositional dip. This pattern of incision and amalgamation of potential reservoir units in a stratigraphically updip direction, that are separated downdip, has important implications for reservoir management. This truncation pattern is common in the Lower Sego. Return to the vehicles and drive to the end of the jeep trail.

99.1 Circle the vehicles at the end of the jeep trail. **STOP FIVE.** At this locality, look across the Mancos to the north to view the edges of two incised valley fills that make up the Lower Sego. This view is illustrated in Figure 1-8b. The tidal deposits of the Lower Sego are encased in open-marine mudstone of the Buck Tongue.

The two valley edges that can be seen at this stop are the edges of the incised valley fills of sequences 5 and 6. As discussed above, sequences 3 and 4 downlap out to the northeast; they are represented by surfaces on top of the lowstand of sequence 2. The red lowstand deposit of sequence 2 can still be seen along the cliff, just below the small nose indicated on Figure 1-8b by number 1. The valley edge for the lowstand of sequence 5 is indicated on the photo (Fig. 1-8b) by the number 1. Even near the edge of the valley, these lowstand deposits are still intertidal and shallow subtidal. The open-marine mudstones of the Buck Tongue laterally encase the valley fill. If the lowstand deposits of sequence 5 were deposited in migrating tidal inlets, they would have to be laterally encased in strata of the beach through which they were migrating. Regionally extensive, migrating tidal-inlet fills do not form out on the shelf, they are tied to barrier and back-barrier processes. The lateral facies relationship observed at this stop can be best explained by incising the Buck Tongue during a sea-level fall and backfilling the incised valleys with tidal deposits during a slow sea-level rise.

Sequence-boundary 5 is expressed in the interfluve to the west of the valley edge as a 1-foot thick zone of sideritic and hematitic concretions. Details of the facies and stratigraphy in this area can be examined on Figure 4 in the Sego paper. In particular, study measured sections 40 through 48.

At the end of the stop retrace the route back to I-70.

115.5 Turn right onto the entrance ramp for I-70 west. Drive west toward Green River, Utah. The road remains on the Mancos all the way to Green River. The road parallels the Book Cliffs. The La Sals, the anticlines and synclines of the Uncompahgre Uplift, and Arches National Park are on the south side of the Interstate. Arches National Park is south of the intersection of I-70 with Utah 191 at Crescent Junction.
173.3 Arrive at the parking lot of the Best Western River Terrace motel in Green River.

REFERENCES

Campbell, M. R., 1922, Guidebook of the western United States, Part E, The Denver and Rio Grande Western route, U. S. Geological Survey Bulletin 707, 186 p.

Erdmann, C. E., 1934, The Book Cliffs coal field in Garfield and Mesa Counties, Colorado, U. S. Geological Survey Bulletin 851, 150 p.

Gannett, H., 1877, Report of topographer of the Grand River division, U. S. Geological and Geographical Survey Territory 9th Annual Report, 346 p.

Heyman, O. G., 1983, Distribution and structural geometry of faults and folds along the northwestern Uncompahgre Uplift, western Colorado and eastern Utah, *in* W. R. Averett, ed., Northern Paradox Basin-Uncompahgre Uplift, Grand Junction Geological Society Field Trip, October 1-2, 1983, p. 45-57.

Young, R. G., 1955, Sedimentary facies and intertonguing in the Upper Cretaceous of the Book Cliffs, Utah-Colorado, Geological Society of America Bulletin, v. 55, p. 177-202.

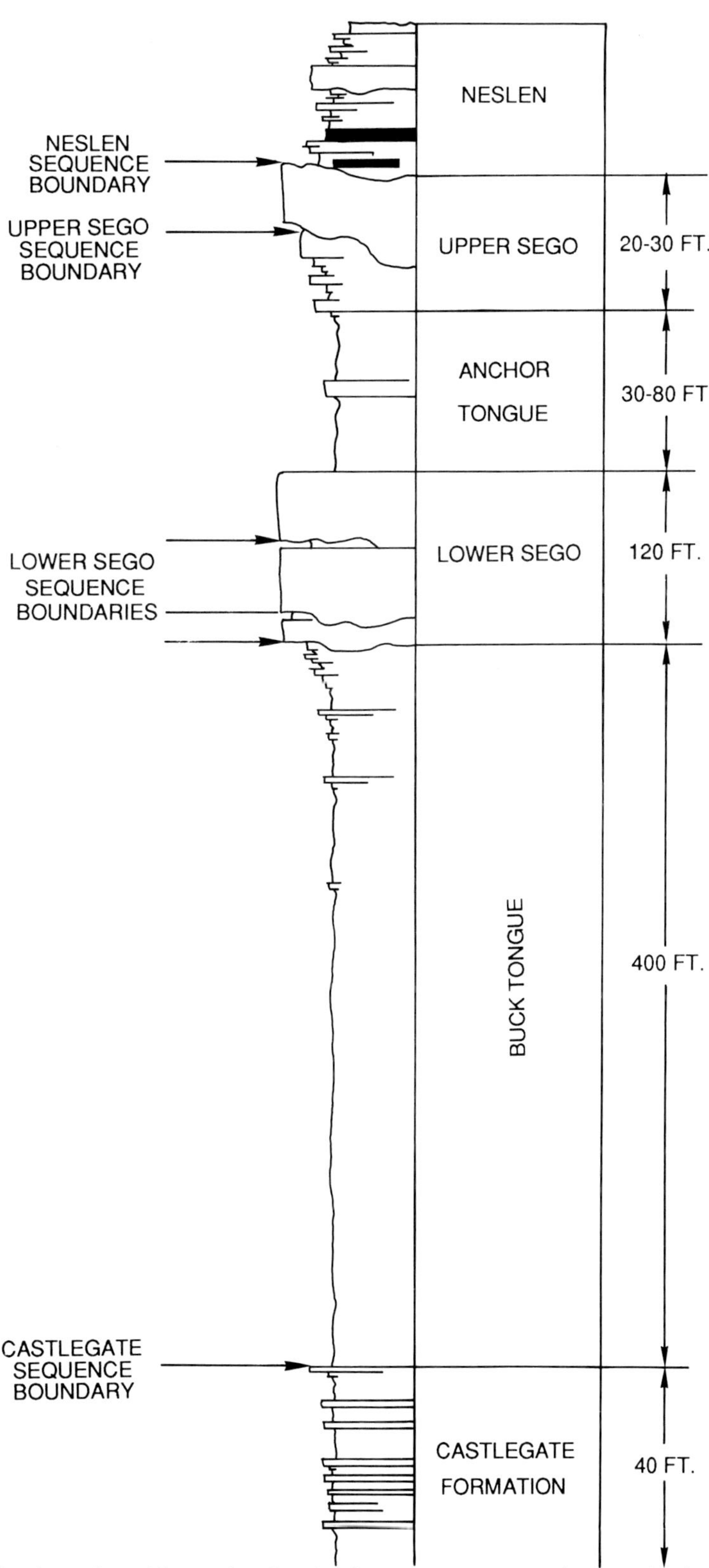

Figure 1-1: Generalized stratigraphic section for the Cretaceous strata at the mouth of Prairie Canyon. In this area, the interval of strata referred to as the Castlegate Formation consists of thin, interbedded, very fine-grained hummocky bedded sandstones and mudstones. These strata were deposited in a distal position on the shelf. The Buck Tongue is a thick, open-marine, dark gray mudstone and siltstone. The Lower Sego truncates the top of the Buck Tongue. The bedding in the Lower Sego is predominantly tidal. The Anchor Tongure is a regionally extensive distal-marine unit. At least one sequence boundary occurs within the Anchor. Strata within the Upper Sego are dominantly tidal. The Neslen is a coal-bearing, marginal marine to nonmarine succession of strata.

Figure 1-2: **A.** This photograph shows the stratigraphy at the mouth of Prairie Canyon located on the Colorado-Utah border in Section 13, T8S R105W. The letters indicate the various stratal units in this Campanian section: Mancos Formation (M), Castlegate Formation (C), Buck Tongue (B), Lower Sego (LS), Anchor Tongue (A), Upper Sego (US), and Neslen Formation (N).

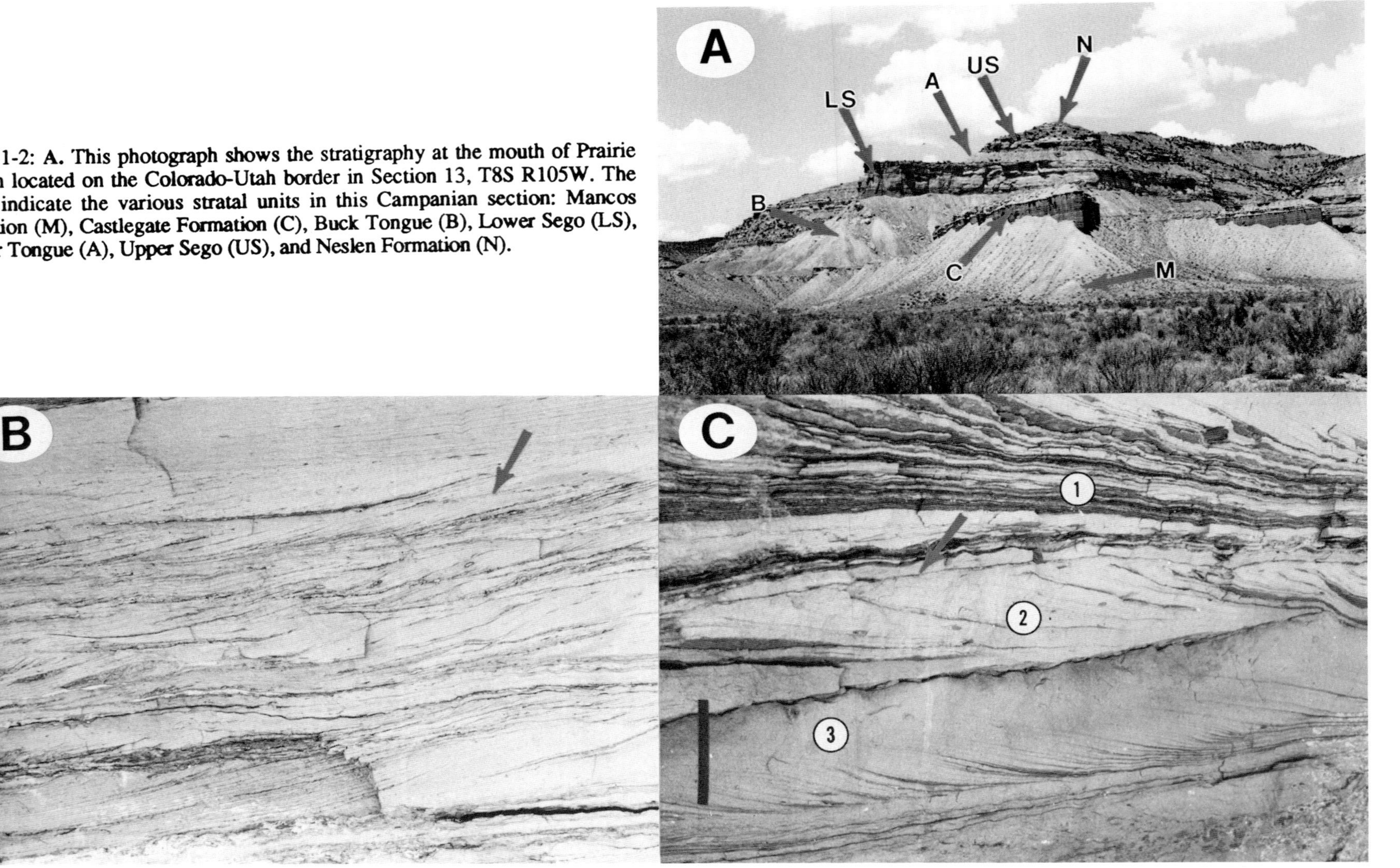

Figure 1-2: **B.** Cross bedding within the Lower Sego at 65 to 75 feet in the measured section shown in Fig. 1-3. Cross beds are two-dimensional with relatively straight crests and tabular geometries in the third dimension. Foresets are extensively draped with clay. Unidirectional reactivation surfaces are common; the arrow in the photo points to one of these surfaces. These features suggest that this sandstone was deposited in a high-energy, shallow subtidal environment. The scale bar is 1-foot high. C. At least five beds within the Lower Sego between 15 and 25 feet in the measured section are shown in Fig. 1-3. The three major beds in the photo are labelled 1, 2, and 3. Bed 1 consists of the mud-draped extremely tangential toes of a large cross bed. Bed 2 is a convex-upward cross bed with well-developed clay drapes on the foreset laminae. Groups of foresets can be subdivided into three bundles based on drape distribution and foreset thicknesses. These bundles are interpreted to reflect neap-spring tidal cycles. The top of bed 2 is a bidirectional reactivation surface indicated by the arrow. Current ripples oriented 180 degrees to the underlying foresets rest on this surface. Bed 3 is a sigmoidal-shaped cross bed also with well-developed clay drapes. Thin mudstone drapes are separated by thin sandstone forming mud couplets. The toes of this cross bed are very tangential. The features in beds 1,2, and 3 strongly indicate deposition in a shallow subtidal, tide-dominated environment. The scale bar is 3-feet high.

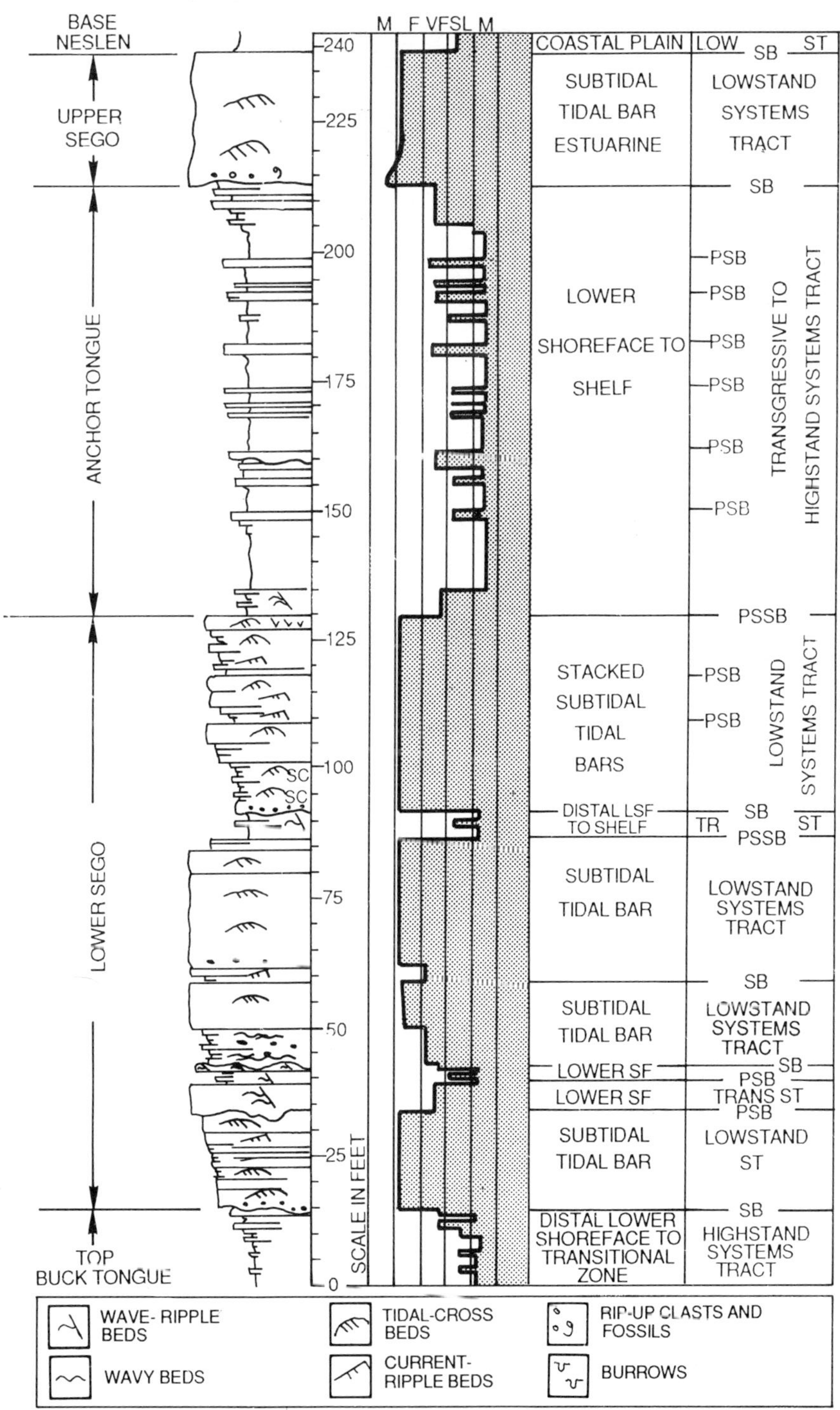

Figure 1-3: Measured section in the Jim Canyon branch of Prairie Canyon. Four high-frequency sequences are present in the Lower Sego at this location. The lowstand systems tract of each high-frequency sequence consists of tide-dominated deltas filling incised valleys. These valleys are up to 10's of miles wide.

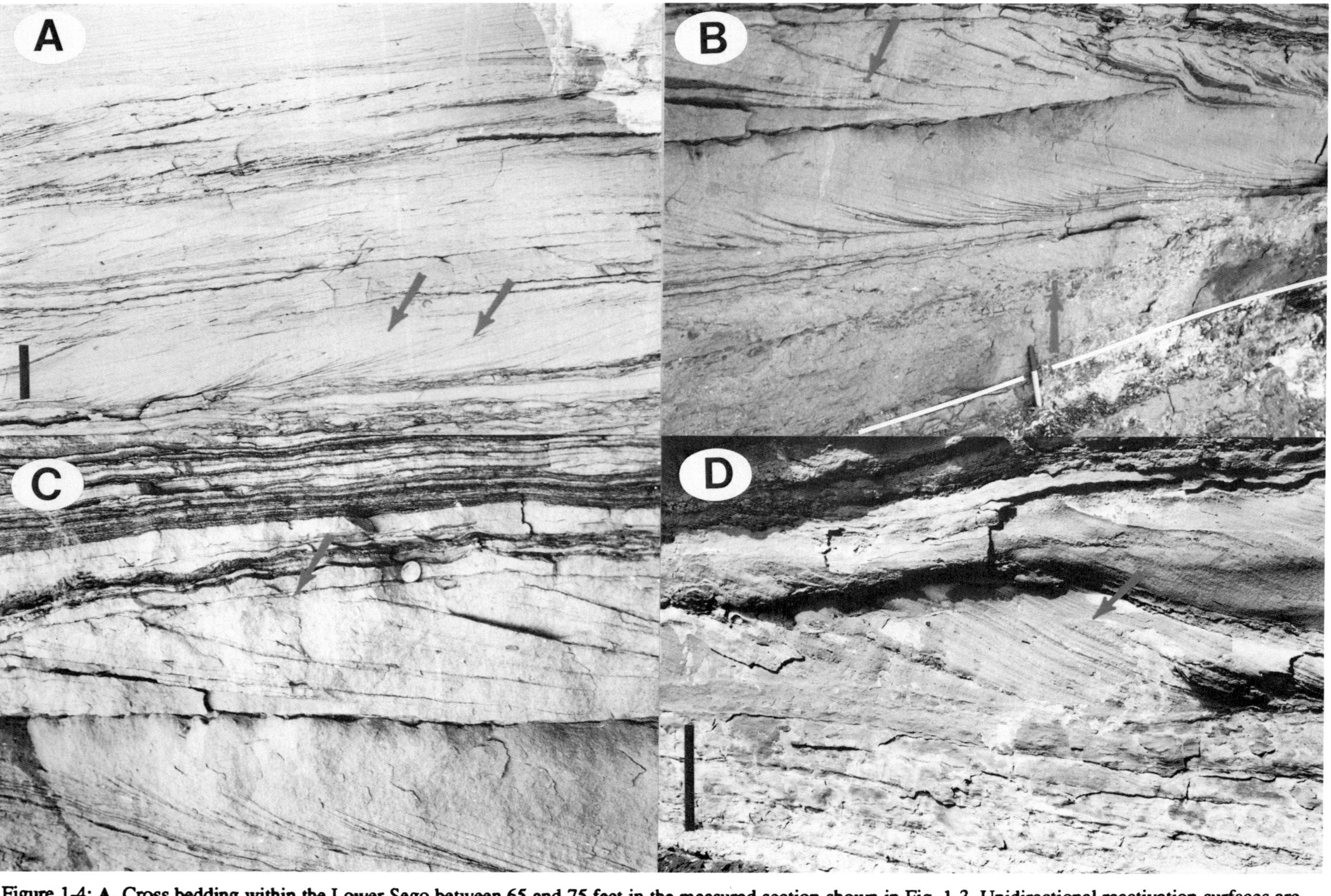

Figure 1-4: **A.** Cross bedding within the Lower Sego between 65 and 75 feet in the measured section shown in Fig. 1-3. Unidirectional reactivation surfaces are shown by the arrows. Clay drapes on the foresets are common. The scale bar is 1-foot high. **B.** Sigmoidal cross beds containing clay drapes, unidirectional reactivation surfaces, siderite clasts shown by the arrows, and tidal bundles in the upper cross bed. The white line indicates the sequence boundary at the base of the Lower Sego shown at 14 feet in Fig. 1-3 . C. A convex-upward cross bed containing clay drapes and siderite clasts overlain by a bidirectional reactivation surface shown by the arrow. Current ripples above the reactivation surface are oriented 180 degrees to the underlying cross bed. **D.** A two-dimensional cross bed with minor clay drapes and unidirectional reactivation surfaces, shown by the arrow occurring at 30 feet in Fig. 1-3. The scale bar is 1-foot high.

Figure 1-5: **A.** Sequence boundary at the base of a lowstand tidal-bar complex occurring at 92 feet in Fig. 1-3. The sequence resting upon this boundary is the youngest sequence within the Lower Sego. Above the boundary, the strata are medium-grained sandstones with siderite clasts and a monospecific trace-fossil assemblege indicating brackish water (Wightman et al., 1987). Below the boundary, the strata are interbedded lower very fine-grained sandstone with wave ripples and interbedded mudstones with a *Cruziana* trace-fossil assemblege. **B.** The lower and middle part of a tidal bar between 95 and 108 feet in Fig. 1-3. This tidal bar rests on the sequence boundary shown in photo A of this Figure. The staff in the photo is 5-feet long.

Figure 1-7: **A.** Sequence boundary 6 truncating distal lower-shoreface hummocky strata at Stop Three. The sequence boundary is overlain by thin-bedded sandstones interpreted to have been deposited in shallow subtidal and intertidal environments. The vertical scale bar is 10-feet high. **B.** Close up of photo A showing additional detail of the sequence boundary and associated strata. Cross bedding above the sequence boundary can be seen. The vertical scale bar is 8-feet high. **C.** This photo shows the outcrop of the Lower Sego at Stop Three. Number 1 identifies the sequence boundary seen in photos A and B. A distal lowstand tidal-bar complex is shown by 2. This reddish bioturbated sandstone is the basinward end of the lowstand tidal-dominated delta in sequence 2. The rectangular box outlines a geologist for scale.

Figure 1-8: **A.** This photo shows the east wall of Middle Canyon located
Number 1, on the point at the south end of the outcrop, points to sequenc
deposits in the lowstand of sequence 2. Number 3 points to marine mudsto
top of the underlying lowstand deposit; these marine strata are 35 feet thick.
or left on the photo, the unconformity truncates progressively deeper into the
marine rocks and cuts completely through the lowstand of sequence 2. T
represent the edge of an incised valley. **B.** The lowstand tidal deposits in th
shows the edge of one of these valleys, the valley for sequence 5. Number
filled with high-energy subtidal and intertidal deposits, the strata are open-
this interfluvial area. The sequence boundary in the interfluvial area is rep
clasts. On the photo, the Buck Tongue is B, the Anchor is A, the Upper Se
feet high.

n NE Sec. 5 and NW Sec. 4, T18S R24 E in eastern Utah.
: boundary 6 in the Lower Sego . Number 2 points to tidal
es and thin hummocky-bedded sandstones lying directly on
As sequence boundary 6 (shown by 1) is traced to the north,
underlying strata. Ultimately, the surface truncates all of the
ıis truncation occurs in a landward direction and does not
: Lower Sego are deposited within incised valleys. This pan
1 points to the valley edge. Outside of the valley, which is
narine to shelf mudstones. Beaches and deltas are absent in
esented by a 1-foot thick horizon of hematitic and sideritic
;o is US, and the Neslen is N. The vertical-scale bar is 200-

HIGH-FREQUENCY SEQUENCE STRATIGRAPHY AND FACIES ARCHITECTURE OF THE SEGO SANDSTONE IN THE BOOK CLIFFS OF WESTERN COLORADO AND EASTERN UTAH

by
John C. Van Wagoner
Exxon Production Research Company
Houston, Texas 77001

INTRODUCTION

The Lower Sego, Anchor Tongue of the Mancos Shale, and the Upper Sego are well exposed along the Book Cliffs in western Colorado and eastern Utah. For the most part these strata crop out on public lands and access is excellent due to the constantly maintained roads leading to the gas fields along the front of the cliffs. Nine sequences and their component systems tracts can be seen in these strata. Because of the high-quality exposure and access, the Lower and Upper Sego are excellent units to study the geometry and expression of sequence boundaries and the facies contained between these regionally extensive surfaces.

This paper discusses the sequence stratigraphy of the Lower and Upper Sego between Prairie Canyon on the Colorado-Utah border and Sulphur Canyon in eastern Utah. The discussion includes criteria for recognizing sequence boundaries in outcrop, the geometry of incised valleys and the nature of their fill, and the facies of the transgressive and highstand systems tracts within the sequence.

LOCATION AND LITHOSTRATIGRAPHY

The Sego Sandstone is composed of two units: the Lower Sego and the Upper Sego. The two are separated by the Anchor Mine Tongue of the Mancos Shale (Fig. 1). The Lower Sego is underlain by the Buck Tongue of the Mancos Shale and the Upper Sego is overlain by the coal-bearing Neslen Formation. The Lower and Upper Sego occur in outcrop along the Book Cliffs from Tuscher Canyon north of Green River, Utah to near Coal Gulch in western Colorado (Fig. 2). The Lower Sego changes facies to the south into the Mancos Shale between East Salt Creek Canyon and Coal Gulch in western Colorado; the Upper Sego changes facies to the south into coal-bearing coastal-plain strata in the same area. This study of the Sego Sandstone in the Book Cliffs includes the outcrop between Hunter Canyon in western Colorado and Thompson and Sego Canyons north of Thompson, Utah. However, in this paper only the work from Prairie Canyon on the Colorado-Utah border to Sulphur Canyon in eastern Utah is reported (Fig. 2).

AGE

Gill and Hail (1975) collected the ammonite *Baculites perplexus* from the upper part of the Buck Tongue in Prairie Canyon, although it is not clear from their work if this is the early or late form, and *Baculites scotti* from the Anchor Tongue in East Canyon. These ammonites place the Sego in the Late Campanian (W.J. Devlin, pers. comm.). Assuming that *B. perplexus* found by Gill and Hail (1975) in the Buck Tongue is the late form, the Sego would roughly span 76.0 ma, about in the middle of the *B. perplexus* range, to 74.6 ma, about at the the top of the *B. scotti* range (ages from W.J. Devlin, pers. comm.), a period of 1.4 ma. Nine sequences have been interpreted in the Lower and Upper Sego and Anchor Tongue during this time period. If all the sequences are of equal duration, then each sequence had a frequency of approximately 150,000 years, the approximate duration of a 4-th order cycle of sea-level change. For this reason, these sequences are referred to as high-frequency sequences (Mitchum and Van Wagoner, 1991).

PREVIOUS WORK

Erdman (1934) and Fisher (1936) studied the coal deposits associated with the Sego Sandstone in Colorado and Utah. Erdmann (1934) recognized a Lower and Upper Sego, separated by a marine unit he called the Anchor Tongue of the Mancos Shale. Young (1955) contributed significantly to our understanding of Book Cliff stratigraphy by publishing a cross section showing the correlation of the major units in the Book Cliffs from Price, Utah to Grand Junction, Colorado. On this cross section Young (1955) showed the general relationships of the lower and upper Sego to the Anchor Tongue and to the Mancos Shale. Gill and Hail (1975) studied the ammonite distribution within the Mancos Shale and Anchor Tongue in western Colorado and eastern Utah. Their data represent the best biostratigraphic control for the Sego available today. Franczyk (1989) published a reconnaissance study to of the

Sego primarily addressing the coal distribution. As part of this work, she recognized upward-coarsening and fining units within the Sego and interpreted them to record deposition in lower- and upper-shoreface, foreshore, and bay/lagoonal environments. Franczyk (1989) ascribed Sego deposition to a relative rise in sea level. Noe (1983) studied the lower-shoreface deposits in the Sego northeast of the study area documented in this paper.

SEQUENCE STRATIGRAPHY

The sequence stratigraphy of the Lower Sego, Anchor Tongue, and the Upper Sego is presented on two cross sections, Figures 3 and 4. Figure 3 covers the cliffs from Prairie Canyon on the Colorado-Utah border to San Arroyo Canyon in eastern Utah. Figure 4 covers the cliffs from San Arroyo Canyon to Sulphur Canyon in eastern Utah. Base maps on each figure show the location of the measured sections used to construct the cross sections. The straight-line distance from Prairie Canyon to Sulphur Canyon is 20 miles.

Six sequences were interpreted in the Lower Sego, one sequence was interpreted in the Anchor Tongue, one sequence was interpreted in the Upper Sego and one sequence boundary was interpreted at the top of the Upper Sego separating it from the overlying Neslen Formation. Stratigraphic analysis ended in the very lower part of the Neslen.

The Lower Sego is separated from the underlying Buck Tongue by a sequence boundary; although, as the cross sections show, it is not always the same sequence boundary that separates the sandstones of the Lower Sego from the sandstones and mudstones of the Buck Tongue. Of the six sequence boundaries in the Lower Sego, sequence-boundary 6 is the best developed. It has the most incision, the most widely developed basinward shift in facies, and the widest incised valleys. Sequence-boundaries 3, 4, and 5 are well developed, although locally sequence-boundary 5 is truncated by 6. Sequence-boundaries 1 and 2 are exposed primarily on the north end of Figure 3 and on the south end of Figure 4. It is difficult to correlate these two sets of sequence boundaries in the outcrop. Because of the similar relationship of both sets of sequence boundaries to sequence-boundary 3, which is widely correlatable as an unconformity or its correlative conformity, both sets of sequence boundaries are assumed to be correlative.

One sequence boundary marked by truncation and a basinward shift in facies is observed within the Anchor Tongue. The lowstand resting on this sequence boundary is well developed on Figure 3 and in sections measured to the north of both cross sections (Figs. 3, 4) up the many canyons that intersect the Book Cliffs. The lowstand is only locally developed along Figure 4, although a zone of siderite nodules in the Anchor up to a foot in diameter correlates with the top of a local incised valley on Figure 4, possibly indicating subaerial exposure on an interfluve.

The lower part of the Upper Sego is locally composed of hummocky beds up to 3-feet thick. They are truncated by a sequence boundary. This sequence boundary is present everywhere along the Book Cliffs in the area studied. The top of the Upper Sego is truncated by a sequence boundary that locally removes all of the Upper Sego resulting in the deposition of the coals in the Neslen Formation directly on top of the Anchor Shale. This boundary is the most pronounced in the Upper Campanian.

Every sequence boundary in the interval studied is marked by regional truncation. Truncation is especially well developed below sequence-boundary 6 in the Lower Sego, below sequence-boundary 8 at the base of the Upper Sego, and below sequence-boundary 9 at the base of the Neslen. In some areas, sequence boundaries are marked by evidence of subaerial exposure developed on open-marine strata. This evidence is present below sequence-boundary 2 in measured section 33 on Figure 4, and below sequence-boundary 5 in measured section 47 on Figure 4.

Each sequence has a well-defined lowstand systems tract resting on the sequence boundary. In every sequence in the interval studied, the lowstand systems tract is an incised valley. The valleys are filled with tidal or mixed tidal and fluvial deposits, with a single exception. That exception is the lowstand of the sequence that rests on top of the Upper Sego. In that sequence, the lowstand systems tract is coal-bearing marginal-marine to nonmarine strata, probably filling a broad, coalesced valley system. In all of the other sequences, the incised valley-fill is tide-dominated deltas consisting of stacked subtidal to intertidal, upward-shallowing tidal-bar parasequences. In a few places, these tidal deposits interfinger along the edges of the incised valleys with fluvial deposits that show evidence of tidal modification. The incised valleys in the Lower and Upper Sego are broad and generally shallow. They range in width from 1 to 15 miles and in maximum thickness from 15 to 70 feet. Incised-valley fills thin toward the edges of the valleys, although the thickest fill can be close to the edge. For example, see the incised-valley fill associated with sequence-boundary 2 at measured section 33 on Figure 4. This fill is 40-feet thick within 100 yards of the valley edge in the adjacent measured section.

Incised valleys truncate into shelf and distal lower-shoreface strata; shallow-marine strata such as upper-shoreface and foreshore sandstones or stream-mouth bar and delta-plain deposits are completely absent adjacent to valley edges. The lack of these shallow-marine to marginal facies adjacent to valley edges is one reason why valley fills cannot be interpreted as tidal-inlet or distributary-channel deposits. Valley edges are best developed in the Lower Sego along the southern end of the cross section in Figure 4. In this area, interfluves developed for each sequence boundary in the Lower Sego over a period of 100,000's of years suggesting a persistent tectonic control on valley distribution.

In addition to valley edges, the cross section in Figure 4 shows basinward facies changes within the lowstand systems tracts of sequences 2, 3, and 4. As the base map on Figure 4 shows, the cliffs turn to the south at measured section 35. North of measured section 35, the cliffs are oriented subparallel to the depositional strike of the lowstand deposits in sequences 2, 3, and 4; south of measured section 35 the cliffs are oriented subparallel to the depositional dip of the lowstand deposits for these sequences. Because of this change in cliff orientation relative to the trend of the lowstand deposits, the valley edge for sequence 2 is observed in measured section 32, downdip facies changes are observed in measured sections 36 to 47, and the final basinward downlap of the lowstand in sequence 2 is observed in measured section 49. The orientations indicate that the lowstand systems in the Lower Sego were flowing to the south, southeast, or south-southwest.

The top of each lowstand systems tract is a flooding surface marked by deeper water rocks resting sharply on the underlying tidal or fluvial deposits. Each lowstand in the Lower Sego and the one lowstand in the Anchor are overlain by one of these flooding surfaces interpreted to be a parasequence set boundary. However, the flooding surface overlying the lowstand of sequence 6 (Figs. 3, 4), separating these tidal deposits from the marine mudstones of the overlying Anchor Tongue, is the most significant parasequence set boundary in the Lower Sego. Due to the quality of the outcrop exposures, these major flooding surfaces can be traced widely; they are typically very planar everywhere they are observed. They are not marked by transgressive lag, nor do they have associated transgressive deposits. They are most commonly overlain by marine mudstones and thin interbedded siltstones containing wave-ripple laminae.

The strata within the transgressive and highstand systems tract of each sequence in the Lower Sego and Anchor Tongue are wave dominated without any indication of tidal influence. Hummocky beds and wave ripples are the predominant bedding types. These strata are interpreted to have been deposited in lower-shoreface, distal lower-shoreface, and shelf environments. Trace fossils include *Asterosoma*, *Teichichnus*, *Chondrites*, *Planolites*, *Paleophycus*, *Thalassinoides*, and *Terebellina*. The strata are arranged in parasequences which stack to form parasequence sets. Clear retrogradational parasequence sets are rare, most parasequence sets observed in these parts of the sequences are progradational. The deepest water mudstones within these systems tracts typically occur directly above the parasequence-set boundary at the top of the lowstand. For example, the lowstand of sequence 4 (Fig.3) is overlain by a black, marine shale that can be correlated visually along the cliffs for 10 miles. For these reasons, the transgressive systems tracts in these sequences are interpreted to be thin and condensed.

In the following sections, details of the sequence boundaries, lowstand system-tract facies, transgressive surfaces on top of the lowstand systems tracts, and the transgressive/highstand systems-tract facies are discussed in more detail. For reference, Figure 5 illustrates a general model of the major surfaces and dominant facies associations found in the sequences in the Sego Sandstone and Anchor Tongue.

SEQUENCE BOUNDARY

The sequence boundaries found in the Lower Sego, Anchor Tongue, and the Upper Sego are continuous, regionally extensive surfaces (Figs. 3, 4). They are marked by a basinward shift in facies, regional truncation, and clast horizons incised valleys and by evidence of subaerial exposure and rip-up clast deposits in interfluves or along the feather edges of incised valleys. Figure 6 shows some examples of basinward shifts in facies so commonly observed in the Sego. Figure 6a shows cross beds with clay drapes, interpreted to have been deposited as part of a tide-dominated delta within an incised valley, resting erosionally on gray-marine mudstones interpreted to have been deposited on the shelf in water depths of 50- to 150-feet. This photo is from 62 feet in measured section 50, Figure 4. 100 yards to the west, this cross-bedded unit is completely absent due to onlap onto the sequence boundary at the incised-valley edge (Fig. 4). As Figure 4 illustrates, the incised-valley fill sandstone in Figure 6a is laterally encased in open-marine mudstones. This facies juxtaposition indicates that the sandstone is not a tidal-inlet or distributary-channel fill cutting through a beach or delta into underlying shelf and prodelta mudstone. If this were so, beach or deltaic deposits should laterally encase the channel, not open-marine strata. The observed lateral-facies juxtaposition requires a relative fall in sea level.

The other photos in Figure 6 (Fig. 6b, c, d) also show basinward shifts in facies marked by high-energy cross beds resting erosionally on lower-shoreface hummocky beds (Fig. 6b, c) or on open-marine mudstones and thin wave-rippled, very fine-grained sandstones (Fig. 6d).

Figure 7a, b, and c show Lower Sego sequence boundaries at the bases of upward-shallowing tidal-bar parasequences within incised valleys. The thicknesses of the beds in the lower parts of the tidal bars are the same as the thicknesses of the thin, wave-rippled sandstones below the sequence boundaries. For this reason, the sequence boundaries are not obvious on the photos or in the field until they are examined closely. Upon close examination, the sequence boundaries are marked by 1) a large grain-size change from medium-grained sandstone above to lower very fine- grained sandstone to mudstone below, 2) changes in bedding types, from current ripples and small scale sigmoidal cross beds with clay drapes above to wave ripples below, and 3) a change in ichnofacies, from a low-diversity assemblage with *Thalassinoides, Teichichnus*, and thin walled, small *Terebellina* above to a diverse Cruziana assemblage below. These sequence boundaries would show clearly in cores and well logs. Figure 7d shows truncation at the base of a sequence boundary. This truncation occurs between measured sections 38 and 37 (Fig.4) along sequence-boundary 6.

Sequence boundaries in the Sego are also marked by clast accumulations. The most common clast type in the Sego and Anchor Tongue is a brick red hematite-rich clay clast up to 4 inches in diameter. This clast type is commonly associated with siderite clasts and nodules of the same dimensions. Figure 8a shows sequence-boundary 6 at a measured section called Swedes Hole located in Sec.2 T20S R22 E just west of Cottonwood Canyon in eastern Utah. At this locality, the sequence boundary is marked by a 4-foot thick sandstone bed with a basal accumulation of clasts, shown with the scale resting on them on the right side of the photo, and by minor truncation. The sandstone bed is only locally present; over much of the cliffs in that area the sequence boundary is marked only by the clast accumulation. Because of regional correlations, these clasts are known to lie on the feather edge of an incised valley. Figure 8b shows a close up of the clasts. The clasts include pieces of bone, sharks teeth, oyster fragments, pelecypods, pieces of coal, siderite-cemented clay clasts, hematite-rich clay rip-up clasts, and limonitic clay clasts all in a medium- to fine-grained sandstone matrix. The unit is heavily cemented and partially replaced with limonite. This deposit is interpreted as a channel lag formed in the base of the incised valley during fluvial incision in response to a relative fall in sea level. The eclectic mixture of particles in the deposit is consistent with a fluvial system incising into the shelf, mixing the larger particles removed by erosion from the open-marine strata with particles transported from updip.

Figure 8c, d also shows clast accumulations on sequence boundaries in the Lower Sego. The clasts are slightly flattened and composed of siderite and hematite. In both of these examples, the clasts are concentrated by fluvial or tidal processes. These same types of clasts are found on interfluves. Figure 9a shows red hematite-rich clay clasts, some up to 2-inches in diameter, mixed with coal and charcoal fragments in a bioturbated gray silty mudstone adjacent to an incised-valley edge. This edge occurs in measured section 32, Figure 4. The origin of the hematite is uncertain but is hypothesized to form as an iron-rich hard pan at the bottom of peat swamps developed on the subaerial interfluves during a sea-level fall. Subsequent sea-level rise removed the thin veneer of subaerial deposition and eroded the hematitic hard pan. Rounded to subangular hematite clasts and coal fragments were left behind as a lag to record the subaerial exposure.

LOWSTAND SYSTEMS TRACT FACIES

Tidal deposits predominate within the lowstand systems tracts of the Sego Sandstone and Anchor Tongue; fluvial deposits are a minor component of the lowstand strata. All of the lowstand facies were deposited within incised valleys. Figure 9b shows the sequence boundary at the base of the incised valley at measured section 32 (Fig. 4). Figure 9c shows the incised valley thinning to the east along the outcrop toward the valley edge. Figure 10 provides additional details about the valley edge, the lowstand facies within the valley, and the expression of the sequence boundary on the interfluve.

Cross bedding ranging in thickness from 4 inches to 2 feet is common within incised valleys. The cross beds are dominantly 2-dimensional with a variety of features that indicate deposition in response to regularly increasing and decreasing flow velocities and flow directions. These features include: bidirectional reactivation surfaces (Nio and Yang, 1989), unidirectional reactivation surfaces, sigmoidal cross-bed geometries (Mutti, 1985), rare mud couplets (Visser, 1980), mud drapes, drapes of finely disarticulated organic material on cross-bed foresets, and rare tidal bundles (Visser, 1980). Figures 1-2b, c and 1-4a to d in the road log for Day One of this trip illustrate many of these features. Figure 11 illustrates some additional examples. In particular, Figure 11a shows bidirectional cross bedding probably developed in association with a bidirectional-reactivation surface. This photo also shows mud drapes and hematite-rich clay clasts on the foresets of the cross bed. Figure 11b shows a bidirectional reactivation surface and mud couplets. Unidirectional reactivation surfaces are shown in figure 11c. A sigmoidal-cross bed (Mutti et al.,1984; Kreisa and Moiola, 1986) is illustrated in Figure 11d.

Tide-dominated deltas fill the incised valleys in the lowstand systems tracts. These deltas are composed of tidal-bar parasequences. Each tidal-bar parasequence is an upward-thickening assemblage of strata recording deposition in the subtidal portion of a valley. Mutti (Mutti et al., 1985) has identified similar tidal bars in the Pyrenees. The parasequences in the Sego have different facies associations and outcrop expressions depending upon where within the tidal parasequence they are observed. Figure 12a shows an upward-thickening parasequence exposed at measured section 27 on Figure 4. Figure 12b shows three upward-thickening tidal-bar parasequences located in Bitter Creek Canyon in SE Sec. 34 T19S R 25E. Figure 12c shows a typical imbricate-stacking pattern for three tidal bars. Each bar builds seaward in a strongly progradational pattern and downlaps onto the sequence boundary farther seaward than the underlying bar (Fig. 14). The facies architecture of the tidal bars seen in the Lower Sego is summarized in Figure 14. This diagram is based on observations of many tidal bars in the Sego. Although it is a summary of the features of many tidal bars, most tidal bars in the Sego closely resemble this model. Figure 15 shows examples of some of the stratification found in tidal bars. Figure 15a and b shows stratification typical of the lower parts of tidal bars. In Figure 15a, continuous beds of mudstone separate current ripples formed in medium to upper fine-grained sandstone. In Figure 15b, the arrows point to current ripples indicating reversing current flow. The current ripple marked by the arrow on the right of the photo is indicating current flow to the right, the current ripple marked by the middle arrow is indicates current flow to the left, and the current ripple marked by the left arrow is indicates to the right also. Figure 15c is an example of distal tidal-bar facies from the interval between 67 and 70 feet in measured section 38, Figure 4. This red, upper fine-grained sandstone is churned by *Asterosoma* and *Thalassinoides*. Bone fragments, charcoal and wood fragments, and hematite rip-up clasts are present in the sandstone. The sandstone is underlain by a 3-inch thick, brick-red bed of hematite or siderite. Figure 4 indicates that this sandstone unit downlaps out to the southwest in a basinward direction; it is gone in measured section 48 (Fig. 4). Nowhere in the tidal-bar system, even in the distal portion, does tide-dominated strata change facies into wave-dominated deposits.

The facies architecture of the lowstand systems tracts in the Sego Sandstone is controlled by the distribution of upward-thickening tidal-bar parasequences confined within incised valleys. No laterally equivalent lowstand-shoreline deposits have been identified outside of the valleys nor have downdip lowstand-shoreline sandstones been found. Interfluves and basinward facies changes within tidal bars have been carefully studied and described above. The apparent lack of lowstand shoreline deposits such as lowstand deltas or beaches is puzzling. Perhaps they were deposited but were destroyed during the subsequent sea-level rise. Perhaps the true lowstand-shoreline deposits are far to the south. The basinward facies changes observed within the tidal bars may occur entirely within the valleys, far updip of the original lowstand shorelines. Additional work must be done to address these issues.

TRANSGRESSIVE SURFACE

The upper bounding surface of the lowstand systems tract is a parasequence set boundary referred to as the transgressive surface. The most pronounced transgressive surface in the Sego Sandstone is the marine-flooding surface that separates the lowstand of sequence 6 from the overlying Anchor Tongue (Figs. 3, 4). In outcrop, this surface is remarkably planar and continuous. There is no observable relief on the surface over distances of 1 to 5 miles. A small part of this transgressive surface is shown in Figure 15d. The upper arrow in this photo points to the transgressive surface at the base of the Anchor Tongue located in a small amphitheater between measured sections 49 and 50 (Fig. 4) in the Sulphur Canyon area. There is a pronounced difference between the geometry of transgressive surfaces and sequence boundaries in the Sego Sandstone. Erosional relief of several feet to 10's of feet over short distances, associated with sequence boundaries, is detectable nearly everywhere the sequence boundaries are observed. In some places, for example between measured sections 39 to 36, 70 feet of relief is visible in a distance of less than 100 yards. No relief is detectable associated with the transgressive surfaces. Even the widespread surface on top of the Lower Sego, which is well exposed almost everywhere from the mouth of Thompson Canyon to East Salt Creek Canyon (Fig. 2), is planar.

Sequence boundaries erode and create accommodation which is commonly filled with reservoir-quality strata. Hydrocarbon traps against valley edges are controlled by sequence boundaries. Transgressive surfaces can erode, producing a regionally persistent, bevelled and planar boundary. They rarely control reservoir; more commonly they control seal distribution. Differentiating between these two types of surfaces is critical to correctly predict, correlate, and map the distribution of reservoir and seal rocks in the subsurface. Identifying a sequence boundary in an interfluve can lead to discovery of a hydrocarbon-filled incised valley. Identifying a transgressive surface aids in correlating but rarely leads directly to more hydrocarbons discovered or produced.

In addition to low relief, these transgressive surfaces are notable for their lack of lag and associated transgressive deposits. Deep-water deposits rest directly on the underlying lowstand strata. In some places, thin, distal beds of backstepping parasequence can be found just above a transgressive surface. These distal deposits are usually siltstones to lower very fine-grained sandstones with wave ripples, thin hummocky beds, and a Cruziana trace-fossil assemblage indicating deposition on the shelf. Deposits like these can be found at 130 to 135 feet in measured section 1, Figure 3. These deposits are interpreted as the distal toes of a backstepping parasequence.

Transgressive surfaces separate two completely different depositional regimes. Below the transgressive surface, the strata are dominated by tidal and fluvial processes. Above the transgressive surface, the strata are dominated by wave processes; the shorelines have been pushed many miles up depositional dip.

TRANSGRESSIVE AND HIGHSTAND SYSTEMS TRACT FACIES

The strata within the sequence between the transgressive surface below and the next overlying sequence boundary in the Sego is wave dominated. If the overlying sequence boundary does not truncate deeply, the upper part of this wave-dominated succession can consist of several upward-shallowing, lower-shoreface parasequences arranged in a progradational parasequence set typical of highstand systems tracts. For example, see the top of the Anchor Tongue below the base of the Upper Sego in measured sections 23 (Fig. 3) and 50 (Fig. 4). Where the overlying sequence boundary erodes into the underlying highstand systems tract, only mudstone and thin, wave-rippled sandstones are preserved. In some places, the overlying sequence boundary erodes deeply through the transgressive and highstand systems tracts juxtaposing one lowstand systems tract on another (see measured sections 18 to 22, Fig. 3).

Transgressive systems tracts are not a significant component of Sego sequences. Wave-dominated parasequences, stacked in a backstepping pattern, are not commonly developed on top of the transgressive surfaces. Where interpreted to be present they range in thickness from 3 to 10 feet. Based on tidal-bar parasequence stacking patterns, some backstepping may occur within the incised valleys during the sea-level rise at the end of the lowstand systems tract.

Figure 16 shows examples of the facies in Sego highstand systems tracts. Figure 16a shows interbedded mudstone and thin hummocky beds up to 1-foot thick overlain by a sequence boundary. The mudstone and hummocky strata are interpreted to have been deposited in a distal lower-shoreface environment. These strata are typical of highstand systems tracts in Sego sequences. The photo illustrates the strata between 120 and 140 feet in measured section 49, Figure 4. Figure 16b shows a hummocky bed within a late highstand parasequence at the top of the Anchor Tongue in measured section 20 (Fig. 3) between 88 and 90 feet. Hummocky stratification is the predominant bedding type within the transgressive and highstand systems tracts. Figures 16c, d show gutter and channel casts commonly found in the lower-shoreface hummocky strata of late highstand systems tracts. These storm-generated structures are very common in the lower shoreface just below sequence boundaries in the Sego and in the Desert and Castlegate. They commonly occur in association with thick, areally extensive, vertically amalgamated hummocky beds (For example, see the discussion of the highstand deposits below the Desert sequence boundary in the road log for Day 3). For these reasons, the gutter casts and channels are interpreted to form in the late highstand when sea-level is slowly rising or just beginning to fall. At these times, accommodation on the shelf is reduced. Lower-shoreface deposits are exposed to storms for longer periods of time resulting in a greater chance for the gutter and channel casts to be formed and preserved. Due to the decrease in accommodation, these beds extend farther across the shelf. Figure 16c is from 30 to 40 feet in a measured section in San Arroyo Canyon at the top of the Buck Tongue. Figure 16d is from 143 to 144 feet in a measured section called Cottonwood 2 located in SW SW Sec. 7 T19S R22E in Cottonwood Canyon south of Figure 4.

RELATIONSHIP OF SYSTEMS TRACTS AND FACIES TO RELATIVE CHANGES IN SEA LEVEL

The facies relationships in the lowstand, transgressive, and highstand systems tracts discussed above can be interpreted in terms of relative changes in sea level. A sequence in the Sego would begin with fluvial incision forming in response to a relative fall in sea level. The amount of incision is a function of the change in slope between the coastal plain and the shelf (Blum, 1991). If the dip of the shelf is greater than the dip of the coastal plain, the rivers will incise more deeply as they cut across the exposed shelf. If the dip of the shelf is the same as the coastal plain, the rivers will erode across the shelf, because rivers are channelized, but will not incise more deeply. In either case, channels will erode into or incise, in the loose sense, into marine deposits. The areal extent of the incised valleys depends on the discharge and the distribution of river systems at the time of sea-level fall. As the rivers cut across marine strata they concentrate lag in the channel bases composed of whatever material was eroded from these marine deposits. Lag material can include phosphate nodules, glauconite, sharks teeth, oyster and ammonite fragments, pieces of bone; quartz, chert, and quartzite pebbles from updip; coal fragments, and clay and sandstone rip-up clasts. These lags have significant marine components although they were deposited in a fluvial system. The feather edges of the valleys can contain lag. In this case, the lag might be interpreted as a transgressive lag and the sequence boundary will not be identified. The presence of clast types that are incompatible with the underlying strata often marks the sequence boundary and can help distinguish the sequence boundary from the transgressive surface. For example, quart, chert, and quartzite pebbles concentrated on a thick marine-mudstone section should be suspected to be the product of a sea-level fall, not solely of a sea-level rise.

As the fluvial systems erode basinward, current thinking predicts that sediment is transported through the valley systems toward the lowstand shoreline. The absence of lowstand-shoreline deposits outside of incised valleys in the Sego can be explained in several ways: 1) The Book Cliffs could be landward of the final lowstand-shoreline position for the Sego. Intermediate lowstand shorelines might be eroded by incised valleys during the fall; only the final shoreline position in another part of the basin would be preserved. 2) Thin lowstand deposits could have been deposited outside of the valleys on the interfluves, but were stripped off during the subsequent sea-level rise. 3) Most of the sediment deposited by the fluvial systems could have been stored in the channels so that very little was actually bypassed. Perhaps all of the above occurred in combination. During the fall of 1991 and in the spring of 1992 additional work will address these hypotheses.

The relative sea-level fall was followed by a slow relative sea-level rise. The valleys flood creating estuaries along what becomes a highly embayed coastline. Strong tidal currents are set up within these estuaries. High velocity tidal currents can scour the incised valleys, further augmenting the erosion associated with the sea level fall. In the Lower Sego, tidal deposits have been correlated for 6 miles updip in West Salt Creek Canyon without encountering fluvial deposits or beds influenced by fluvial processes. None-the-less, based on observations made in the Castlegate along the Book Cliffs between Bull Canyon and the Horsepastures (see the guidebook for Day Four) where the fluvial deposits can be traced into tide-dominated estuarine deposits, the Sego tidal deposits are interpreted to have updip fluvial equivalents, probably braided-stream sandstones. Franczyk (1989) states that the Sego Sandstone is temporally equivalent to the middle part of the braided-stream Castlegate Sandstone. During the initial sea-level rise some tidal bars may backstep up the valleys but the interfluves remain above sea level. Swamps, peat bogs with hematitic hard pans, and lakes could form on interfluves between flooded valleys. Soil horizons might develop.

As sea-level rise accelerates, the interfluves also flood. The sea-level rise is probably energetic enough to strip off much of the evidence of subaerial exposure leaving behind indirect evidence of the exposure in the form of concentrations of hematite nodules and coal and charcoal fragments. A sharp surface is created by the flooding with no significant deposits marking the transgression. When the interfluves flood, the embayments which focused the tidal currents are eliminated. Wave processes dominate the shelf until the next sea-level fall.

PALEOGEOGRAPHY AND IMPLICATIONS FOR BASIN-FILL PATTERNS IN EPICONTINENTAL SEAS

The incised valleys in the Lower Sego are oriented north-south to southeast-northwest. The interpretation of this orientation is based on orientation of the valley edges, the direction of the majority of facies changes in the tidal bars, and the direction of imbrication within the tidal bars, for example in San Arroyo Canyon. This transport direction is at nearly right angles to the presumed predominant easterly progradation direction in Cretaceous strata in the western interior seaway (Mallory, 1972). The transport direction is similar to that observed in the Shannon in the Powder River basin (Tillman and Martinsen, 1987), and to the Tocito observed in the San Juan basin (Jennette and others, in this guidebook). In the past, these paleotransport directions have been used to interpret cross-bedded sandstones like the Sego, as offshore bars (Tillman and Martinsen, 1987; Nummedal et al., 1989).

We propose another model for the origin of the "offshore bars" in the Cretaceous seaway based on the Sego Sandstone. This model relies on sea-level falls to periodically reduce the area of the Cretaceous epicontinental basin covered by the seas. During minor high-frequency sea-level falls, only the center of the basin would contain water. During major high-frequency falls the seaway might be completely drained. In this model, a fall in sea level results in a realignment of fluvial-drainage patterns with major, basinwide tectonic elements such as basement-involved faults. In the western interior of the United States these faults trend northwest-southeast and northeast-southwest (Stevenson and Baars, 1986). Fluvial systems incising the shelf would flow in a southerly or northerly direction, parallel to the trend of the dominant fault system, toward the lowstand position of sea level. Because the sea level in the basin has dropped, the fluvial systems would incise widely across strata originally deposited in open-marine environments.

Many of the incised, lowstand fluvial systems would parallel the trend of the old highstand shoreline. Additionally, because the fluvial systems follow fault trends, many of these systems would flow across now exposed shelf strata 10's of miles "basinward" of the old highstand shoreline trend.

As sea level rises, extensive lower-shoreface deposits would form along the margin of the basin. During the sea-level rise, the shoreline would gradually rotate from an east-west orientation during the lowstand to a north-south orientation during the new highstand systems tract. Ultimately, the new highstand position would subparallel the old shoreline. On the shelf, now under water, a sandstone body with the following characteristics would be preserved: 1. elongate geometry, oriented north-south to northwest-southeast, 2. composed of cross beds with abundant mud drapes and siderite nodules, 3. possibly containing marine constituents in lag deposits, especially glauconite, phosphate, and marine mollusk fragments, 4. composed of upward-coarsening successions of beds and bedsets, 5. surrounded by opoen-marine deposits, apparently 10's of miles seaward of correlative shoreline sandstones.

During a fall in sea level, the interaction between tectonics and sequence-boundary formation is enhanced. Fluvial systems incising and bevelling the shelf reduce slight differences in elevation in an attempt to reach a new equilibrium profile. Uplifted strata tend to be "planed off" producing angular discordance beneath sequence boundaries if the timing between the sea-level fall and uplift is correct. Additionally, as mentioned above, incised-valley positions on the shelf commonly are controlled by structures.

The interactions between tectonics and sea level are documented in the paper on the Tocito by Jennette and others (this guidebook) and in the Castlegate road log by Van Wagoner.

CONCLUSIONS

1) Nine high-frequency sequence boundaries are present in the Lower Sego, Anchor Tongue, and Upper Sego along the Book Cliffs in eastern Colorado and western Utah. These sequence boundaries are regional surfaces that have been traced widely along the outcrop. Sequence boundaries are marked by truncation and a basinward shift in facies.

2) Each sequence defined by the sequence boundaries can be subdivided into lowstand, transgressive, and highstand systems tracts. The lowstand systems tracts contain tidal deposits arranged in upward-coarsening tidal-bar parasequences that build tide-dominated deltas. The tidal deposits are associated with minor fluvial deposits which show tidal modification. Both the tidal and fluvial strata are filling incised valleys. Wave-dominated deposits are rare to absent in this systems tract. The lowstand system-tract deposits are bounded above by regional-flooding surfaces termed transgressive surfaces. These surfaces are overlain by the transgressive and highstand system tracts. The strata within these systems tracts are wave dominated. In the Sego, they are arranged in coarsening-upward, lower-shoreface parasequences.

3) Interfluves between incised valleys show indications of subaerial exposure. Hematite-rich clay rip-up clast units are interpreted to record iron-rich deposition in peat swamps.

4) Incised-valley orientation and tidal-bar progradation to the south, southwest, and southeast are consistent with similar orientations observed in the Shannon and Sussex Sandstones in the Powder River Basin, the Tocito Sandstone in the San Juan Basin, and some transport in the Muddy Sandstone in the Denver Basin (Weimer, 1983). Each of these units have been or could be interpreted as incised-valley complexes filled with tidal or fluvial deposits. The north-south to northwest-southeast orientation of incised valleys in the Sego reflects realignment of "normal" easterly paleodrainage patterns developed during highstand systems tract time with loci of minor subsidence controlled by regional basement-involved faults during times of sea-level fall.

5) Incised valleys contain the highest quality reservoir within each sequence. Recognition of high-frequency sequence boundaries and their associated lowstand facies is essential to understand the distribution of potential reservoir within the Sego and to define all of the potential hydrocarbon traps. Once the sequence boundaries have been identified using high-frequency sequence stratigraphy, it is critical to correctly identify, correlate, and map the individual incised valleys as well as incised-valley interactions to predict reservoir distribution in the subsurface. Finally, once a correlation of incised valleys is established, it is necessary to map the distribution of individual tidal-bar parasequences within each valley system to understand differences in reservoir quality.

ACKNOWLEDGEMENTS

The author gratefully acknowledges the assistance of the many geologists who accompanied him in the field to study the Sego over the years. Art Donovan and Fred Zelt assisted during the first field work in September, 1986. It rained hard. Since that time the author worked closely with Norm Corbett, Elijah White, John Suter, Jeff Larson, Dave Jennette, Clive Jones, Richard Lovell, Kim Miskell-Gearhardt, Charles Harris, Christi Harrison, and Barbara Faulkner. Their efforts to help carefully collect data during long hours in the field are deeply appreciated. Ted Lukas spent many days in Sego Canyon collecting the cores from the Exxon Production Research Co. Sego Canyon no. 2. His attention to detail allowed us to acquire one of the finest cores Exxon has ever owned. The author thanks Exxon Production Research Co. for providing the support to conduct Sego field work and to ultimately publish the results. Hundreds of Exxon and Esso geologists have visited the Sego on Exxon schools over the years. Their insightful questions have contributed to the author's understanding of the Sego.

REFERENCES

Blum, M.D., 1991, Systematic controls on genesis and architecture of alluvial stratigraphic sequences: a late Quaternary example, abstract, in D.A. Leckie, H.W. Posamentier, R.W.W. Lovell, 1991 NUNA Conference of High-Resolution Sequence Stratigraphy, August 23-30, 1991, Banff, Canada, pgs. 7-8.

Erdmann, C. E., 1934, The Book Cliffs coal field in Garfield and Mesa Counties, Colorado, U. S. Geological Survey Bulletin 851, 150 p.

Fisher, D. J., 1936, The Book Cliffs coal field in Emery and Grand Counties, Utah, U. S. Geological Survey Bulletin 852, 104 p.

Franczyk, K. J., 1989, Depositional controls on the Late Campanian Sego Sandstone and implications for associated coal-forming environments in the Uinta and Piceance Basins, U. S. Geological Survey Bulletin 1787-F, 17 p., 2 plates.

Gill, J.R., W. J. Hail, Jr., 1975, Stratigraphic sections across Upper Cretaceous Manco Shale-Mesaverde Group boundary, eastern Utah and western Colorado, U.S.Geological Survey Oil and Gas Investigations Chart OC-68, 1 sheet.

Kreisa, R. D., R. J. Moiola, Sigmoidal tidal bundles and other tide-generated sedimentary structures of the Curtis Formation, Utah, Geological Society of America Bulletin, v. 97, p. 381-387.

Mallory, W. M., ed., 1972, Atlas of the Rocky Mountain region: Rocky Mountain Association of Geologists 331p.

Mitchum R.M.,Jr., J. C. Van Wagoner, 1991, High-frequency sequences and their stacking patterns: sequence-stratigraphic evidence of high-frequency eustatic cycles, Sedimentary Geology, v. 70, p. 131-160.

Mutti, E., G. P. Allen, J. Rosell, 1984, Sigmoidal cross stratification and sigmoidal bars: depositional features diagnostic of tidal sandstones, abstract, 5th European Regional Meeting of Sedimentology: International Association of Sedimentologists, Marseille, p. 312-313.

Mutti, E., J. Rosell, G. P. Allen, F. Fonknesu, M. Sgavetti, 1985, The Eocene Baronia tide-dominated delta-shelf system in the Ager basin, *in* Field trip guidebook of the VI European meeting of the International Association of Sedimentologists, Lerida, Spain, excursion 13, p.579-600.

Nio, S. D., C. S. Yang, 1989, Recognition of tidally influenced facies and environments, Short Course Note Series 01, prepared for the Second International Research Symposium on Clastic Tidal Deposits, Calgary, Canada, August 22-25, 1989, International Geoservices B. V., 230 p.

Noe, D. C., 1983, Storm-dominated shoreface deposits, Sego Sandstone (Campanian), northwestern Colorado (abs.), AAPG Bulletin, v. 67, no. 3, p. 454.

Nummedal, D. R., R. Wright, D. J. P. Swift, R. W. Tillman, R. W. Wolter, 1989, Depositional systems architecture of shallow marine sequences, *in* D. Nummedal and R. Wright (eds.), Cretaceous shelf Sandstones and Shelf Depositional Sequences, 28th International Geological Congress field Trip, p. 35-73.

Stevenson, G. M., and D. L. Baars, 1986, The Paradox: A pull-apart basin of Pennsylvanian age: *in* J. A. Peterson (ed.), Paleotectonics and Sedimentation in the Rocky Mountain Region, United States, AAPG Memoir 41, p. 513-539.

Tillman, R. W., R. S. Martinsen, 1987, Sedimentologic model and production characteristics of Hartzog Draw Field, Wyoming, A Shannon shelf-ridge sandstone *in* R. W. Tillman and K. J. Weber (eds.), Reservoir Sedimentology, Society of Economic Paleontologists and Mineralogists Special Publication 40, p. 15-112.

Visser, M. J., 1989, Neap-spring cycles reflected in Holocene subtidal large-scale bedform deposits: a preliminary note, Geology, v. 8, p. 543-546.

Weimer, R. J., 1983, Relation of unconformities, tectonism, and sea-level changes, Cretaceous of the Denver basin and adjacent areas, *in* M. W. Reynolds and E. D. Dolly, (eds.), Mesozoic paleogeography of west-central United States: Rocky Mountain Section , Society of Economic Paleontologists and Mineralogists Rocky Mountain Paleogeography Symposium 2, p. 359-376.

Young, R. G., 1955, Sedimentary facies and intertonguing in the Upper Cretaceous of the Book Cliffs, Utah-Colorado, Geological Society of America Bulletin, v. 55, p. 177-202.

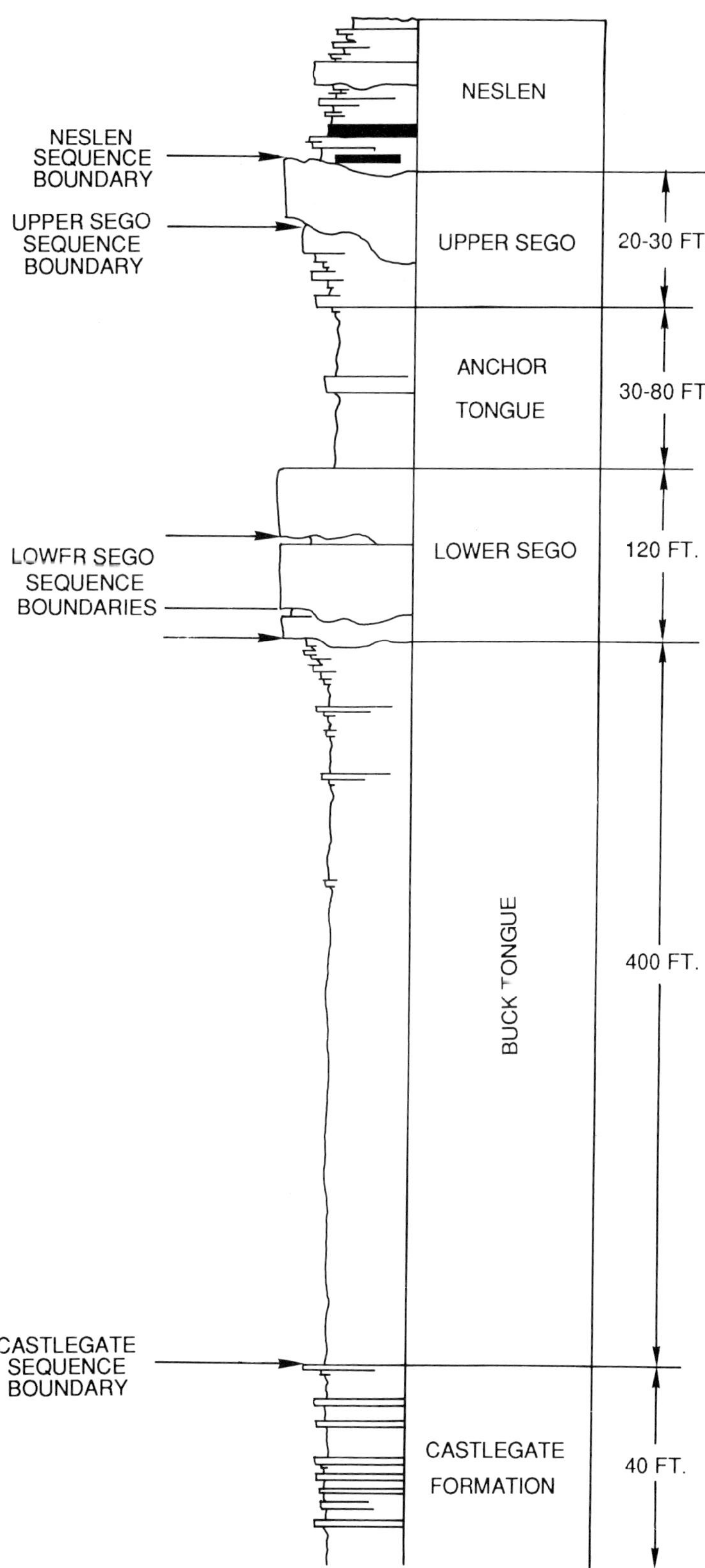

Figure 1: Generalized stratigraphic section through the Lower Sego, Anchor Tongue, and Upper Sego showing the relationship of these stratal units to one another, their thicknesses, and the position of the sequence boundaries in this Campanian section.

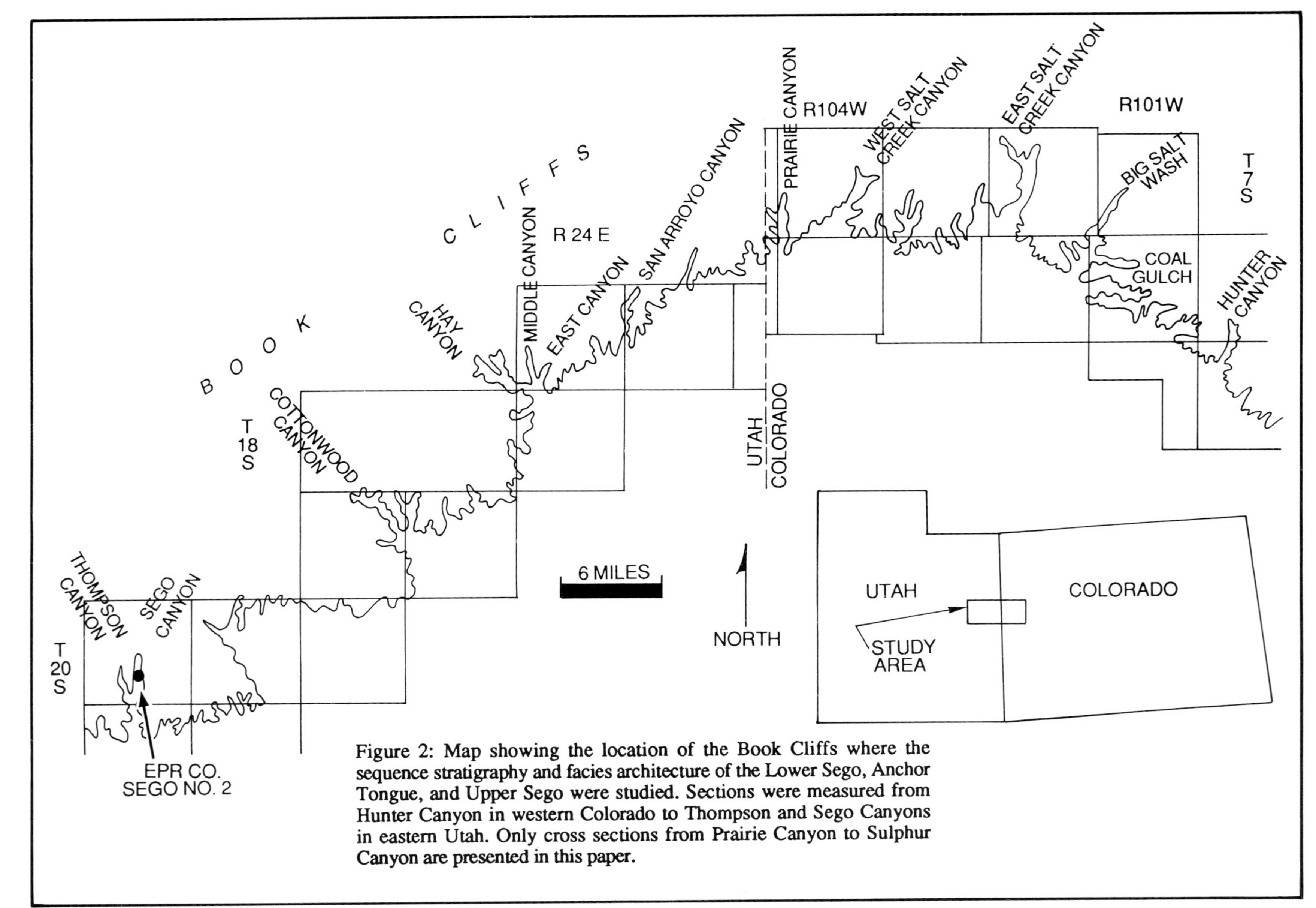

Figure 2: Map showing the location of the Book Cliffs where the sequence stratigraphy and facies architecture of the Lower Sego, Anchor Tongue, and Upper Sego were studied. Sections were measured from Hunter Canyon in western Colorado to Thompson and Sego Canyons in eastern Utah. Only cross sections from Prairie Canyon to Sulphur Canyon are presented in this paper.

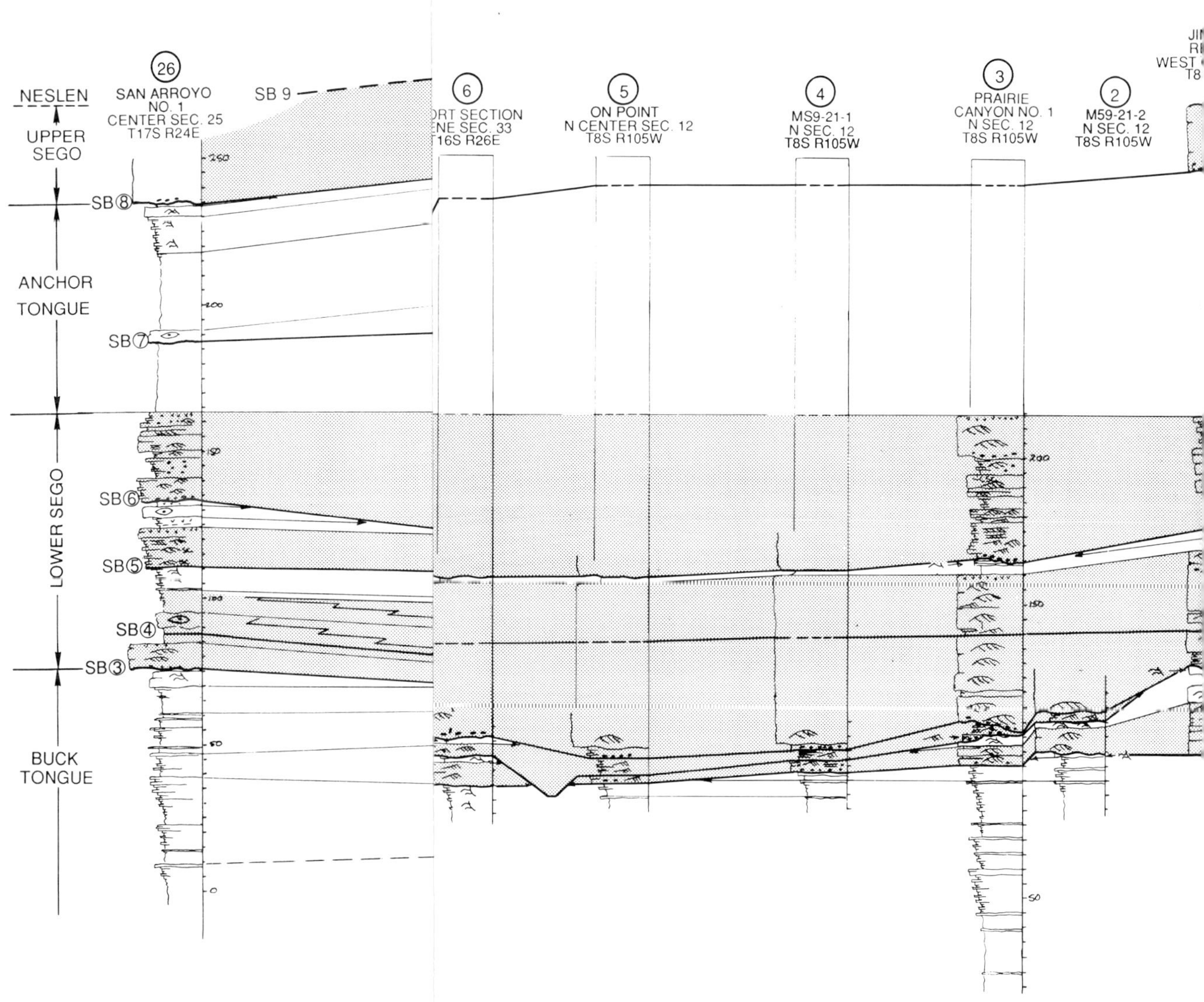

NESLEN
UPPER SEGO
ANCHOR TONGUE
LOWER SEGO
BUCK TONGUE
SB⑧
SB⑦
SB⑥
SB⑤
SB④
SB③
SB 9
26
SAN ARROYO NO. 1
CENTER SEC. 25
T17S R24E
6
ORT SECTION
ENE SEC. 33
T16S R26E
5
ON POINT
N CENTER SEC. 12
T8S R105W
4
MS9-21-1
N SEC. 12
T8S R105W
3
PRAIRIE CANYON NO. 1
N SEC. 12
T8S R105W
2
M59-21-2
N SEC. 12
T8S R105W

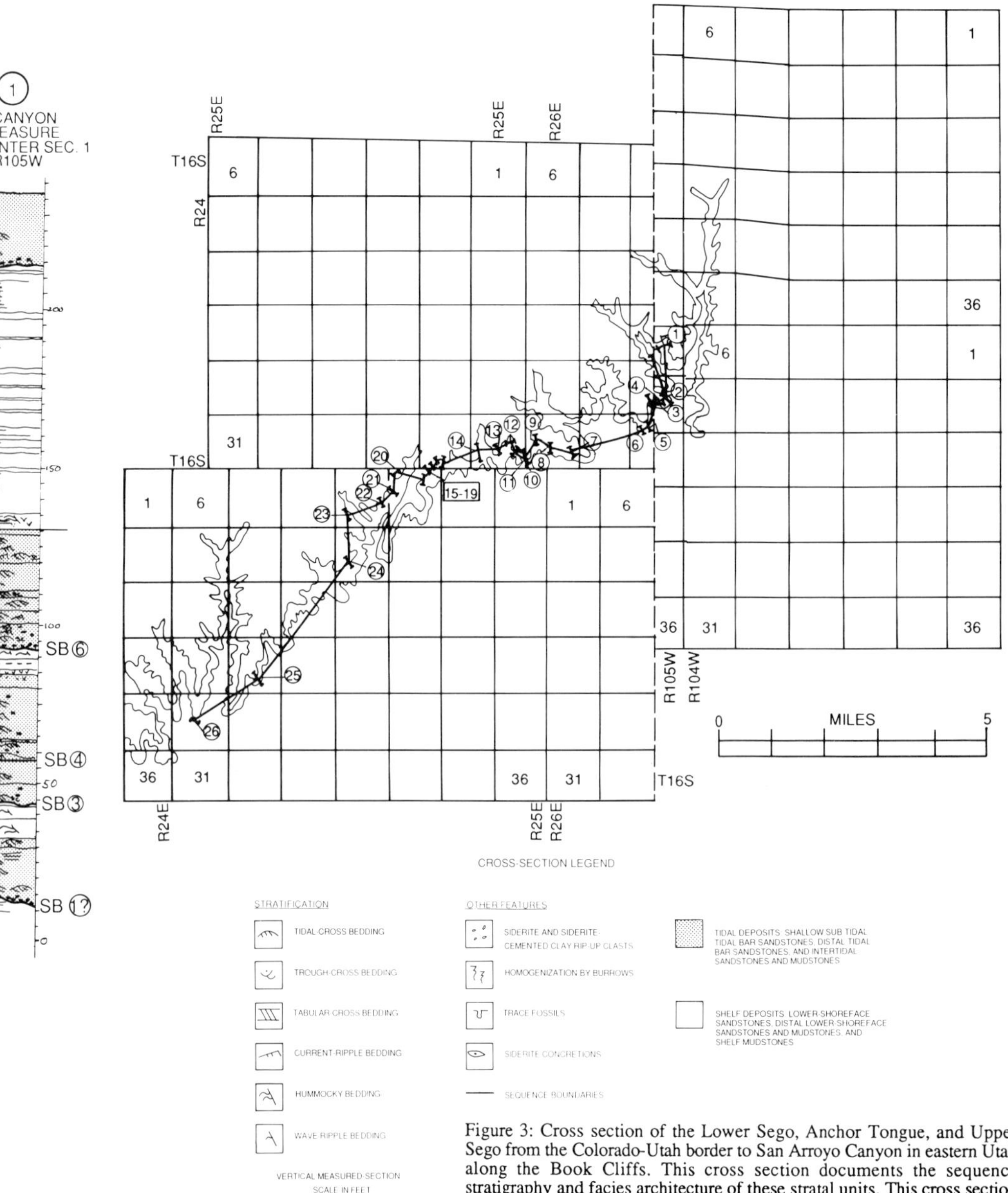

Figure 3: Cross section of the Lower Sego, Anchor Tongue, and Upper Sego from the Colorado-Utah border to San Arroyo Canyon in eastern Utah along the Book Cliffs. This cross section documents the sequence stratigraphy and facies architecture of these stratal units. This cross section is continued on Figure 4.

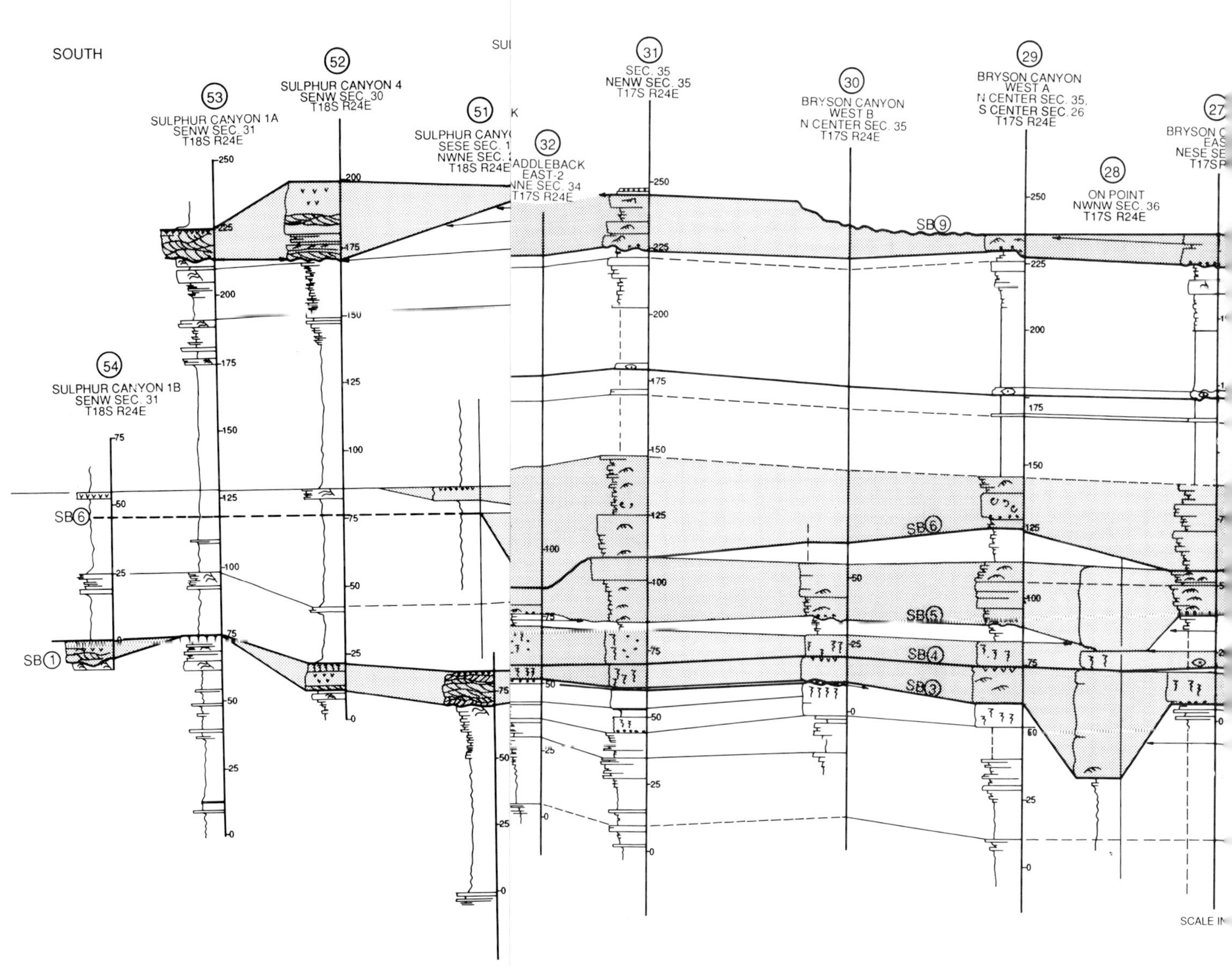

SOUTH
53
SULPHUR CANYON 1A
SENW SEC. 31
T18S R24E
52
SULPHUR CANYON 4
SENW SEC. 30
T18S R24E
54
SULPHUR CANYON 1B
SENW SEC. 31
T18S R24E
31
SEC. 35
NENW SEC. 35
T17S R24E
30
BRYSON CANYON
WEST B
N CENTER SEC. 35
T17S R24E
29
BRYSON CANYON
WEST A
N CENTER SEC. 35,
S CENTER SEC. 26
T17S R24E
28
ON POINT
NWNW SEC. 36
T17S R24E
SB⑨
SB⑥
SB⑤
SB④
SB③
SB①
SCALE IN

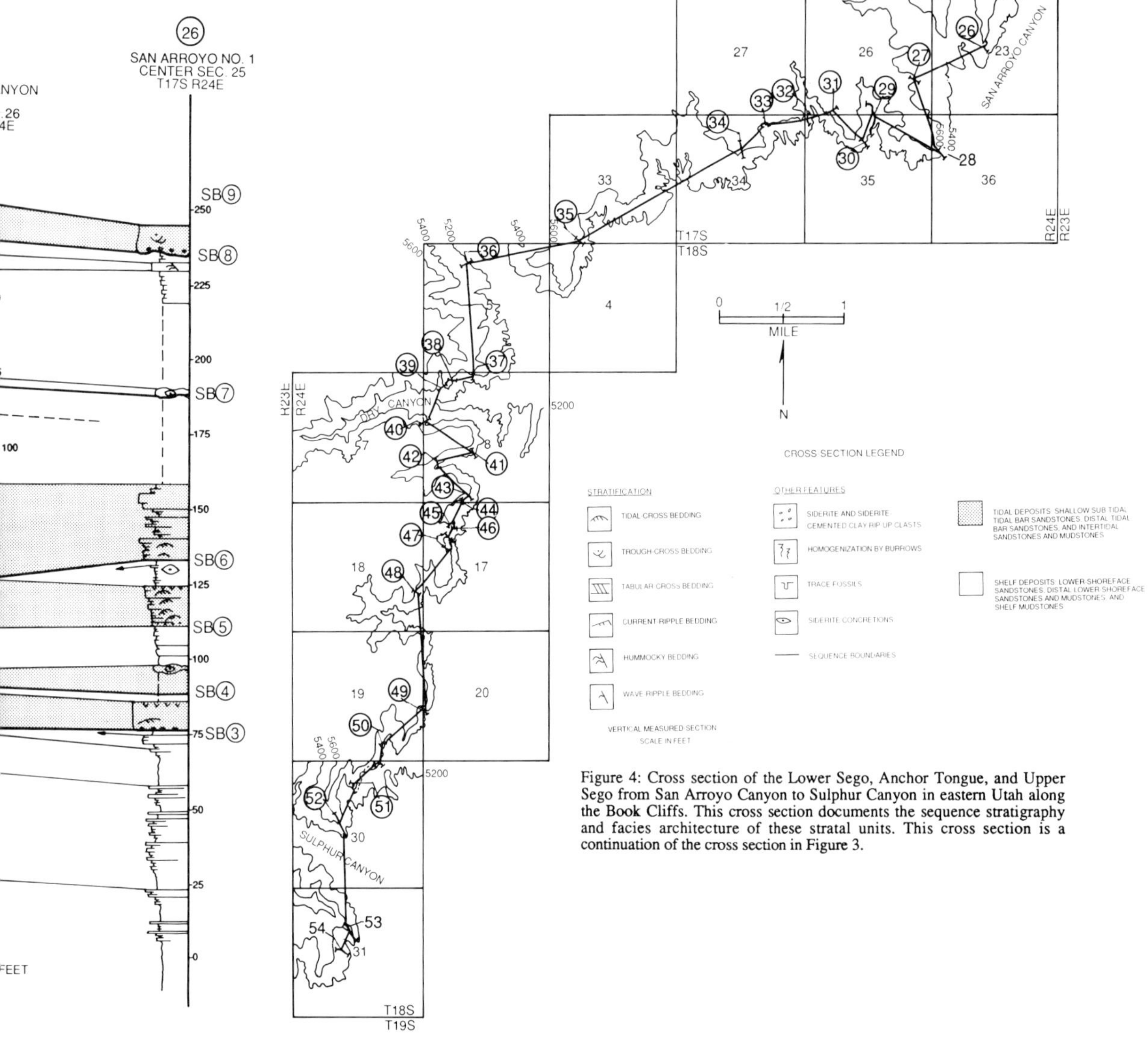

Figure 4: Cross section of the Lower Sego, Anchor Tongue, and Upper Sego from San Arroyo Canyon to Sulphur Canyon in eastern Utah along the Book Cliffs. This cross section documents the sequence stratigraphy and facies architecture of these stratal units. This cross section is a continuation of the cross section in Figure 3.

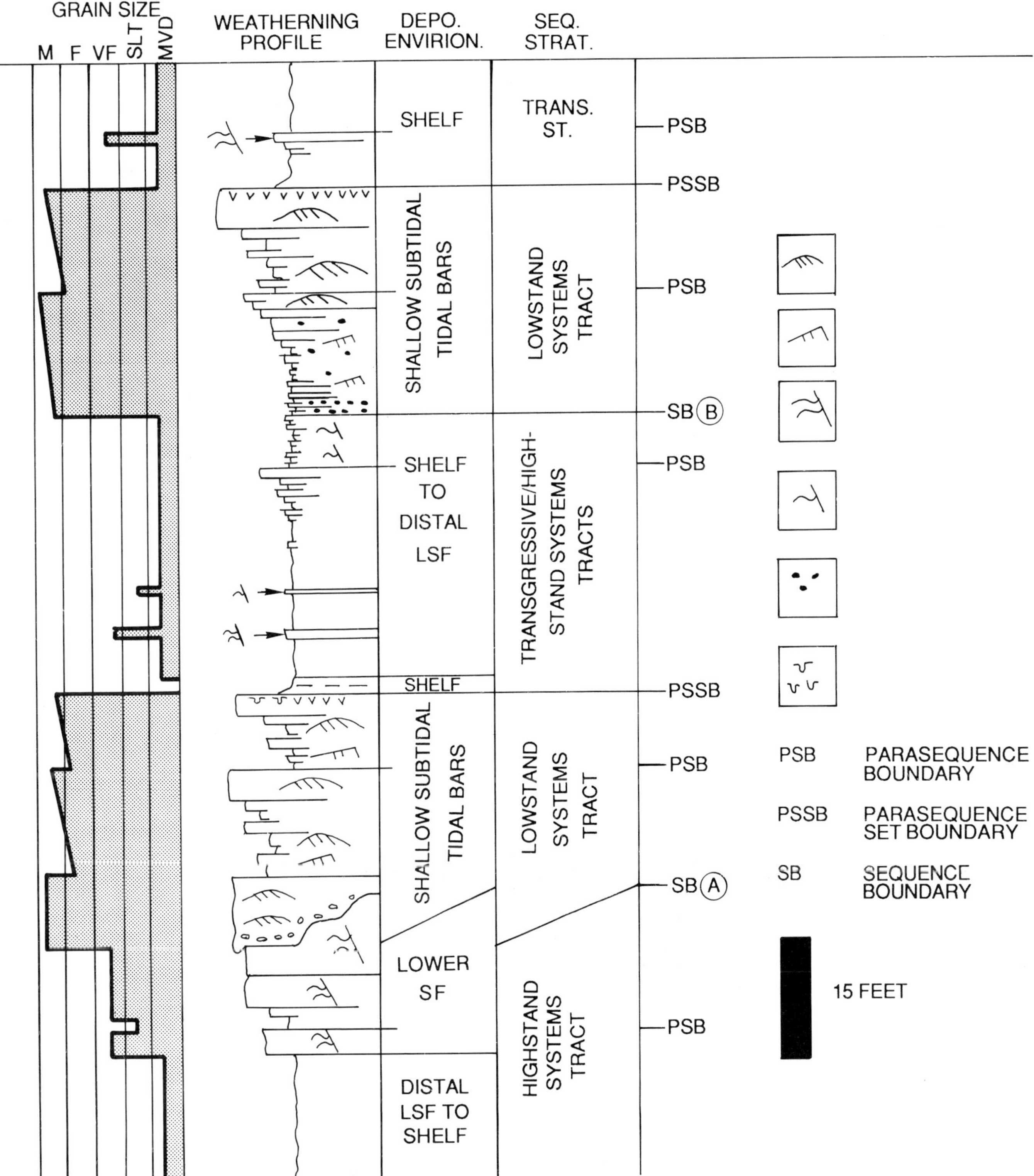

Figure 5: General model of the major surfaces and vertical-facies associations found in the sequences observed in the Lower Sego, Anchor Tongue, and Upper Sego along the Book Cliffs in eastern Utah and western Colorado.

Figure 6: **A.** High-energy cross beds interpreted as tidal deposits rest erosionally on gray-marine mudstones. This contact is a basinward shift in facies and marks a sequence boundary shown by the arrow. A parasequence boundary exists at the top of the continuous 6-inch thick sandstone bed behind the geologists head.The staff next to the geologist is 5-feet long. The sequence boundary is at 133 feet in measured section 49 (Fig. 4). **B.** Large-scale cross beds resting erosionally on lower-shoreface hummocky strata. This contact is a basinward shift in facies marking a sequence boundary shown by the arrow. The staff in the geologists hand is 5-feet long. The sequence boundary is at 45 feet in measured section 15 (Fig. 3). **C., D.** Two examples of basinward shifts in facies marking sequence boundaries. In both cases, large-scale cross beds interpreted as tidal deposits erode distal lower-shoreface, thin sandstones and mudstones. The arrow in C marks the base of the Upper Sego in San Arroyo Canyon. The arrow in D marks sequence-boundary 6 in Bull

Figure 7: **A to C.** Sequence-boundary expression at the bases of tidal-bar parasequences indicated by the arrows. The sequence boundaries are marked by a grain size increase from very fine-sandstone below to upper fine-grained sandstone above. The surfaces defined by these contacts can be traced regionally. The sequence boundary in A is at 90 feet in measured section 11 (Fig. 3). The arrow in B marks sequence-boundary 5 at 111 feet in measured section 26 (Fig. 3 or 4). The arrow in C marks sequence-boundary 6 in a section measured in NE SE Sec. 4 T20S R21E in eastern Utah. **D.** Truncation at the base of sequence-boundary 6 at 90 feet in measured section 37 (Fig. 4).

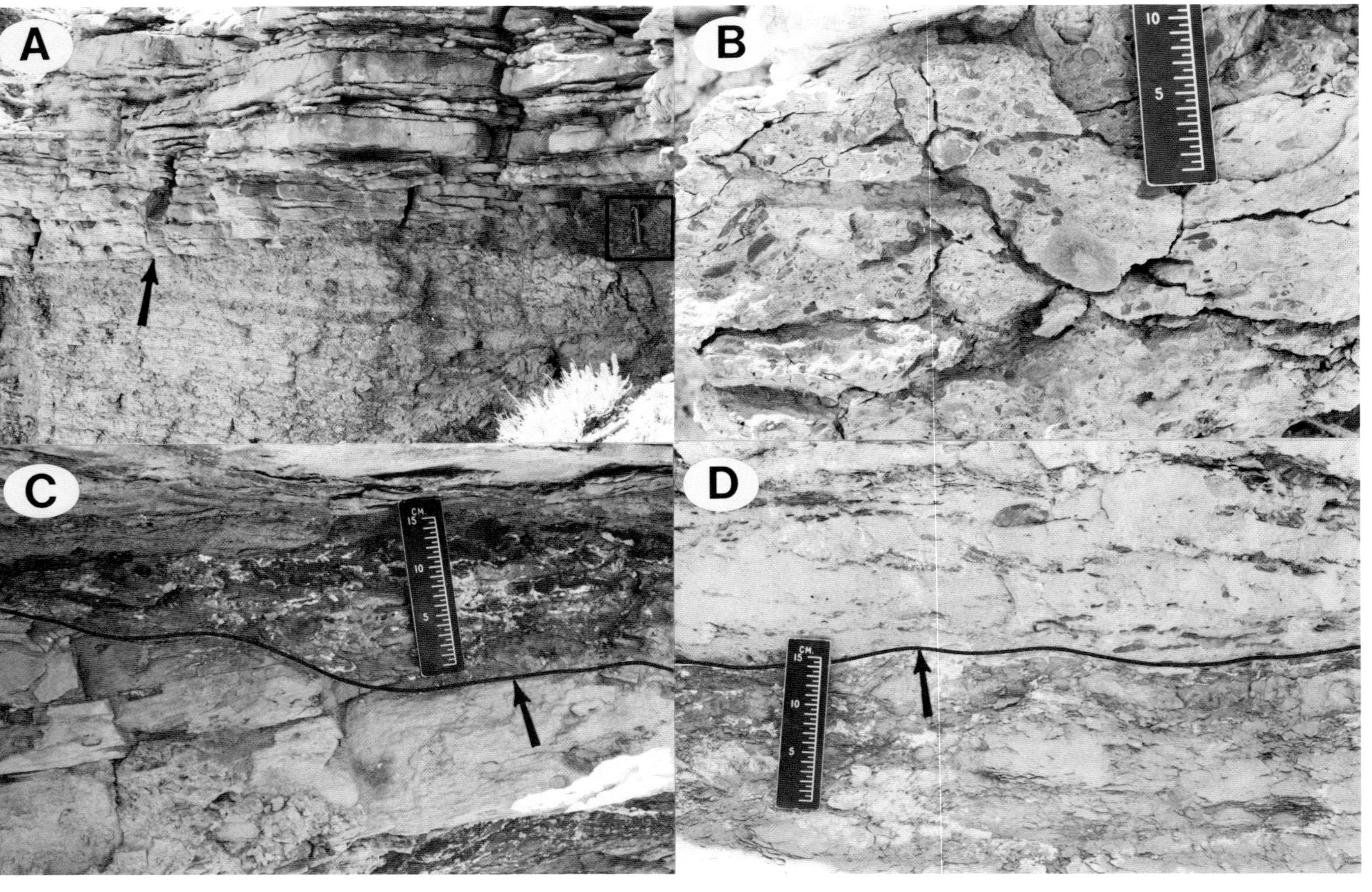

Figure 8: **A.** Sequence-boundary 6, indicated by the arrow, marked by truncation, a basinward shift in facies, and a basal-channel lag located in a measured section at NE NE Sec. 2 T20S R22E. **B.** Close up of the lag marking the sequence boundary in the previous photo. Bone framents, coal and charcoal clasts, oyster and ammonite fragments, and pelecypods are some of the clast types found in this lag. **C.** Truncation and a red clay-clast lag resting on a burrowed lower-shoreface siltstone marking a sequence boundary, indicated by the arrow and the black line. The sequence boundary occurs at 116 feet in measured section 38 (Fig. 4). **D.** Sequence-boundary 6 at 92 feet in measured section 1 (Fig. 3). The sequence boundary in the photo is marked by a basinward shift in facies and red clay rip-up clasts. It is indicated by the arrow and the black line.

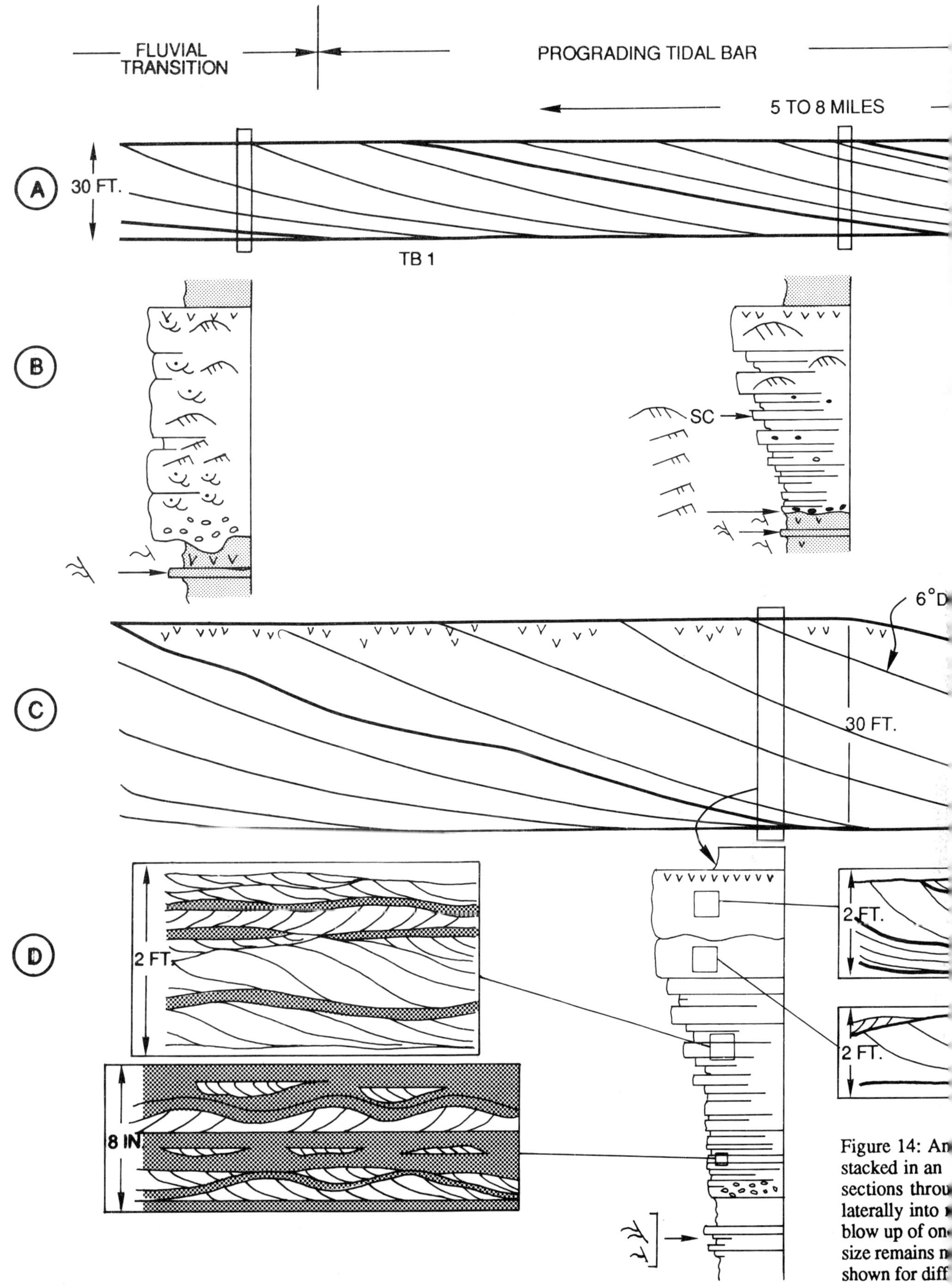

Figure 14: An
stacked in an
sections throu
laterally into
blow up of on
size remains n
shown for diff

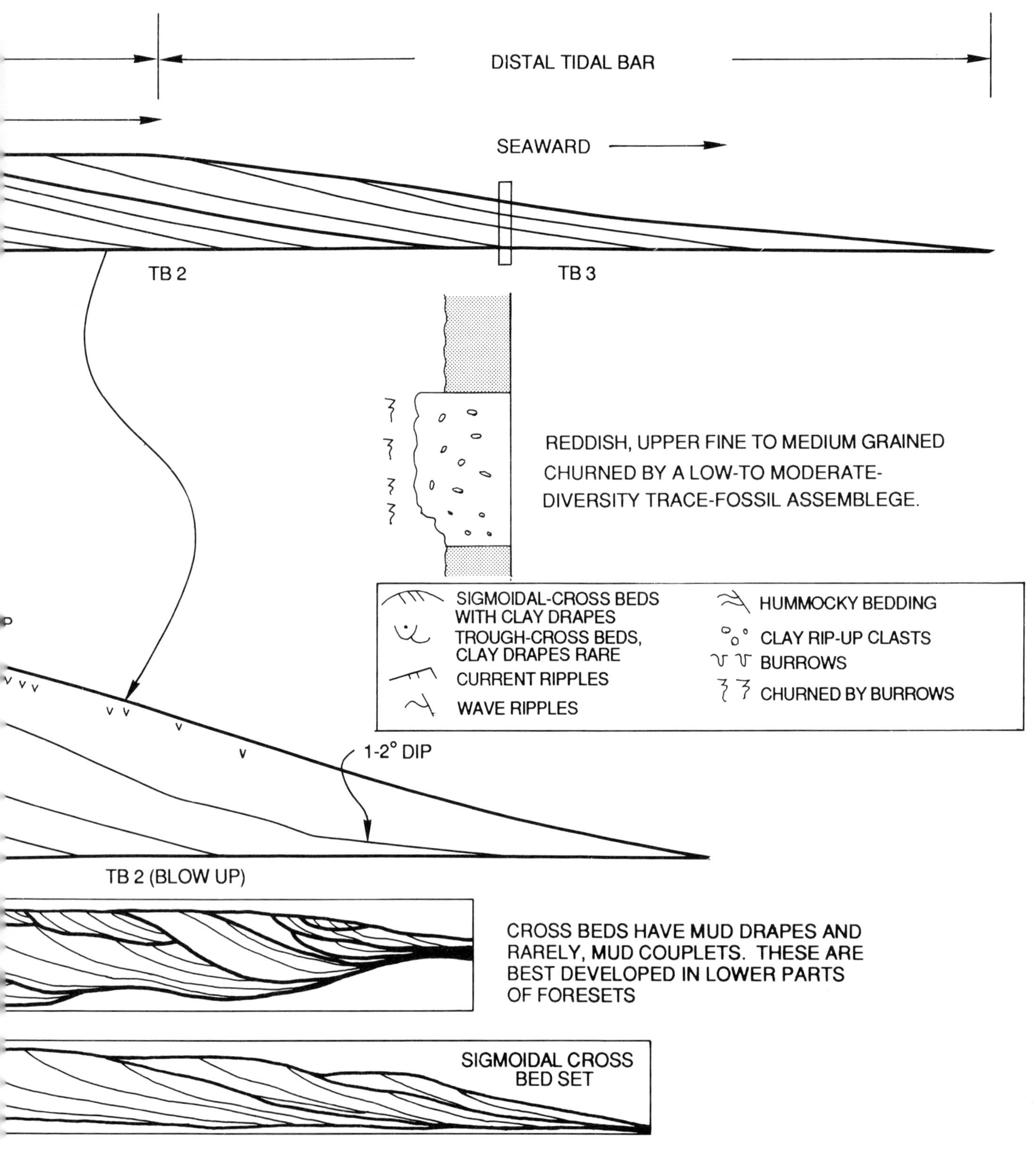

ıtomy of a typical tidal-bar parasequence in the Lower Sego. **A.** Geometry of 3 tidal bars ımbricate pattern. Clinoforms are well developed with each tidal bar. **B.** 3 different vertical ;h the tidal-bar complex. The most distal facies thins basinward by downlap and grades ıarine mudstones. No wave-dominated beds are found in the tidal bar, even downdip. **C.** A tidal bar showing the clinoform pattern. **D.** A blow up of the tidal-bar parasequnence. Grain edium grained from bottom to top, mudstone content decreases upward. Typical bed types are erent parts of the tidal bar.

Figure 9: A. Part of a 2-foot thick bed of dark red, hematite-rich clay clasts, indicated by the arrows, and charcoal fragments in bioturbated marine silty mudstones. These clasts are interpreted to be the remnants of a soil horizon developed at sequence-boundary 2 on an interfluve. These clasts are at 50 feet in measured section 32 (Fig. 4). B. Low-angle truncation of thin, hummocky-bedded sandstones by an incised valley filled with fluvial strata. The surface indicated by the arrow is sequence-boundary 2 at 26 feet on measured section 33 (Fig. 4). C. Outcrop in the Saddleback East area, where sections 32 and 33 were measured(Fig. 4). The photo shows an incised valley resting on sequence-boundary 2 . Number 1 indicates the sequence boundary shown in B at the base of the valley fill. Number 2 points to the valley edge. Number 3 points to the transgressvie surface marking the flooding over the top of the valley, creating a parasequence set boundary. This location is documented in more detail in Figure 10.

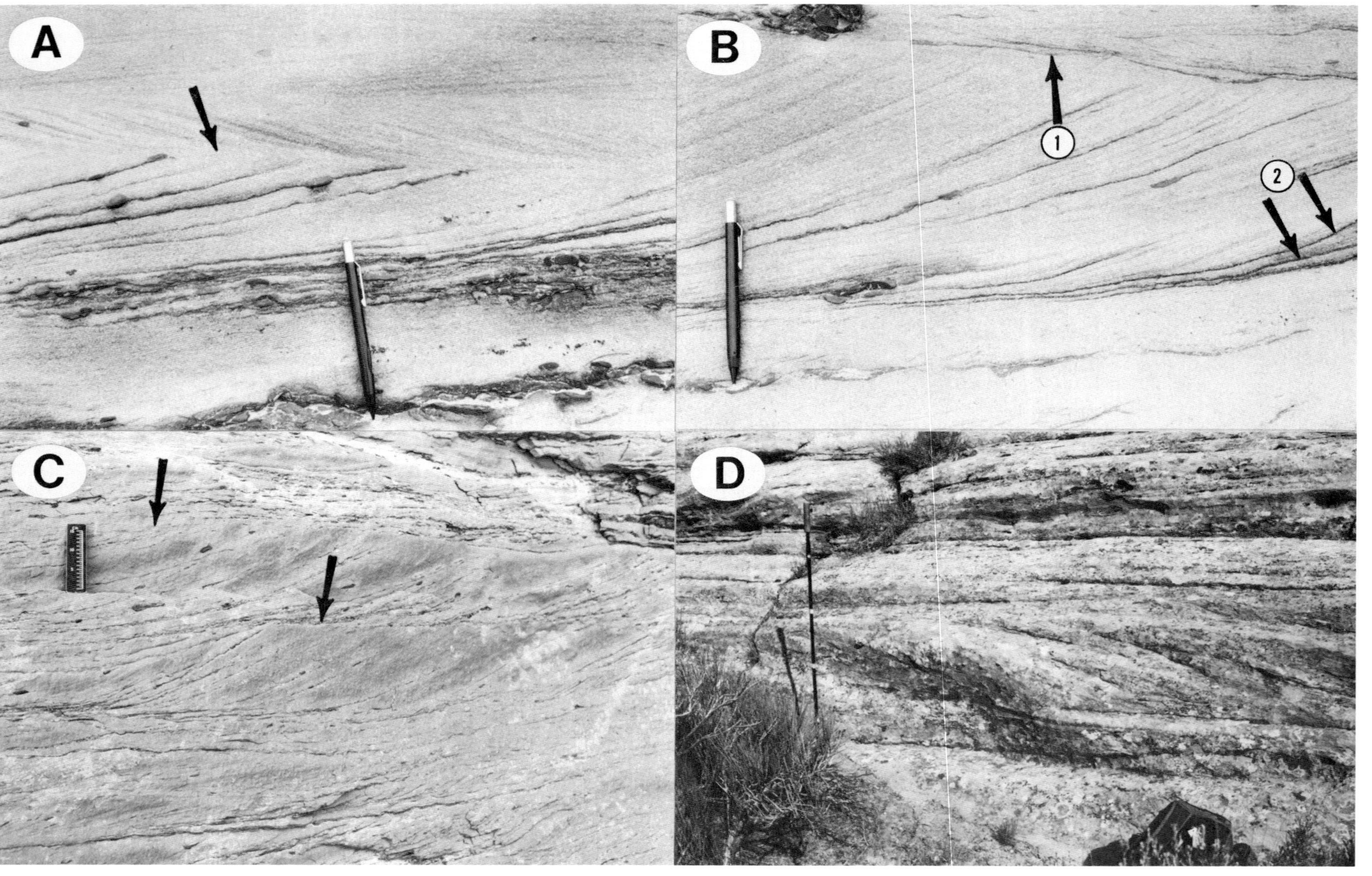

Figure 11: Cross beds within the Lower Sego. **A.** Bidirectional cross beds. The surface indicated by the arrow is interpreted to be a bidirectional reactivation surface. Clay drapes are common on foreset laminae. The toes of the lowest cross bed at the middle of the pencil are very tangential, draped with mud and organic material, and contain siderite- and hematite-rich clay clasts. The pencil is 6-inches long. **B.** Cross beds showing mud couplets (number 2) and a bidirectional reactivation surface (number 1). A unidirectional reactivation surface is present just below the number 2. **C.** Unidirectional reactivation surfaces in cross bedsets. These cross beds also exhibit clay drapes on the foresets and clay clasts. **D.** Sigmoidal cross beds. The staff is 5-feet long.

Figure 12: **A.** An upward-coarsening tidal-bar parasequence located between 38.5 to 50 feet in measured section 27. **B.** Three upward-coarsening tidal-bar parasequences in the Lower Sego in Bitter Creek Canyon, SE Sec. 34 T16S R25E. The vertical-scale bar is 10 high. **C.** Three tidal bars exhibiting an imbricate stacking pattern forming a tidal-bar complex. Basinward is to the right. Tidal-bar 1 is the oldest. It dips to the left and is overlain by tidal-bar 2. Above the number-1 arrow, tidal-bar 2 is at the top of the complex. Tidal-bar 2 also dips to the left and is located in the middle part of the complex below the number 3. Tidal-bar 3 thins to zero at the top of the tidal-bar complex by the tree to the left of the number-2 arrow and thickens to the left. At the tip of arrow number 3, tidal-bar 3 represents about half of the total tidal-bar complex thickness. The tidal-bar complex is 35-feet thick. It is located between 20 and 55 feet in measured section 13 on Figure 3.

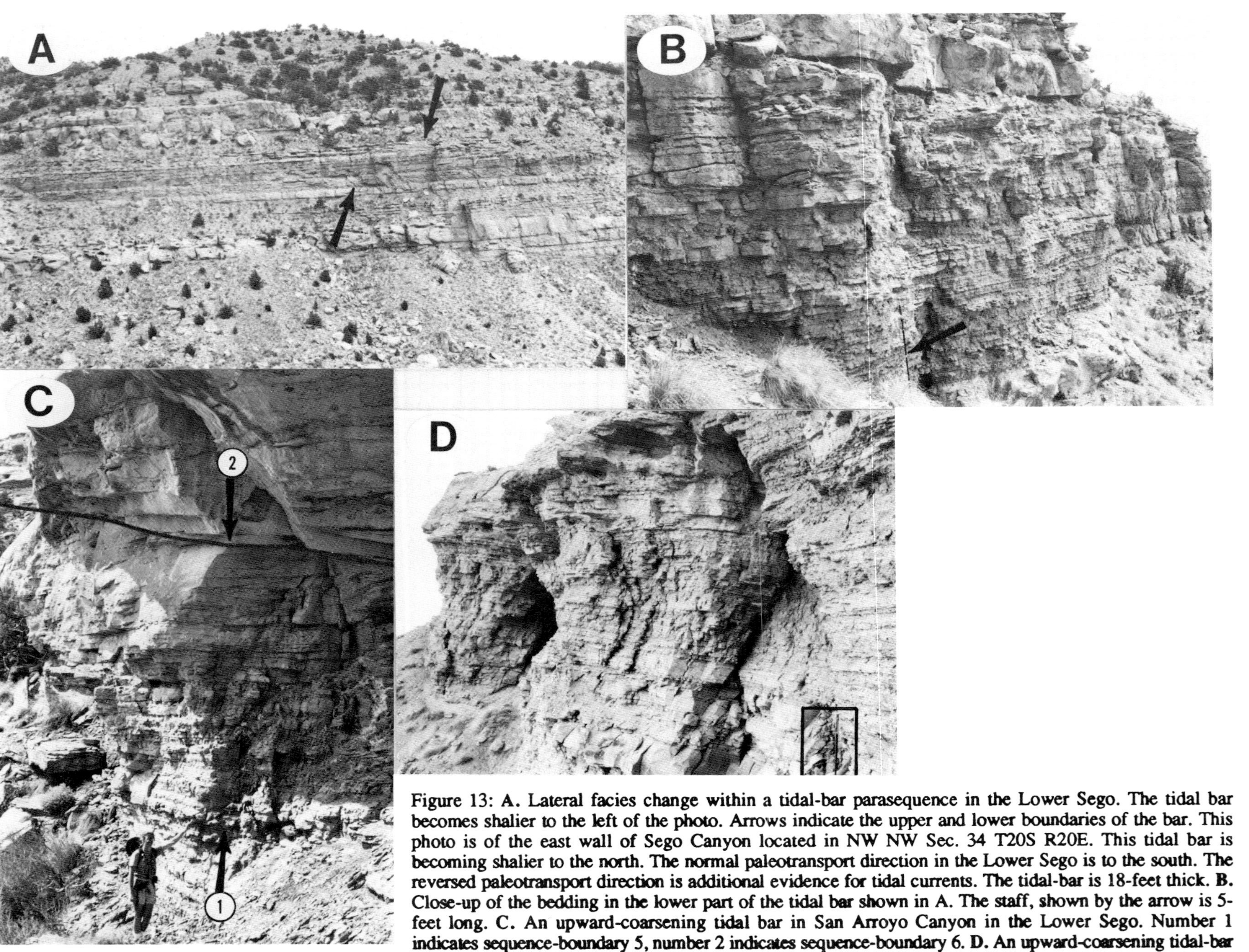

Figure 13: **A.** Lateral facies change within a tidal-bar parasequence in the Lower Sego. The tidal bar becomes shalier to the left of the photo. Arrows indicate the upper and lower boundaries of the bar. This photo is of the east wall of Sego Canyon located in NW NW Sec. 34 T20S R20E. This tidal bar is becoming shalier to the north. The normal paleotransport direction in the Lower Sego is to the south. The reversed paleotransport direction is additional evidence for tidal currents. The tidal-bar is 18-feet thick. **B.** Close-up of the bedding in the lower part of the tidal bar shown in A. The staff, shown by the arrow is 5-feet long. **C.** An upward-coarsening tidal bar in San Arroyo Canyon in the Lower Sego. Number 1 indicates sequence-boundary 5, number 2 indicates sequence-boundary 6. **D.** An upward-coarsening tidal-bar parasequence in the Upper Sego at 220 feet in measured section 47, Figure 4. The geologist in the box is

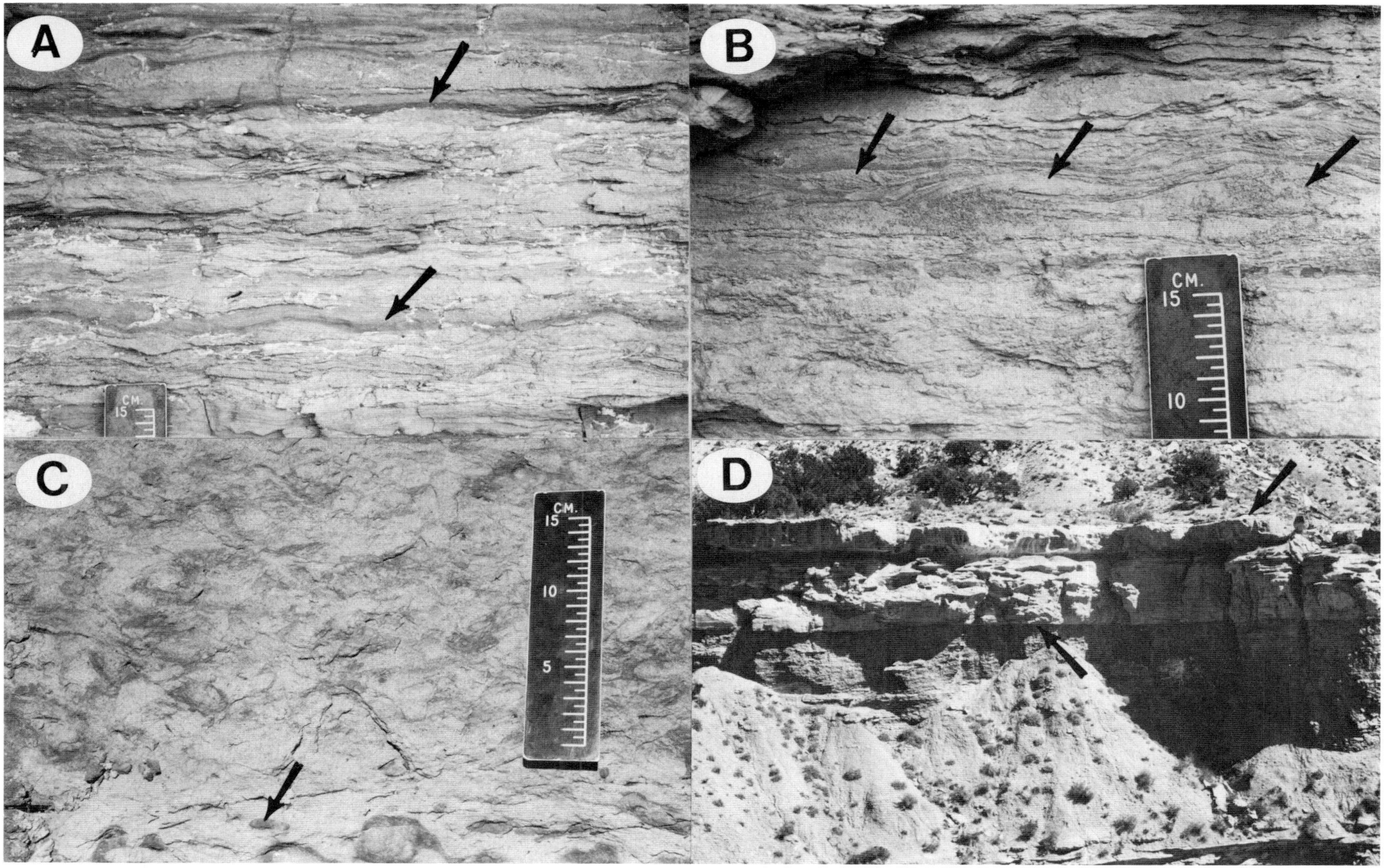

Figure 15: **A.** Bedding in the lower part of the tidal-bar parasequence illustrated in Figure 13d. The mud drapes, indicated by the arrows are continuous. Upper fine-grained current ripples are interbedded with the drapes. **B.** Current ripples in the lower part of the tidal-bar paraseqence shown in Figure 13d. These ripples indicate reverse flow. The ripples are highlighted by the arrows. **C.** The facies typically found in the distal portions of tidal bars. This bioturbated, upper fine-grained sandstone is red and contains hematite-rich clay clasts as well as clasts of charcoal and bones. **D.** The incised-valley fill between 63 and 95 feet in measured section 50 (fig. 4) bounded above by the transgressive surface at the base of the Anchor Tongue indicated by the upper arrow. The lower arrow indicates sequence-boundary 6.

Figure 16: **A.** Highstand distal lower-shoreface strata beneath sequence-boundary 6 in measured section 50. The highstand strata in this area consist of silty mudstones and thin, interbedded very fine-grained, hummocky-bedded sandstones. The staff is 5-feet long. **B.** Hummocky bedding in the late highstand systems tract located between 88 and 90 feet in measured section 20, Figure 4, just below the Upper Sego sequence boundary. **C.** Late highstand, lower-shoreface hummocky strata showing gutter channels. A large gutter channel is indicated by the arrow. These beds are at the top of the Buck Tongue in San Arroyo Canyon. Staff in the geologist's hand is five-feet long. The black tape and upper arrow indicate sequence-boundary 2. **D.** A gutter channel filled with hummocky strata in the late highstand systems tract. This photo is located in a measured section at Cottonwood Canyon at SW SE Sec. 7 T19S R22E near the top of the Anchor Tongue, 3 feet below the Upper Sego sequence boundary.

DAY TWO

Central Utah

ROAD LOG DAY 2: HIGH-FREQUENCY SEQUENCE STRATIGRAPHY AND FACIES ARCHITECTURE OF THE KENILWORTH MEMBER OF THE BLACKHAWK FORMATION, BOOK CLIFFS, UTAH

by
David R. Taylor
Esso Resources Canada Limited, Calgary, Alberta

OBJECTIVE: During Day 2 of the field trip we will examine the stratigraphic architecture of the Kenilworth Member of the Blackhawk Formation and demonstrate how high-frequency cyclicity is expressed in wave-dominated shoreline deposits. Participants will be shown sedimentary facies, parasequence stacking patterns, regional truncation surfaces and lateral changes in depositional environments. Discussion will focus on how these elements are interpreted within a sequence stratigraphic framework.

The Kenilworth Member is Campanian in age (Speiker, 1931; Young, 1957; 1955) and occurs near the middle of the Blackhawk Formation (Fig. 2-1). During the day we will make five stops (Fig. 2-2), beginning approximately half-way along the lateral extent of the Kenilworth exposures. The objective will be to trace the sandstones in a basinward direction, observing sequence stratigraphic and facies relationships.

0.0 - Depart Green River, Utah from the parking lot of the West Winds Rodeway Inn Motel and drive west along Main St. (Loop 70). Turn west on Interstate 70.
4.3 - Turn north on Hwy 6/50 towards Price, Utah. The Kenilworth Member is the lowest cliff-forming sandstone succession exposed on the east side of the highway.
26.9 - Pass Price River Canyon access road (Stop 3) on the right-hand side of Hwy 6/50. There is an old abandoned gas station with a blue and red sign that says "Food and Fuel" on the west side of the highway just before the access road.
49.1 - Turn east on Hwy 123 at Sunnyside Junction and drive toward the town of Sunnyside.
59.0 - STOP ONE, A: Sunnyside Municipal Park, Sunnyside, Utah. Turn right and stop in the parking lot. Looking towards the north across the road, the Kenilworth Member is the lowermost series of sandstone ledges above the houses. The purpose of this stop is to give an overview of the parasequences in the Kenilworth Member.

Description:

At Stop 1a, the Kenilworth can be seen above the houses north of Hwy 123 as a series of sandstones at the base of the cliffs (Fig. 2-3). Five parasequences are present which can be traced along the cliff-face. The lowermost parasequence (PS 1) is a thin, resistive sandstone ledge within the shale-dominated slope. Above that ledge is Parasequence 2 which is a coarsening-upward succession with a thick, blocky sandstone at the top. A recessive shale overlies PS 2 and marks the marine-flooding surface at the base of PS 3. Parasequences 3 and 4 are present in the uppermost sandstone ledge of the Kenilworth and are best examined on the promontory to the left of the cliff-face. Looking closely, PS 3 can be seen to coarsen-upward from a shale into a blocky sandstone. There is a thin "whitecap" within the sandstone succession which represents the upper shoreface/foreshore, and above this is a light yellowish-brown recessive-weathering sandstone overlain by a second whitecap sandstone. The flooding surface between PS 3 and PS 4 is at the top of the lower whitecap. At the base of the large recessive slope above PS 4 are some coal-bearing beds of the coastal plain and these are overlain by light brown marine sandstones and shales within PS 5. The Sunnyside Member overlies these light brown beds and forms a coarsening-upward succession capped by the the prominent cliff-forming reddish colour sandstone.

Exit the parking lot and turn right (east) on Hwy 123. Note the Kenilworth dipping toward the highway.
60.3 - STOP ONE, B: Whitmore Canyon near Sunnyside, Utah. Turn left and stop in the parking area of "Hard Bodies" aerobics center. The Kenilworth is the thick sandstone succession exposed

behind the houses on the north side of the road. By climbing the knoll to the west of the parking area you can look south across the road to the Sunnyside Mine property. The Kenilworth is the lowermost series of sandstones above the roadway on the mine property. The purpose of this stop is to examine the facies relationships in a measured outcrop section of the Kenilworth.

Description:

At Stop 1b we will examine the sedimentary facies and parasequence stacking patterns of PS 3, 4 and 5 which were observed in the cliff-face at Stop 1a. In addition, there is clear evidence for the presence of an erosional surface within the Kenilworth which truncates PS 4. A measured section for this locality is shown in Figure 2-4, which represents a composite vertical profile of the facies present in the immediate area of Whitmore Canyon. The lowermost part of the measured section is exposed in the tributary canyon to the north and cannot be examined behind "Hard Bodies". The section starts in PS 3 at about 100-120 feet on the measured section with a thick, amalgamated hummocky cross-stratified (HCS) lower shoreface sandstone. The interval coarsens-upward into a sandstone characterized by large, gently curved, low-angle laminations followed vertically by a trough cross-bedded upper shoreface sandstone with a few *Ophiomorpha* burrows. The top of this sandstone is heavily bioturbated to churned by *Ophiomorpha* and marks the flooding surface at the top of PS 3. Parasequence 4 comprises HCS sandstone and bioturbated sandstone of the lower shoreface which are overlain by a thin, trough cross-bedded upper shoreface sandstone. Erosively overlying this upper shoreface sandstone is a fine- to medium-grained, large-scale trough cross-bedded sandstone with mudclasts at the base. This sandstone is interpreted as a fluvial channel-fill which cuts into the lower shoreface of Parasequence 4 (Fig. 2-5).

The uppermost part of the measured section is exposed in the gully behind the abandoned day care center. Although the exposure is poor, the sandstones and shales overlying the cross-bedded fluvial sandstone are highly carbonaceous, and have prominent root traces and *Planolites* burrows. This succession is interpreted as having been deposited in a coastal plain/lagoonal environment. There is a thin coal at the top of the coastal plain succession which is overlain by interbedded wave rippled and bioturbated sandstones, with small-scale HCS occurring in some places.

To further examine the incision at the top of PS 4, walk to the top of the knoll to the west of the entrance to "Hard Bodies" and look south across Hwy 123 to the mine property. In the cliff-face, truncation of PS 4 can be observed along the entire exposure (Fig. 2-6).

Sequence Stratigraphic Interpretation of Stops 1a and 1b:

The outcrops in the vicinity of Whitmore Canyon show the progradational stacking pattern of Parasequences 1 through 4. These parasequences are interpreted as part of the highstand systems tract, with successively younger parasequences stepping further basinward due to decreased accommodation in the latter part of the highstand. The erosional surface at the top of PS 4 cuts down into lower shoreface deposits and is interpreted as a sequence boundary that formed during a fall in relative sea-level. A basinward shift of facies occurs above this surface, with fluvial and coastal plain/lagoonal sediments resting on lower shoreface deposits. The facies overlying the sequence boundary are interpreted as part of the lowstand systems tract which back-filled an incised valley. The transgressive surface overlies the coal bed and Parasequence 5 is stacked retrogradationally near the top of the Kenilworth Member. Parasequence 5 is interpreted to have been deposited in the transgressive systems tract and it is overlain by a maximum flooding surface at the base of the Sunnyside Member. Parasequence 5 can be traced to the west (paleo-landward) where it becomes a thick, coarsening-upward beach parasequence in the vicinity of Rock Creek Canyon. To the east of Whitmore Canyon, PS 5 downlaps out on top of PS 4 and can no longer be recognized.

Leave Whitmore Canyon (Stop 1b) and turn right (east) along Hwy 123 to Hwy 6/50.
71.4- Turn left (south) on Hwy 6/50.
91.2- Follow milepost signs along Hwy 6/50 to Milepost 276; drive south another 0.1 miles to a gate on the west side of the highway, turn left and follow the dirt track.
91.7- STOP TWO: Overview of cliff exposures near Woodside, Utah. Stop near rise in dirt track. The purpose of Stop 2 is to examine the erosional surface near the top of the Kenilworth and the

facies changes in the underlying parasequences.

Description:

The Kenilworth Member is continuously exposed in this area of the Book Cliffs for about 6-8 miles. Several normal and reverse faults offset the section, but the Kenilworth can easily be traced with binoculars as the lowermost series of sandstone ledges. Parasequences 1, 2, and 3 are poorly developed at this locality as it is basinward of Stop 1. The main cliff-forming sandstone in the Kenilworth is PS 4.

From right to left (south to north) along the cliffs, the massive shoreface sandstone of PS 4 changes facies and becomes less sand-prone. An erosional surface truncates the top of PS 4 and is overlain by a sandstone displaying large-scale lateral accretion surfaces (Fig. 2-7a). This sandstone is up to 60 feet in thickness and is interpreted as a (fluvial) channel-fill that rests on offshore transition/distal delta front deposits. To the north along the outcrop, channel-forms can be recognized which appear to be more shale-dominated (Fig. 2-7b). The stratigraphic position of the erosional surface and channel-fills is the same as at Stop 1b.

Sequence Stratigraphic Interpretation:

The erosional surface at the top of PS 4 in this locality is the sequence boundary that occurs at the base of the fluvial channel-fill sandstone at Stop 1b. This surface has regional extent, both along the cliffs at Stop 2 as well as over the 17 miles between Stops 1 and 2. Channels are not ubiquitous along this surface as we will see at the next stop. A basinward shift of facies occurs above the sequence boundary, with fluvial sandstones resting directly on offshore transition/distal delta front deposits. The observations at Stops 1 and 2 indicate that the erosional relief on the sequence boundary is caused primarily by fluvial incision.

The facies change observed in PS 4 reflects basinward progradation during the highstand. Paleocurrent data and facies mapping show an easterly/east-northeasterly paleoslope; however, changes in the orientation of the Book Cliffs with respect to the paleoshoreline orientation expose the basinward facies change at this stop. Not considering the orientation of the outcrops with respect to the paleoshoreline commonly results in the appearance of thin sandstones being "isolated" in marine shale. These sandstones have often been interpreted as "offshore bars" in the literature but paleogeographic mapping in this study suggests that these sandstones in the Kenilworth are shoreface-attached (Figs 2-8a and 2-8b).

Drive back to Hwy 6/50; turn left and drive south towards Green River.
94.0- Price River Canyon access road is on the east side of the highway; turn left and follow the dirt road. If you pass the abandoned gas station on the east side of the road with the blue and red sign, you've gone too far. Drive along the dirt road; the Kenilworth is the lowermost series of sandstone ledges along the canyon.
98.3- STOP THREE: North Price River Canyon near Woodside, Utah. After passing a small knoll on the left side of the road, turn left and park in the open area. The measured section is in a tributary canyon north of the dirt road. Walk about 0.25 miles to the measured section. The purpose of Stop 3 is to look at the sedimentary facies, parasequence stacking pattern, and grit lag at the top of the Kenilworth.

Description:

The measured section from North Price River Canyon (Fig. 2-9) shows a series of coarsening-upward parasequences, with PS 2 having "split" into PS 2a, b, and c. This "split" in PS 2 can be traced in a landward direction where it forms the thick parasequence in the cliffs at Stop 1a. Parasequence 3 is a coarsening-upward succession capped by interbedded sandstones and shales while PS 4 is the thick, ridge-forming sandstone that occurs at the top of the Kenilworth.

The base of the measured section begins with bioturbated shales and thin sandstones containing wave and current ripples. A massive sandstone about 4 feet in thickness that pinches and swells and contains HCS, occurs at the top of PS 2a. Above PS 2a are bioturbated shales that change facies upward into HCS and wave rippled sandstones interbedded with bioturbated shales of PS 2b. These are overlain by a succession of very thin, wave rippled, bioturbated sandstones and interbedded shales of PS 2c. The facies of PS 2a, b and c are interpreted as having been deposited in an offshore marine/offshore transition zone environment.

The lowermost part of PS 3 is a thick churned offshore/offshore transition silty mudstone which begins to coarsen-upward into wave rippled and HCS sandstones. Wave ripple crest orientations are approximately north-south to northwest-southeast in this interval. A churned mudstone overlies PS 3 and marks the flooding surface at the base of Parasequence 4.

In this locality, PS 4 has also "split" into PS 4a and 4b, which are separated by a minor flooding surface. PS 4a is thin and consists of a sandier-upward succession, whereas PS 4b is a thick, coarsening-upward succession. The offshore transition zone in the lowermost part of PS 4b is characterized by HCS sandstones with *Ophiomorpha* burrows, interbedded with thin silty sandstones. The lower shoreface is dominated by amalgamated HCS sandstones that are beautifully exposed in this canyon (Fig. 2-10a). The uppermost part of PS 4b is a whitish, medium-grained sandstone that is characterized by trough cross-bedding and is interpreted to have been deposited in the upper shoreface. Trough cross-bed orientations in this unit are predominantly toward the south-southeast.

The top of PS 4b is intensely churned and contains large *Ophiomorpha, Thallasinoides* and *Skolithos* burrows (Fig. 2-10b). Some coarse sand grains have been incorporated into this churned interval, and the top of PS 4b is capped with a grit lag containing shark's teeth, bone fragments, broken oyster shells, pebbles, and abundant black chert grains (Fig. 2-10c). This surface is overlain by offshore marine shales at the base of the Sunnyside Member.

Sequence Stratigraphic Interpretation:

The parasequences observed at North Price River Canyon display a progradational stacking pattern and are interpreted as part of the highstand systems tract. One of the differences between Stop 1 and Stop 3 is that in this more basinward setting, Parasequences 2 and 4 have split due to the intertonguing of offshore transition zone and lower shoreface environments.

The coarse-grained lag occurring at the top of PS 4 is interpreted to mark the location of the sequence boundary seen at Stops 1 and 2. At this stop it is coincident with the flooding surface at the base of the Sunnyside Member and represents an interfluve area between incised valleys. The fluvial channels observed in the cliff-face at Stops 1 and 2 incise from this stratigraphic position. The subsequent transgression resulted in marine reworking of this surface, forming a lag. The transgressive systems tract cannot be identified at this location because PS 5 has downlapped between Stop 3 and Whitmore Canyon.

Depart North Price River Canyon, returning along the dirt road toward Hwy 6/50.

103.6- Turn south on Hwy 6/50 to Interstate 70.

129.8- Travel east on Interstate 70 toward Green River. Take the Main St. exit into town.

130.6- Turn left (north) at Long St. near the center of town. This road heads out of the townsite and has a couple of sharp 90° turns.

137.3 Turn left onto a dirt track that angles up a small hill.

137.8- STOP FOUR: Bluecastle Butte, Middle Mountain and Gunnison Butte overview near Green River, Utah. Stop vehicle and view exposures in surrounding cliff-faces. The purpose of this stop is to examine the basinward facies changes in the Kenilworth Member.

Description:

At Stop 4, the sandstone at the top of the Kenilworth is Parasequence 4, which is well-developed on Bluecastle Butte to the west. Figure 2-11 is a pan of Bluecastle Butte that shows clinoforming geometries in the shoreface. The base of the shoreface on the left-hand side of the photo is sharp and sits abruptly on marine shale. It can be traced to the right (east) where the sandstone begins to change facies into interbedded sandstones and shales. On the right-hand side of the photo near the end of Bluecastle Butte, massive shoreface sandstones also rest abruptly on marine shale; however, the base of this sandstone is younger than the one to the left. Tracing the base of the younger sandstone back to the left indicates that it merges updip and disappears into the body of the thick, cliff-forming sandstone of PS 4. The geometry and the sharp, diachronous nature of the base of the massive sandstones suggests rapid easterly clinoform progradation of this parasequence.

The basinward facies change in Parasequence 4 can be seen along the Middle Mountain/Gunnison Butte exposures to the north. The thick, clinoforming sandstone observed earlier on Bluecastle Butte has "split" in a basinward (easterly) direction and displays an intertonguing of sandstone and shale lithologies (Fig. 2-12). Measured sections from Middle Mountain and Gunnison Butte show a facies change from thick, amalgamated HCS sandstone of the lower shoreface to thin, wave rippled offshore transition zone deposits, respectively. Both measured sections have a thin bioturbated lag present at or near the top of PS 4.

Sequence Stratigraphic Interpretation:

At this location, the maximum progradational extent of the shoreface in the highstand systems tract and the associated basinward change of facies can be observed. The clinoforming geometry and sharp-based nature of the shoreface at Bluecastle Butte suggest rapid progradation during the late highstand when accommodation potential was negligible. The lag observed in measured sections at the top of PS 4 is interpreted as the sequence boundary; however, at this locality there is no evidence of downcutting, suggesting that this may have been an interfluve area during the lowstand.

Return to gravel road and turn right (south) toward Green River.
144.6- Turn left on Main St. and travel east toward I-70.
147.5- At the I-70 junction, head east toward Crescent Junction.
158.4- Exit I-70 at Exit 173, which has a sign indicating "Ranch Exit". Turn left at the "STOP" sign and follow the old highway.
159.3 Turn right on a small gravel road with a sign indicating "Floy Canyon Road".
159.6- Cross the railroad tracks.
160.0- STOP FIVE: Hatch Mesa, near Crescent Junction, Utah. Park vehicle on the side of the road. The outcrop is about 1/4 mile east of the gravel road. The purpose of this stop is to examine coarse-grained lag deposits and deltaic facies in basinal Kenilworth outcrops.

Description:

The succession of strata exposed at Stop 5 is interpreted to be the basinal equivalent of the Kenilworth Member. This interpretation presents a potential correlation problem due to the lack of exposures between Tusher Canyon and Hatch Mesa (a distance of about 10 miles). Data from this study, as well as previous work by Newman (1985) and Swift *et al.* (1987), suggest that the Hatch Mesa succession is approximately time-equivalent to the Kenilworth in the vicinity of Green River, Utah. Palynological samples collected from shales below and above PS 4 at Stop 4 (Gunnison Butte) and from the shales at Hatch Mesa all yielded consistent dates of Early Campanian, probably no younger than 78 ma (R.W. Harris Jr., Exxon Production Research Co., pers. comm., 1990). Fouche (pers. comm., 1991) has suggested that the succession at Hatch Mesa may be older and time-equivalent to the Santonian Emery Member based on ammonite data. Since these paleontological data appear to be conflicting and equivocal, and because the succession cannot be directly traced between Tusher Canyon and Hatch Mesa, we have interpreted the Hatch Mesa sandstones at Stop 5 to be the basinward continuation of the Kenilworth Member.

A measured section from Stop 5 is shown in Figure 2-13, which is a composite section from this locality summarizing the main stratigraphic and sedimentary features of the outcrop. Paleontological data indicates that the base of the section consists of nearshore marine shales (R.W. Harris Jr., Exxon Production Research Co., pers. comm., 1990) that are overlain in the northern part of the outcrop by a discontinuous, coarse-grained lag deposit. Here the lag is only a few feet in thickness and is characterized by chert, mudclasts, bone fragments, shell debris and shark's teeth. Sedimentary structures are difficult to recognize but include low angle dipping laminations which may be cross-bedding.

In the southern part of the outcrop, the base of the succession is characterized by oolites which are present in the axis of a small syncline. The oolites are trough cross-bedded, contain mudclasts, and rest abruptly on marine shales. Thin section analysis indicates that the oolites have nucleated on chert and quartz grains or shark's teeth fragments. The oolite beds are interpreted as shoal-water deposits that have a northwest-southeast orientation and a patchy, en-echelon distribution.

The overlying coarse-grained lag is lithologically similar to the lag observed in the northern exposures at Stop 5. Here it contains large-scale, sigmoidally-shaped cross-beds with well-developed clay drapes (Fig. 2-14a), as well as trough cross-beds, lateral accretion surfaces (Fig. 2-14b), current ripples and some burrows. The lag is interpreted as having been deposited in a tidal channel/tidal inlet environment.

Above the oolites and grits are about 60 feet of marine shales (R.W. Harris Jr., Exxon Production Research Co., pers. comm., 1990) which are overlain by a succession of very fine-grained sandstones and interbedded shales. The sandstones are parallel laminated and current rippled, with the tops of some beds displaying rib-and-furrow features. Bioturbation is rare to absent in these deposits. Paleocurrents from the sandstones indicate flow toward the east. The interbedded sandstone and shale succession at Hatch Mesa is interpreted as having been deposited in a delta front environment.

Sequence Stratigraphic Interpretation:

The succession at Hatch Mesa is interpreted as the lowstand systems tract of the Kenilworth. The oolites at the base of the measured section mark the initial relative sea-level fall in the early part of lowstand, with wave agitation of the sea-floor resulting in the development of oolite shoals. The base of the coarse-grained lag is interpreted as the sequence boundary which formed during the maximum rate of relative sea-level fall (lowstand fan *time* on the eustatic sea-level curve). The shoreline position shifted abruptly basinward during this fall, from the area of Bluecastle Butte to the location of Hatch Mesa (a distance of about 12 miles). Tidally-influenced deposition of the coarse-grained lag occurred at this time. The overlying marine shales were deposited following the initial relative sea-level rise in the early part of the lowstand and the deltaic deposits subsequently prograded into the basin in the latter part of the lowstand.

The coarse-grained lag observed at Stop 5 is petrographically similar to the lag observed at North Price River Canyon. The schematic correlation of the lag deposits between these two stops is shown in the schematic sequence stratigraphic model of Figure 2-15. At North Price River Canyon (Stop 3), the lag marks the sequence boundary at the top of the highstand systems tract in an interfluve area with little erosional relief. At Hatch Mesa, the base of the lag is interpreted as the sequence boundary, with the overlying coarse-grained sandstones and deltaic deposits occurring in the lowstand systems tract.

References

Newman S.L., 1985, Facies interpretations and lateral relationships of the Blackhawk Formation and Mancos Shale, East-central Utah. *In:* SEPM Mid-year Meeting Field Guide 10, p. 69-114.

Speiker E.M., 1931, The Wasatch Plateau coal fields, Utah. USGS Bulletin 819, 206p.

Swift D.J.P., Hudelson P.M., Brenner R.L. and Thompson P., 1987, Shelf construction in a foreland basin: storm beds, shelf sandbodies, and shelf-slope depositional sequences in the Upper Cretaceous Mesaverde Group, Book Cliffs, Utah. Sedimentology, V. 34, p. 423-457.

Young R.G., 1955, Sedimentary facies and intertonguing in the Upper Cretaceous of the Book Cliffs, Utah-Colorado. GSA Bulletin, V. 66, p.177-202.

Young R.G., 1957, Late Cretaceous Cyclic Deposits, Book Cliffs, Eastern Utah. AAPG Bulletin, V. 41, p.1760-1774.

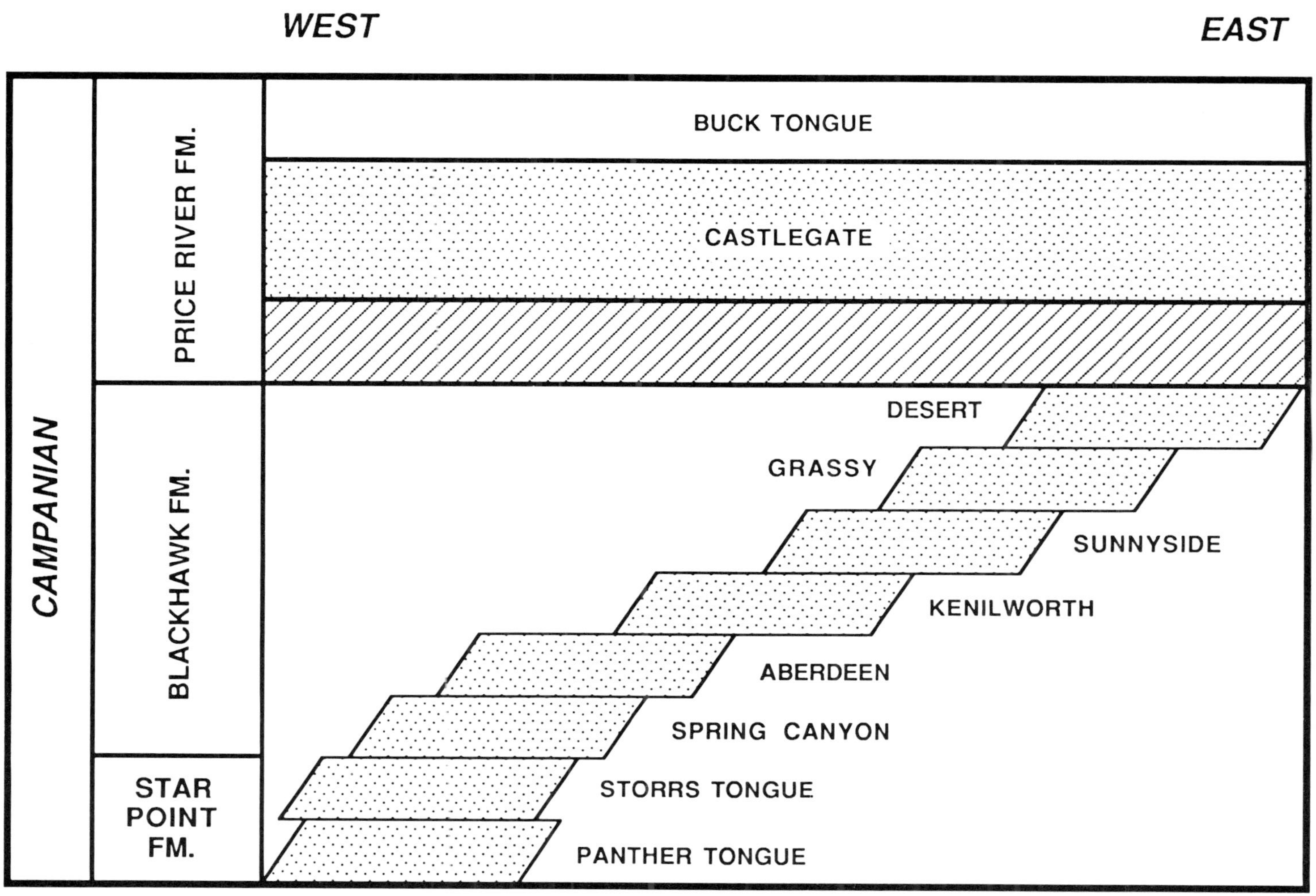

Figure 2-1 Stratigraphic chart of Upper Cretaceous rocks in the Book Cliffs area (modified from Young, 1955).

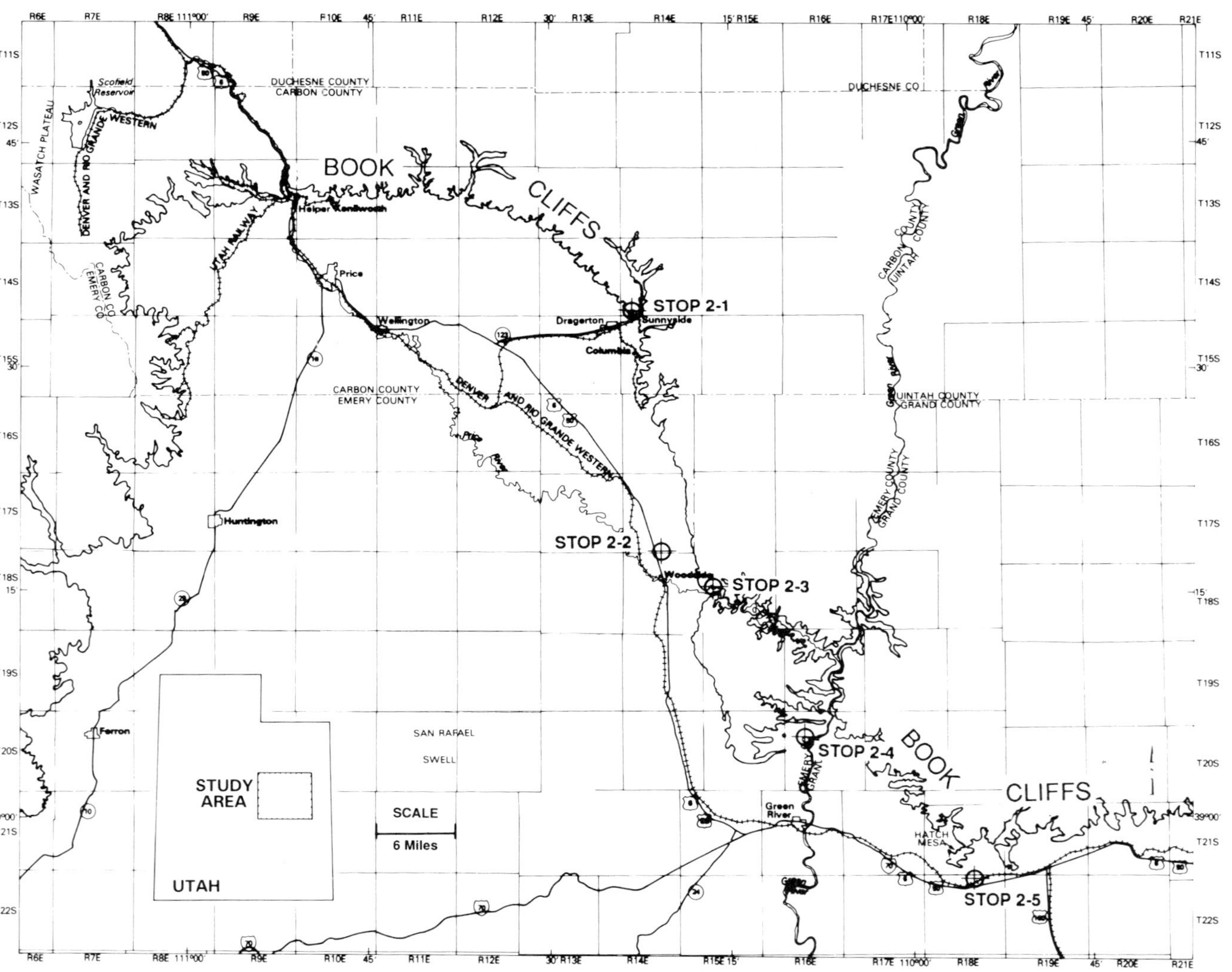

Figure 2-2 Location map indicating the field trip stops to be examined for the Kenilworth Member.

Figure 2-3 Photo of the cliff-face to the north of Stop 2-1a showing parasequences in the Kenilworth Member.

STOP 2-1: WHITMORE CANYON

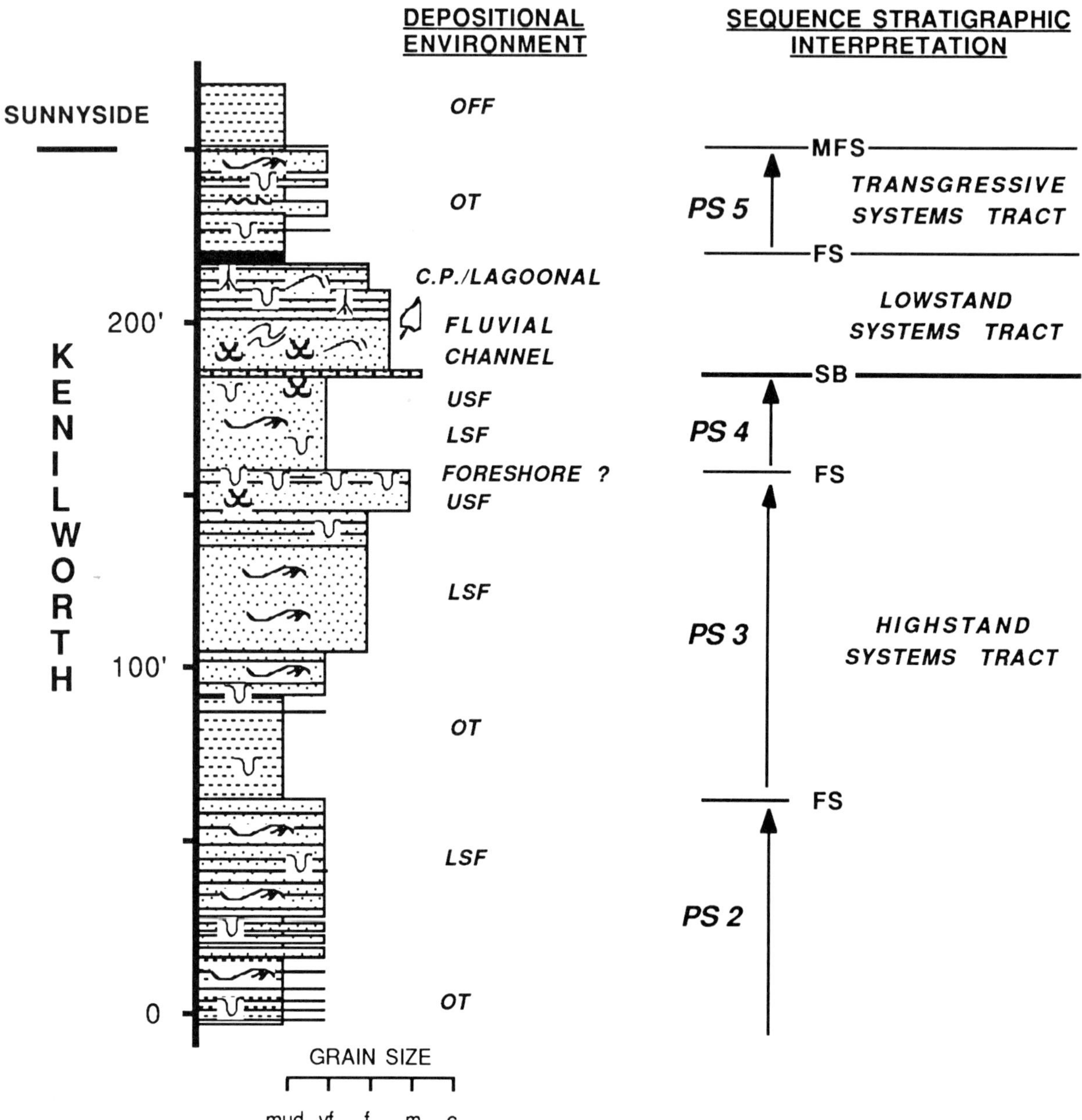

Figure 2-4. Measured section of the Kenilworth Member in the vicinity of Whitmore Canyon near Sunnyside, Utah. The measured section shows a progradational parasequence set (PS 2 - PS 4) in the highstand systems tract which is truncated by a sequence boundary. This surface is overlain by fluvial and coastal plain/lagoonal deposits of the lowstand systems tract. The transgressive surface overlies the lowstand deposits and there is one parasequence at the top of the Kenilworth (PS 5) that is part of the transgressive systems tract. The top of Parasequence 5 is a maximum flooding surface which marks the base of the Sunnyside Member. See Figure 2-13 for legend of sedimentary structures.

Figure 2-5 Panorama showing of fluvial channelization into lower shoreface deposits of Parasequence 4 at Stop 2-lb.

Figure 2-6 Panorama across Hwy 123 to Sunnyside Mine showing erosional truncation along the sequence boundary at the top of PS 4.

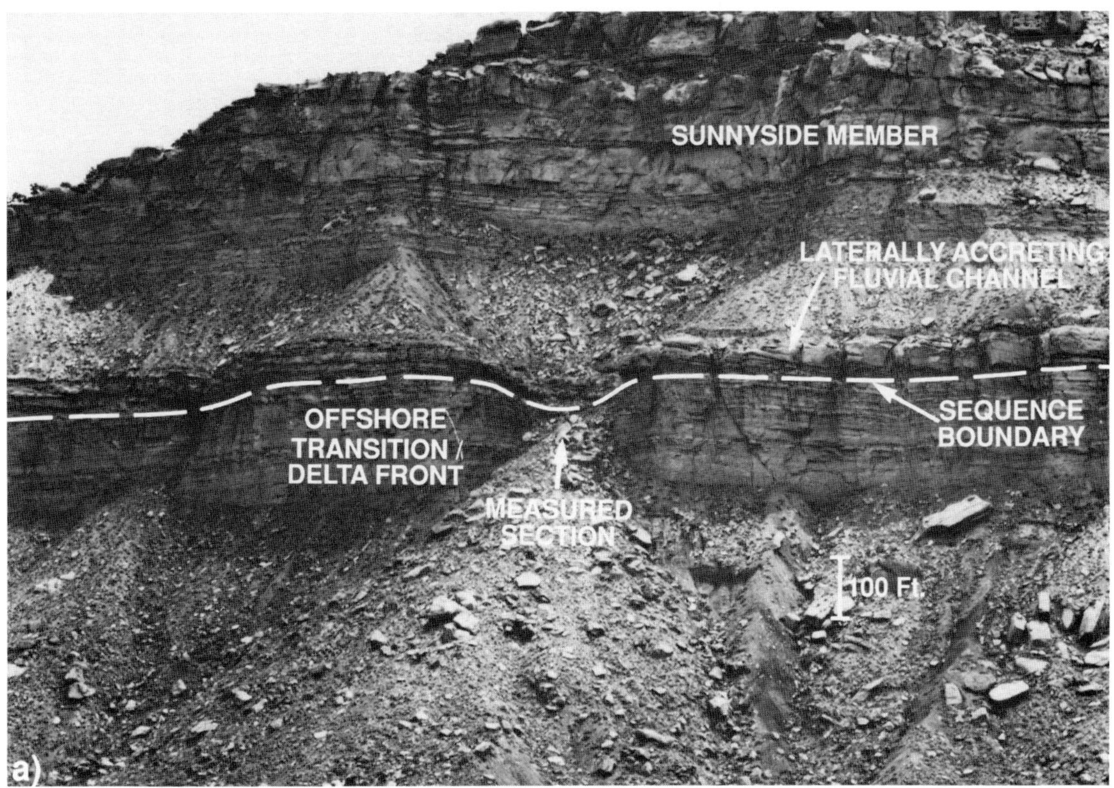

Figure 2-7 a) Laterally accreting fluvial channel above the sequence boundary at the top of Parasequence 4. The channel cuts into offshore transition/distal delta front deposits;
b) Shale-dominated channel-fill along the outcrop to the north.

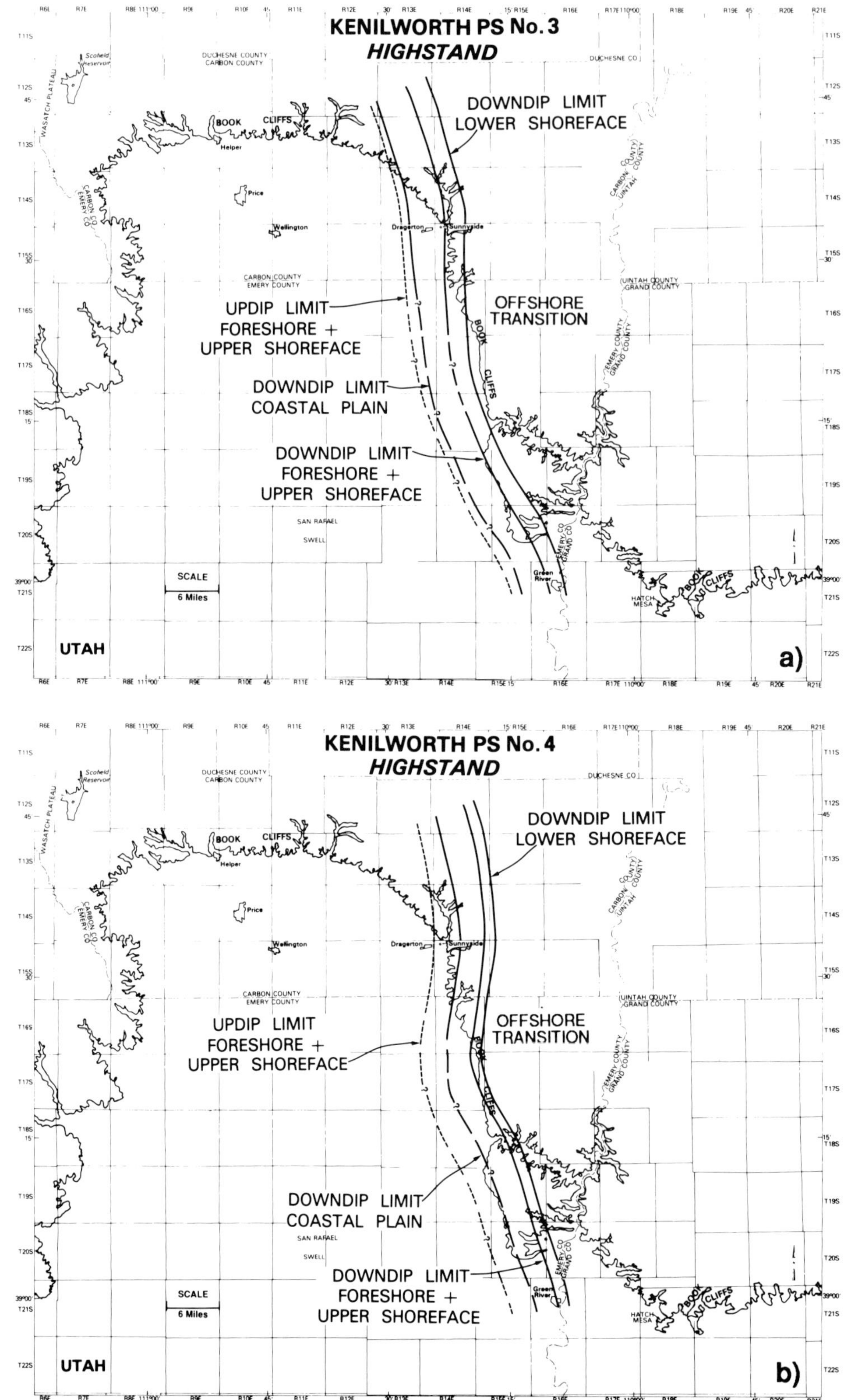

Figure 2-8 Paleogeographic maps showing the distribution of mapped depositional environments:
a) Parasequence 3;
b) Parasequence 4.

STOP 2-3: NORTH PRICE R. CANYON

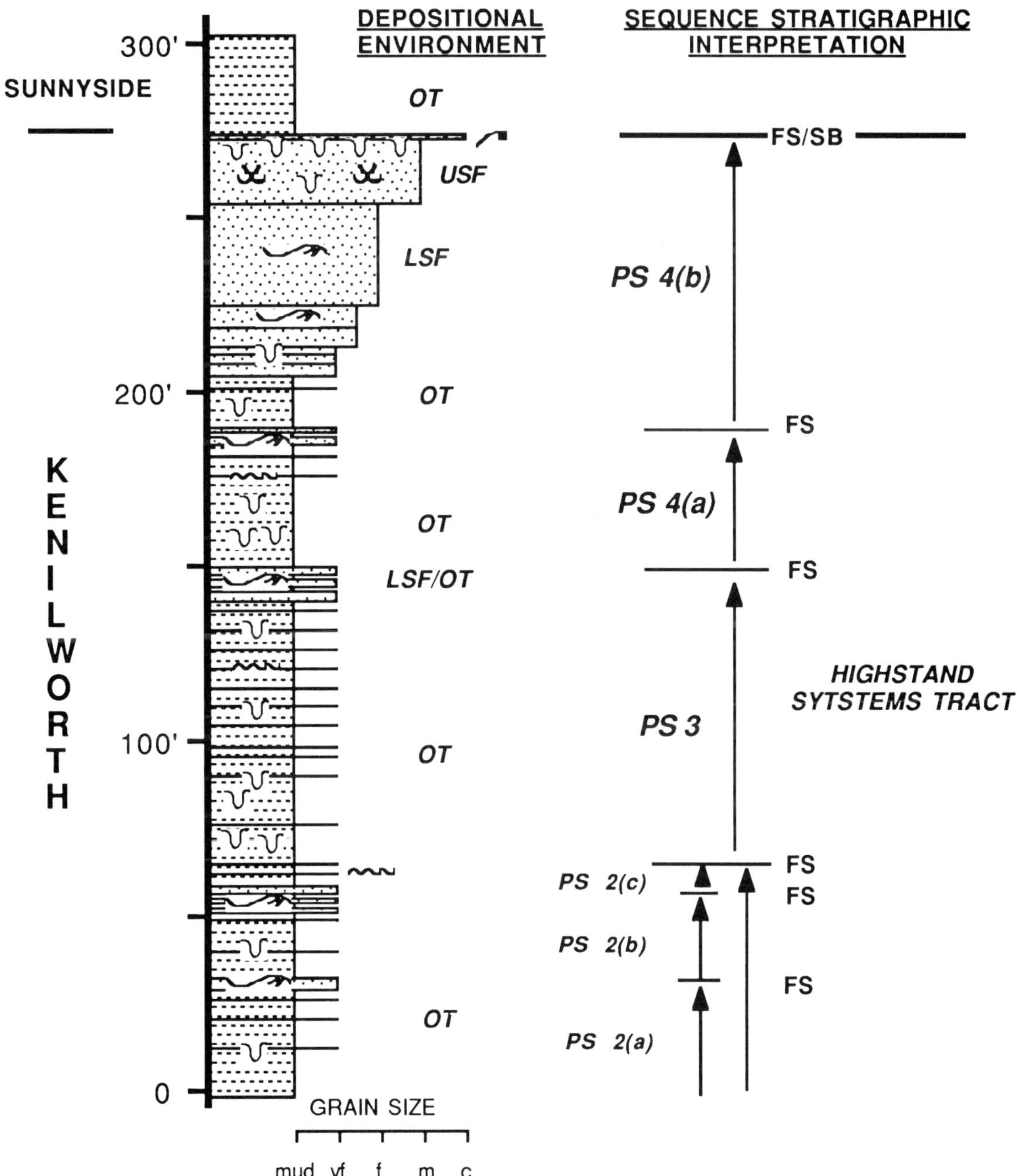

Figure 2-9. Measured section of the Kenilworth Member in a tributary canyon of North Price River Canyon. The measured section shows a progradational parasequence set (PS 2 - PS 4) in the highstand systems tract, which is capped by a coarse-grained lag deposit marking the location of the sequence boundary. This surface is overlain by marine shales at the base of the Sunnyside Member and is therefore also a flooding surface (FS/SB). See Figure 2-13 for legend of sedimentary structures.

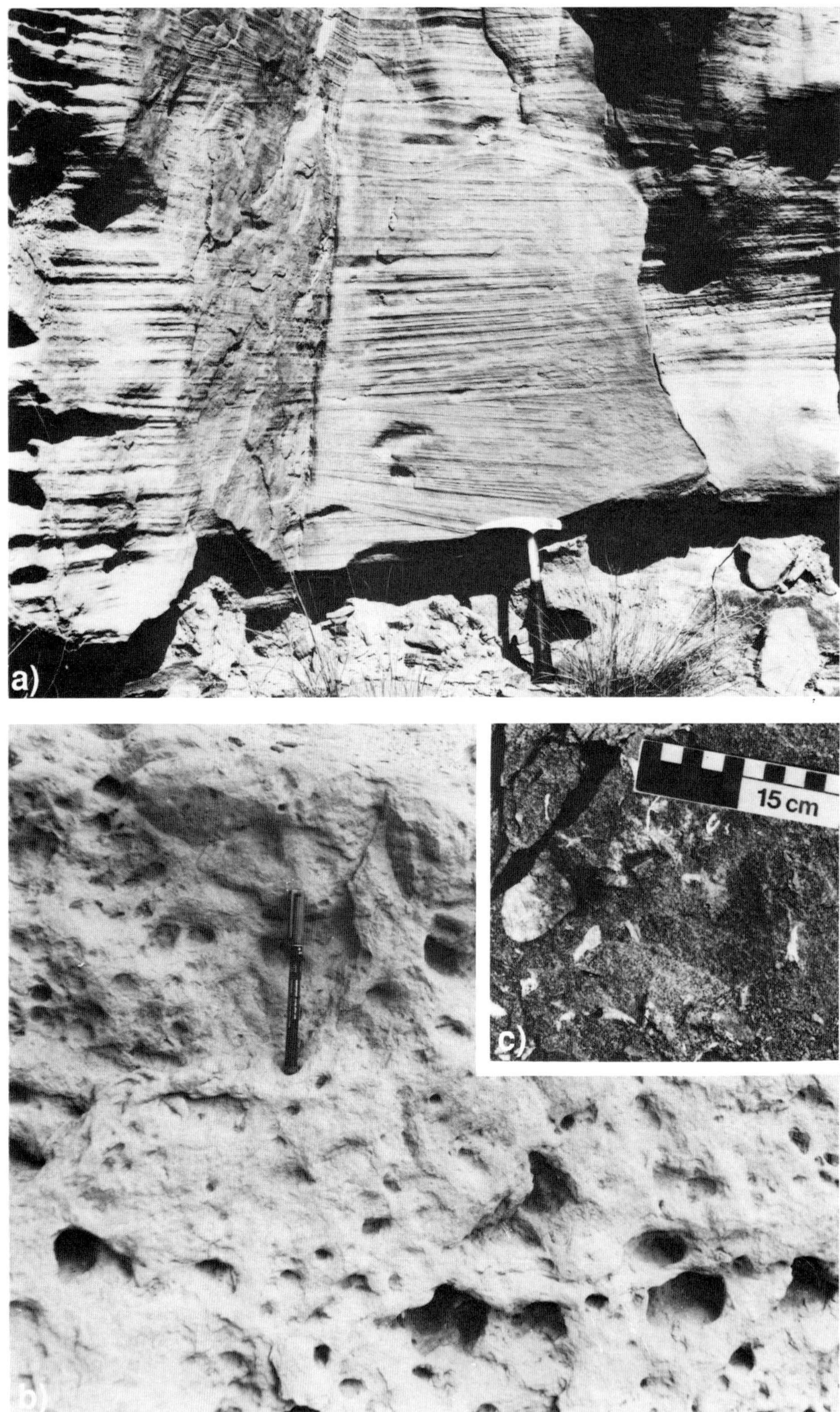

Figure 2-10 a) Amalgamated HCS sandstone from lower shoreface deposits of PS 4;
b) Bioturbated interval near the top of PS 4;
c) Coarse-grained lag deposit capping PS 4.

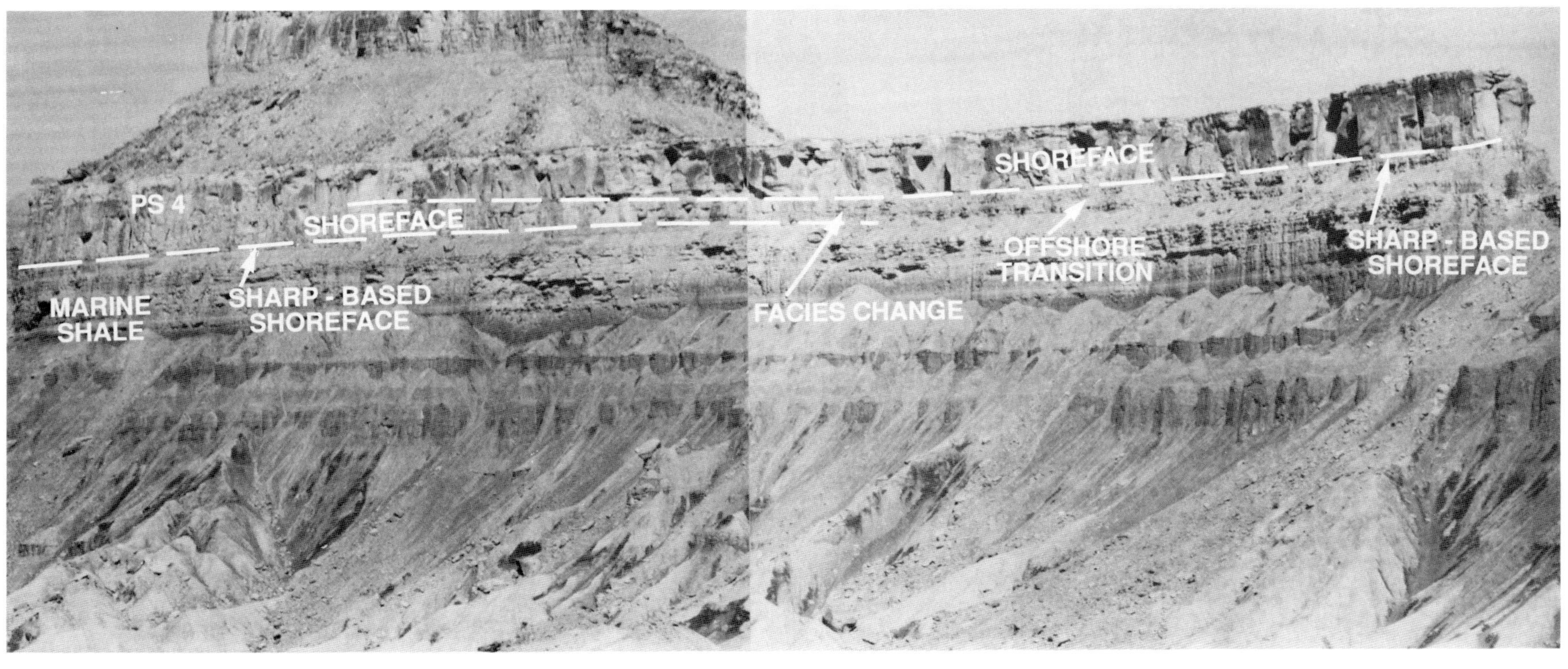

Figure 2-11 Panorama of Bluecastle Butte showing the clinoforming and sharp-based diachronous nature of the shoreface in the late highstand systems tract.

Figure 2-12 Panorama of Gunnison Butte showing the basinward facies change in PS 4. Note the intertonguing of distal lower shoreface sandstones and marine shales.

STOP 2-5: HATCH MESA

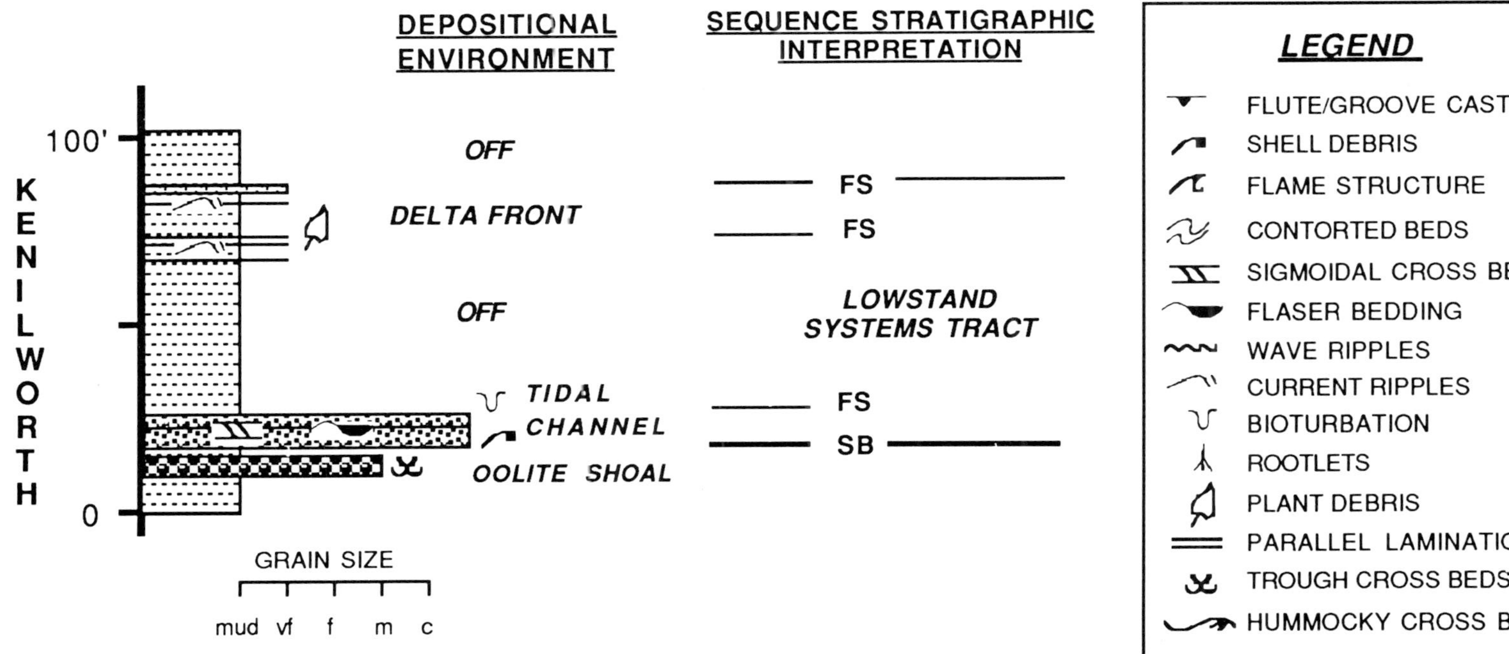

Figure 2-13. Measured section of the Kenilworth Member at Hatch Mesa near Crescent Junction, Utah. The measured section shows a thick oolitc sandstone at the base which is overlain by a sequence boundary and coarse-grained tidal channel deposits. These are in turn overlain by marine shale and lowstand delta front sandstones and shales.

Figure 2-14 a) Sigmoidal cross-bedding and associated clay drapes from coarse-grained tidal deposits above the sequence boundary at Hatch Mesa;
b) Lateral accretion surfaces within the trough cross-bedded lag.

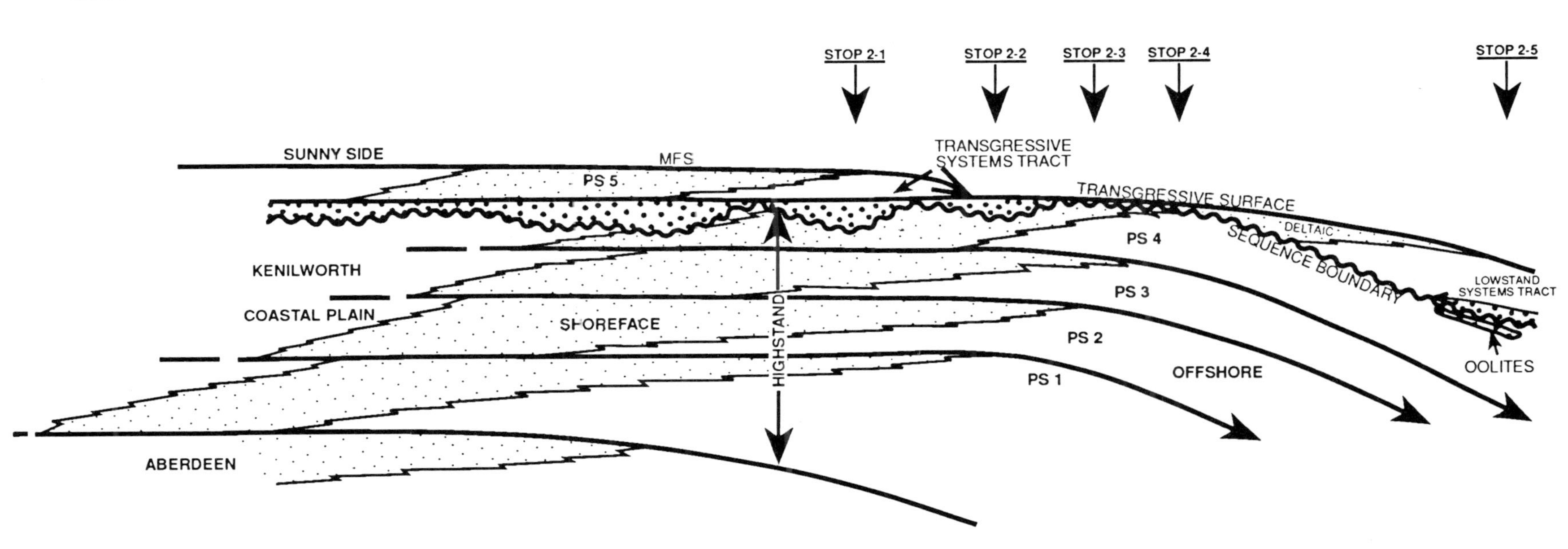

Figure 2-15. Schematic sequence stratigraphic model for the Kenilworth Member.
The approximate location of the field trip stops is indicated along the top of the diagram.

RECOGNITION OF HIGH-FREQUENCY SEQUENCES IN THE KENILWORTH MEMBER OF THE BLACKHAWK FORMATION, BOOK CLIFFS, UTAH

DAVID R. TAYLOR and RICHARD W.W. LOVELL
Esso Resources Canada Limited
Calgary, Alberta

ABSTRACT

Shallow marine strata of the Campanian Kenilworth Member, cropping out in the Book Cliffs of east-central Utah, were examined from their updip to downdip depositional limits. Thirty-one outcrop sections were measured recording facies, stratal surfaces, and paleoflow indicators. These data were used to interpret the depositional environments and to develop a chronostratigraphic framework for the Kenilworth. The geometry of stratal surfaces and the continuity of sandstones were traced between measured sections using binoculars and photographic panoramas of cliff exposures.

A variety of depositional environments were identified in the Kenilworth including fluvial channels, coastal plain, foreshore, shoreface, offshore transition/offshore marine and deltaic. The vertical and lateral associations of these depositional environments indicate the presence of several parasequences separated by regionally correlative marine-flooding surfaces.

Five wave-dominated shoreline parasequences were recognized, each with a north-south to northwest-southeast paleoshoreline orientation. The oldest four parasequences are stacked as a progradational parasequence set and are interpreted to be part of the highstand systems tract. A regional erosional surface overlies the highstand systems tract and east-west trending fluvial channel systems incise into the underlying shoreface deposits. The erosional surface was traced basinward where deltaic deposits occur above a coarse-grained lag. The extensive erosional surface is interpreted to be a sequence boundary which formed in response to a relative sea-level fall. The coarse-grained lag and deltaic deposits in the basin are interpreted to be part of the lowstand systems tract. A major flooding surface occurs above the lowstand and was traced updip beneath a backstepped wave-dominated shoreline parasequence of the transgressive systems tract.

Sandstones of the Kenilworth Member are thus interpreted to comprise parts of two high-frequency sequences. A sequence boundary occurs within the Kenilworth and separates the highstand systems tract of an older sequence from the lowstand and transgressive systems tract of a younger sequence. The sequence boundary can be recognized by changes in parasequence stacking patterns, regionally extensive erosional truncation, and a basinward shift of facies. The magnitude of the relative sea-level fall that occurred during deposition of the Kenilworth is estimated to be at least 60 feet, based on the amount of fluvial channel incision observed at the sequence boundary. This magnitude of relative sea-level fall resulted in a basinward shift in facies of about 10 miles. The resultant paleoslope is interpreted to be 0.07 degrees, which is comparable to depositional slopes on the present day Gulf of Mexico shelf.

INTRODUCTION

The existence of shallow-marine deposits in sandstone members of the Blackhawk Formation exposed in the Book Cliffs of east-central Utah has been recognized many years. Previous workers have described these shoreline successions in terms of their sedimentary facies and associated depositional environments, but have given limited discussion to lateral facies variability and detailed stratigraphic architecture (e.g. Swift *et al.*, 1987; Newman, 1985). Regional studies by Young (1957, 1955) and Balsley and Horne (1980) examined most of the Campanian section in the Book Cliffs. These are important studies that established the stratigraphic framework; however, studies which integrate facies interpretations and chronostratigraphic correlations are currently lacking. The Kenilworth Member of the Blackhawk Formation is well-exposed in outcrop from its updip to downdip depositional limits and provides an excellent opportunity to investigate these relationships.

Recent developments in sequence stratigraphy suggest that within siliciclastic stratigraphic successions, high-frequency depositional sequences are common (Mitchum and Van Wagoner, 1991; Van Wagoner *et al.*, 1990). High-frequency sequences and their associated sequence boundaries can be identified in the rock record by changes in parasequence stacking patterns, abrupt basinward shifts in depositional environments, and evidence of erosional truncation and/or subaerial exposure. The ability to recognize these criteria and interpret the existence of high-frequency cyclicity within stratigraphic successions is improved when sedimentary facies relationships are thoroughly documented and well understood.

The purpose of this paper is to discuss the depositional environments and interpret the chronostratigraphic framework of the Kenilworth Member of the Blackhawk Formation. The observations and interpretations from these well-exposed outcrops show that the three-dimensional distribution of facies, when integrated with high-resolution sequence stratigraphic correlations, provides a better understanding of the spatial relationships of depositional elements within high-frequency sequences.

DATA BASE AND METHODS

The data base for this study consists of 31 outcrop sections exposed in canyons of the Book Cliffs between Price and Crescent Junction, Utah (Fig. 1). Lithology, grain size, sedimentary and biogenic structures, and paleocurrent data were recorded in each measured section. Several shale samples were also collected and analyzed for palynology. Excellent outcrop exposures in the study area allowed for the correlation of flooding surfaces and sequence boundaries between canyons using binoculars and photographic panoramas. Also, facies relationships could be traced along the cliff-faces using weathering profiles and surficial weathering features such as "whitecaps". The observed facies relationships were plotted on small-scale topographic maps in the field and transposed on to the large-scale maps presented in this paper.

STRATIGRAPHY

The Blackhawk Formation is Campanian in age (Speiker, 1931) and is part of the Mesa Verde Group. The generalized stratigraphy and nomenclature has been addressed by Young (1957, 1955) and Kamola *et al.* (1985) and will not be discussed in detail here. The Kenilworth Member is one of six members of the Blackhawk Formation (Fig. 2). It occurs in the middle part of the Blackhawk and consists of massive, cliff-forming, fine- to medium-grained sandstones. The sandstones are sublitharenites to litharenites consisting of quartz, chert and carbonate rock fragments (B. Pruett, pers. comm., 1989). These sandstones are intercalated with coal-bearing strata to the west and the marine Mancos Shale to the east. The thickness of the Kenilworth is estimated to average about 300 feet along the outcrop exposures of the Book Cliffs.

Previous Interpretations

Early interpretations of the Blackhawk Formation in the Book Cliffs recognized the cyclic nature of sandstone members and suggested they comprised "prominent littoral sandstone tongues and many lesser ones, all projecting eastward into the Mancos where they lose their identity by grading into shale" (Young, 1955). The littoral sandstones were considered to be "offshore bars" that changed facies toward the west into "lagoonal" deposits and toward the east into marine shales. The Kenilworth Member was further subdivided by Young (1957, 1955) into four "bar sandstones".

These studies by Young (1957, 1955) were amongst the first to recognize different orders of cyclicity within siliciclastic rocks of the Book Cliffs. Two orders of cyclicity were interpreted and were called "megacyclothems" and "cyclothems". These corresponded to formation members and tongues within members, respectively. Based on the diagrams presented in these studies, "megacyclothems" were defined from the major flooding surface at the base of a sandstone member to the next major flooding surface at the base of the overlying sandstone member.

Recent studies of the Kenilworth have focussed on the sedimentary facies and depositional environments. Balsley and Horne (1980) considered all the Blackhawk members to have been deposited in a wave-dominated deltaic environment, while Swift *et al.* (1987) considered the Kenilworth to represent a "delta-prodelta shelf depositional system". Swift *et al.* (1987) also expounded upon the nature of shelf sediment transport and attributed it to "downwelling coastal storm flows". They also speculated on large-scale depositional sequences in foreland basins, based on partial examination of the Kenilworth Member. In addition to these studies, Newman (1985) published an interpretation of an interesting succession of strata at Hatch Mesa and suggested that these deposits were age-equivalent to the Kenilworth and represented delta lobes and oolitic "offshore marine bars".

The interpretation of the Hatch Mesa succession as the basinward equivalent of the Kenilworth Member presents a potential correlation problem due to the lack of exposures between Tusher Canyon and Hatch Mesa (a distance of about 10 miles). Data from this study, as well as previous work by Newman (1985) and Swift *et al.* (1987), suggest that the Hatch Mesa succession may be time-equivalent to the Kenilworth in the vicinity of Green River, Utah. Palynological samples collected from shales within the Kenilworth Member at Gunnison Butte and from the shales at Hatch Mesa all yielded consistent dates of Early Campanian, probably no younger than 78 ma (R.W. Harris Jr., Exxon Production Research Co., pers. comm., 1990). Fouche (pers. comm., 1991) has suggested that the succession at Hatch Mesa may be older and time-equivalent to the Santonian Emery Member based on ammonite data. Since these paleontological data appear to be conflicting and equivocal, and since the succession cannot be directly traced between Tusher Canyon and Hatch Mesa, we believe that the Hatch Mesa succession is the basinward equivalent of the Kenilworth Member.

Finally, an important point to note is that the recent studies discussed above did not examine the entire Kenilworth Member from its updip to downdip limit. Also, the earlier studies of Young (1957, 1955), which recognized the cyclicity within the Blackhawk Formation and correctly correlated many of the sandstone members, were done when the discipline of clastic sedimentology was in its infancy.

DEPOSITIONAL ENVIRONMENTS

Detailed descriptions of measured outcrop sections indicate that the siliciclastic facies present in the Kenilworth Member were deposited in several different environments. The following part of the paper will briefly describe the sedimentary attributes used to interpret the depositional environments, thus setting the stage for a discussion of the paleogeography and chronostratigraphy. The interpretations of depositional environments made in this paper are generally consistent with previous interpretations in the literature.

Offshore Marine/Offshore Transition Zone

Offshore marine and offshore transition zone deposits consist predominantly of shales, mudstones, siltstones, and interbedded very fine-grained sandstones. This succession of strata can be up to hundreds of feet in thickness. Mudstones, shales and silty shales are commonly bioturbated and may contain thin, wave rippled, silty to very fine-grained silty sandstone lenses (Fig. 3a). Burrow types are generally not identifiable due to poor preservability in the shales, but *Terebellina* was recognized in many of the outcrops. Offshore transition zone sandstones are clean, very fine-grained and a few inches in thickness. Internally the sandstone beds contain parallel laminae, wave ripples and current ripples, and individual beds may have tool marks and flute casts on their bases (Fig. 3b).

Lower Shoreface

Lower shoreface deposits are dominated by sandstones containing hummocky cross-stratification (HCS). These sandstones are a few feet to tens of feet in thickness and are interbedded with bioturbated silty shales (Fig 3c). Individual beds show varying degrees of burrowing, from completely churned to non-bioturbated, but the uppermost part of each bed generally has some burrowing associated with it. Trace fossils identified from lower shoreface deposits include; *Ophiomorpha nodosa*, *Ophiomorpha sp.*, *Terebellina*, *Paleophycus*, *Thalassinoides*, *Planolites*, *Skolithos* and *Cylindrichnus*.

Upward in the lower shoreface succession, the sandstone:shale ratio increases and the beds begin to amalgamate, forming amalgamated HCS (Fig. 3d). In these sandstones, the wavelength of the hummocks increases and may be up to a few tens of feet. The grain size also increases from very fine to fine, and the density of bioturbation decreases. Overlying the amalgamated HCS beds, a medium-grained, parallel-laminated sandstone is commonly observed.

Although the exact hydrodynamic conditions under which hummocky cross-stratification forms is still under debate, it is generally believed to be the result of storm-enhanced wave action below fairweather wave base (Walker, 1984). The ubiquity of HCS in many of the measured sections suggests that most of the shorelines were wave-dominated in this area during deposition of the Kenilworth Member. The parallel-laminated sandstones that are present within the lower shoreface succession of some Kenilworth outcrops may be related to very long wavelength hummocks and swales. Alternatively, they may indicate that the grain size of the sands being moved and reworked on the sea-floor below fairweather wave-base was too large to form HCS under storm wave influence, thus forming upper flow regime, parallel-laminated sands instead.

Upper Shoreface and Foreshore

Upper shoreface deposits are characterized by trough cross-bedded, fine- to medium-grained sandstones (Fig. 4a) and range from a few feet to tens of feet in thickness. Interbedded shales and siltstones are absent and bioturbation is sparse, although *Ophiomorpha sp.*, *Ophiomorpha nodosa*, *Skolithos* and

Cylindrichnus burrows were observed. In some outcrops, carbonaceous debris was found along trough cross-bed foresets. Higher concentrations of plant debris in these localities may indicate proximity to distributary systems. Fluid escape structures were noted throughout upper shoreface sandstones.

Cross-bedded upper shoreface sandstones are interpreted to have formed in response to fairweather reworking of sediment by nearshore currents. As will be discussed later in the paper, a dominant component of sediment transport sub-parallel to the shoreline can be demonstrated by comparing paleogeographic maps to paleocurrent measurements from plan views of trough cross-bed sets.

Foreshore deposits comprise medium- and/or medium- to coarse-grained, parallel to sub-horizontally laminated sandstones (Fig. 4b). They range from less than one foot to a maximum of eight feet in thickness. In some localities, interbedded scour-and-fill structures are present. Bioturbation is sparse, with only a few *Ophiomorpha nodosa*, *Ophiomorpha sp.* and *Macronichnus* occurring within this interval. Root traces were commonly observed at the top of the foreshore.

Upper shoreface and foreshore deposits in the Kenilworth commonly have a characteristic surficial "whitecap" weathering appearance in the cliffs which is helpful for tracing the lateral extent of this facies.

Coastal Plain

Coastal plain strata comprise a relatively complex association of lithologies that include interbedded very fine- to fine-grained sandstones, siltstones, carbonaceous shales and coals. Sandstones range from a few inches to a few feet in thickness. Interbedded sandstones, siltstones and shales are commonly rooted and contain abundant comminuted carbonaceous material. Minor amounts of burrowing, pelecypod debris, and oyster shell debris were observed in some localities. Individual sandstone beds are very fine- to fine-grained and contain current ripples. Coals range from a few inches in thickness to a maximum of 8 feet and in some cases can be traced for several miles along the outcrop.

Interbedded, fine-grained coastal plain successions are interpreted as crevasse splay and overbank deposits associated with fluvial channels, and as thick, coal-bearing, coastal swamp and minor lagoonal sediments landward of the shoreline. Thin coals that are present throughout the succession may be associated with fluvial point bars or crevasse splays and are not as laterally continuous as those deposited in coastal swamp environments (McCabe, 1988).

Channels

Two different scales of channels were observed in the Kenilworth Member in the study area. Large channels up to 60 feet thick were recorded in several outcrops, with channel-fills consisting of very fine- to medium-grained sandstone. These have a sharp, erosional base which is overlain by mudclast lags and coal debris. The sandstone bodies fine-upward and may contain large-scale lateral accretion surfaces. The lowermost part of the sandstones is trough cross-bedded, while the uppermost part is current rippled and contains comminuted carbonaceous debris and root traces.

Large channelized sandstones are interpreted to have been deposited by meandering fluvial systems. This interpretation is based on the upward-fining grain size, the presence of lateral accretion surfaces and the lack of burrowing. All of the large channels that were identified in the Kenilworth appear to incise deeply into underlying shoreface sandstones and shales. A series of channel-form sandstones were observed in the cliff-face at the East of Grassy measured section and these had a significant amount of erosion associated with them that could be traced along the entire cliff face for 3 to 4 miles. The existence of these large, laterally extensive channels that incise deeply into shoreface deposits has important stratigraphic implications that will be discussed later in the paper.

Several thinner channelized sandstones up to 20 feet in thickness were also observed in the Kenilworth. These sandstones have similar sedimentary features to the large channels, but are laterally discontinuous. They are also interpreted as meandering fluvial deposits, but were only found in association with fine-grained coastal plain sediments (Fig. 4c). This suggests that these thin, discontinuous channels are an element of the normal coastal plain succession.

Delta Front and Distributary Mouth Bar

Delta front and distributary mouth bar deposits are generally found in the eastern part of the study area near Hatch Mesa. A typical vertical succession is shown in Figure 4d. Delta front sediments comprise very fine- to fine-grained sandstones interbedded with carbonaceous shales. Sandstone beds are a few inches in thickness and commonly form a thickening- and coarsening-upward succession. Sedimentary structures within individual sandstone beds include abundant current ripples (Fig. 5a) and "b-c" or "b" Bouma sequences. The bases of these beds have sole marks that include flute casts, prod marks and grooves. Flame structures were noted in some places and burrowing was very sparse to absent.

Distributary mouth bar sandstones range from 1 to 8 feet in thickness and are very fine- to fine-grained. They commonly pinch-and-swell dramatically (Fig. 5b) and have minor shale interbeds. Sedimentary features include massive to parallel laminated intervals that are characteristic of "a-b" Bouma sequences (Fig. 5c). Current ripples, comminuted carbonaceous debris and minor bioturbation were observed locally within these sandstones.

Lag Deposits and Oolites

In several outcrops, lag deposits were observed to occur within marine shales or at the top of the upper/lower shoreface. These were also noted by other workers (e.g. Swift *et al.*, 1987; Balsley and Horne, 1980; Newman,1985). The lags comprise medium- to coarse-grained quartz and chert-rich sandstones containing abundant shark's teeth, bone fragments, shell debris (commonly oysters and pelecypods), shale clasts, and scattered chert pebbles (Fig. 5d). In outcrops where the lag occurs at the top of the upper shoreface (e.g. North Price River Canyon), it is commonly churned by *Skolithos, Ophiomorpha* and *Thalassinoides* (Fig. 6a). This churning is likely the result of biogenic reworking subsequent to transgression, as these lags are overlain by marine shales.

In localities where the lag is isolated within marine shales (e.g. Hatch Mesa 2), it has a sharp base and locally rests in small channels on a surface marked by angular erosional truncation (Fig. 6b). At Hatch Mesa 6 (Fig. 6c) the shales immediately underlying the channelized lag contain abundant marine dinoflagellates and were deposited in nearshore, well-aerated marine waters (R.W. Harris, pers. comm., 1990). An erosional surface separates the lag from the shales. The lag deposits at Hatch Mesa 6 have large-scale, sigmoidally-shaped cross-beds with preserved clay drapes (Fig. 6d), as well as trough cross-beds, lateral accretion surfaces, current ripples and some burrows. These sharp-based lags resting on marine shales are interpreted as tidal channels and/or tidal inlet deposits.

Oolites were only observed in the most basinward measured section at Hatch Mesa 6 (Fig. 7a). Here they occur below the channelized lag described above. The oolites are nucleated on quartz and chert grains or shark's teeth fragments, and oolite beds contain a significant proportion of associated bioclastic and plant debris. They sit sharply on underlying silty shales that contain *Glossifungites* traces. Locally, a thin grit layer was found at the base of the oolites. Internal scouring occurs throughout the oolite beds, although this is not readily apparent due to minor tectonic folding of the strata. Primary sedimentary structures are not well preserved but appear to be predominantly trough cross-beds. Oolite deposits are patchy and have a linear, en-echelon distribution. They have been interpreted by Newman (1985) as "shelf bar forms" and may represent oolitic shoals.

KENILWORTH PARASEQUENCES

Five parasequences that are interpreted to have been deposited in a wave-dominated shoreline environment were identified in the Kenilworth Member. The parasequences recognized in this study are similar to those described by Van Wagoner *et al.* (1990), Van Wagoner (1987), and Kamola *et al.* (1985) from the Spring Canyon Member of the Blackhawk Formation. A measured section through a series of wave-dominated parasequences is shown in Figure 8. Each parasequence corresponds to a cyclothem of Young (1957) and varies in thickness, averaging about 75 to 100 feet.

Wave-dominated shoreline parasequences have predictable vertical and lateral facies associations. An idealized parasequence is bounded at the base by a marine-flooding surface that is successively overlain by offshore marine/offshore transition zone, lower shoreface, upper shoreface/foreshore and coastal plain environments upward in the vertical profile. Complete parasequences are not commonly observed due to non-deposition of some facies and environments, particularly near the basinward limit of progradation.

Deltaic deposits were observed locally in some parasequences, but most of them occur near Hatch Mesa. A typical measured section from this locality is shown in Figure 9. An idealized deltaic parasequence at Hatch Mesa shoals-upward from offshore marine (prodelta) shales into delta front sandstones and shales, then into distributary mouth bar sandstones at the top of the parasequence.

DISCUSSION

Sequence Stratigraphic Architecture

The continuous outcrop exposure along the Book Cliffs allows for direct observations of stratal patterns and stratigraphic architecture within the Kenilworth Member. Flooding surfaces and sequence boundaries were the key stratal surfaces observed in the field. Flooding surfaces were easily recognized because of recessive intervals in the cliff-faces. Candidates for sequence boundaries were more difficult to identify because they commonly exist in the sand-prone part of the sedimentary section. This results in sand-on-sand contacts that can easily be missed in the field unless careful outcrop observations are made. Criteria used to identify possible sequence boundaries included: 1) evidence of erosional truncation of regional extent; 2) basinward shifts of facies and environments; and 3) changes in parasequence stacking patterns, from progradational to retrogradational.

The cross sections in Figures 10 and 11 show the stratal surfaces, systems tracts, parasequences, and facies relationships that were traced along the cliff-faces between the measured sections. The base of the Kenilworth Member is a maximum flooding surface and parasequence set boundary that can be followed updip to its maximum landward extent near the town of Helper, Utah. Four parasequences (PS1 through PS4) overlie this parasequence set boundary and each successively younger parasequence progrades further basinward forming a progradational parasequence set.

The upper surface of PS4 is marked by erosional truncation and channelization throughout the study area. Channelization is best exhibited near the East of Grassy section, where fluvial channels displaying lateral accretion are cutting into offshore transition zone, delta front, and lower shoreface deposits (Fig. 7b). In areas where channels do not occur at the top of PS4, medium- to coarse-grained, burrow-churned lags or coastal plain deposits are commonly found.

Because the top of PS4 is marked by regional erosional truncation, incision through older shoreface deposits, and a basinward shift of facies, it is interpreted as a sequence boundary. Up to 60 feet of incision is observed in outcrop along this surface. The sequence boundary occurs at the top of a progradational parasequence set and is overlain to the west (in a landward direction) by a backstepping parasequence (PS5; Fig. 10). This change in parasequence stacking pattern from progradational to retrogradational also suggests that the erosional surface at the top of PS4 is a sequence boundary.

The parasequence stacking patterns and the position of the sequence boundary can be used to interpret systems tracts. Parasequences 1 through 4 form a progradational parasequence set and are interpreted as the highstand systems tract. The incised valleys overlying the sequence boundary at the top of PS4 are filled with fluvial channel deposits and are interpreted as the lowstand systems tract. Interfluve areas between incised valleys are marked by bioturbated lags. PS5 backsteps over the lowstand deposits (or the sequence boundary in interfluve areas) and is interpreted as the transgressive systems tract. The top of PS5 is the maximum flooding surface that denotes the base of the Sunnyside Member.

In a basinward direction at Hatch Mesa, the lowermost part of the succession is characterized by lag deposits and oolites. The lag deposits are lithologically similar to those found on the sequence boundary in the interfluve areas at the top of PS4. This suggests that the base of the lag at Hatch Mesa represents the basinward expression of the sequence boundary identified in the updip areas. The underlying oolite deposits are interpreted as the initial basinal response to a relative fall in sea-level, with wave reworking of the sea-floor resulting in the development of oolitic shoals. The overlying coarse, channelized lag is interpreted to rest on the sequence boundary. This surface represents the point in time at which the shoreline shifted to its maximum basinward position. The lags coinciding with the sequence boundary at Hatch Mesa are tidally-influenced and were deposited during the early part of the lowstand systems tract. The marine shales and deltaic sediments overlying the lags are interpreted to have been deposited in the later part of the lowstand systems tract.

The Kenilworth Member is therefore interpreted to contain parts of two high-frequency sequences. A schematic sequence stratigraphic model is shown in Figure 12. The lowermost part of the Kenilworth, consisting of PS1 through PS4, is the highstand systems tract of one high-frequency sequence. This systems tract is truncated by the sequence boundary at the top of PS4. The uppermost part of the Kenilworth, consisting of incised fluvial channels, lag deposits, deltaic deposits at Hatch Mesa, and Parasequence 5, comprises the lowstand and transgressive systems tracts, respectively, of a second high-frequency sequence.

The Kenilworth Member as defined by lithostratigraphy is correlated from maximum flooding surface to maximum flooding surface and corresponds to a "megacyclothem" of Young (1957). Swift *et al.*, (1987) considered each of the members of the Blackhawk a single parasequence, thus missing the importance of the detailed architecture within the Kenilworth. By focussing on the maximum flooding surfaces, a significant chronostratigraphic erosional surface (i.e. the sequence boundary) was not recognized by these authors.

Paleogeography

Facies Mapping from Outcrop Exposures

Paleogeographic maps for each parasequence were created directly from field observations. In most cases, stratal surfaces and characteristic whitecap weathering profiles from closely spaced outcrop sections could be traced laterally between canyons or along the cliff-faces using binoculars to detect lateral facies changes. The areal distribution of interpreted environments was plotted on topographic maps from which summary paleogeographic maps were constructed. The elements that were mapped included: 1) the updip limit of marine sandstone, which represents the change from shallow marine shoreface sandstone to coastal plain in a landward direction; 2) the downdip limit of coastal plain, which could be mapped using the seaward extent of coastal plain coals immediately overlying the foreshore; 3) the downdip limit of upper shoreface, identified by the disappearance of whitecaps in the cliff-face; 4) the downdip limit of lower shoreface, recognized by the change from massive, cliff-forming sandstones to interbedded sandstone and shale; 5) the location and orientation of channels; and 6) the presence and distribution of deltaic deposits. The paleogeographic maps constructed for each parasequence show the updip to downdip distribution of depositional environments. Paleoshoreline orientations were determined from these maps and paleocurrent data was compared to the mapped shoreline trends (see the following section).

The paleogeography for highstand parasequences in the Kenilworth (PS1 through PS4) is shown in Figures 13 and 14. These parasequences consist primarily of wave-dominated shorelines, although deltaic deposits were observed locally, as shown on the maps. Shorelines are oriented approximately north-northwest to south-southeast in all parasequences. The amount of progradation within each parasequence can be estimated from the maps by measuring the distance from the updip limit of the foreshore to the downdip limit of the upper shoreface. The average distance of progradation is 8.5 miles, although this is somewhat skewed by the large progradational extent of PS1 (19 miles). The paleogeographic maps suggest that the lower shoreface zone averaged 2-3 miles in width, at least at the maximum progradational extent of individual parasequences.

The observed distribution of fluvial and distributary channels within incised valleys along the cliff-faces was used to construct the lowstand paleogeographic map shown in Figure 15a. Lowstand deltaic deposits were only observed at Hatch Mesa. The east-west channel trends were interpreted from field mapping and paleocurrent data. The amount that sea-level fell during the lowstand is estimated to be about 60-to-100 feet, based on the depth of incision at the top of PS4. From the paleogeographic map of the lowstand systems tract, the distance into the basin that the shoreline shifted during the lowstand was measured from the downdip limit of the foreshore in PS4 to the position of the lowstand delta at Hatch Mesa (Fig. 15a). This basinward shift of facies is approximately 10-to-12 miles and yields a paleoshelf gradient of 0.07°, which is consistent with modern shelf gradients in parts of the Gulf of Mexico.

Following lowstand deposition, wave-dominated shorelines once again predominated in the transgressive systems tract (Fig. 15b). The mapped distribution of depositional environments suggests that the amount of shoreline progradation in this parasequence was more areally restricted than those of earlier highstand parasequences. North-northwest/south-southeast paleoshoreline trends existed during this time and no deltaic deposits were observed.

Paleocurrents

Paleocurrent measurements were taken from a variety of directional indicators in sandstones of the highstand, lowstand and transgressive systems tracts. In lower shoreface deposits, the strikes of wave ripple crests were recorded, while in upper shoreface deposits orientations of trough cross-bed axes were measured. In channel deposits, trough cross-bedding and the strike of the channel walls were estimated, as were the strike and dip of lateral accretion surfaces. Paleocurrent indicators from delta front sandstones were predominantly current ripple foresets and sole marks. The data recorded from each of the measured sections are summarized in Figures 16a and 16b.

In the highstand and transgressive systems tracts of the Kenilworth, paleocurrent data from wave ripple crests indicate that the shoreline orientations were approximately northwest-southeast, which compares favorably with the paleogeographic maps. The strikes of wave ripple crests in wave-dominated deposits have been shown by Leckie and Krystinik (1989) to closely approximate local paleoshoreline trends. Trough cross-bed orientations have a dominant southerly component and suggest that there were strong shore-parallel currents.

Paleocurrent data from the fluvial/deltaic lowstand deposits show predominantly easterly paleoflow directions (Fig. 16b). This indicates that paleoslope was to the east and suggests that during periods of lowered sea level, fluvial and deltaic systems shifted basinward across the pre-existing highstand paleoslope.

Regional studies indicate that the shoreline trend in the study area during the Campanian was approximately north-south (Fouche *et al.*, 1983; Balsley and Horne, 1980; Young, 1957). Paleocurrent data published by other workers from the Kenilworth Member (e.g. Swift *et al.*, 1987; Newman, 1985) generally support the north-south strandline trend, although Swift *et al.*, (1987) suggested that in the area of Hatch Mesa the shoreline was northeast-southwest. Observations made during this study do not support that conclusion. Data from Leckie and Krystinik (1989) show that the shoreline orientation approximated by wave ripple crests was north-northwest/south-southeast (340-160°) and that offshore was to the east-northeast (070°). These observations closely agree with the data presented in this paper.

Speculations on Large-Scale Sequence Stacking Patterns in the Blackhawk Formation

Observations and interpretations from this paper suggest that high-frequency sea-level fluctuations were occurring during deposition of the Blackhawk Formation. Previous studies by Balsley and Horne (1980), and Young (1957, 1955) have shown that the Blackhawk Formation is an overall progradational succession, with each successively younger sandstone member prograding further basinward and changing facies into the Mancos Shale. The oldest member, the Spring Canyon, comprises a thick aggradational to progradational parasequence set (Van Wagoner *et al.*, 1990; Kamola *et al.*,1985), while the youngest Desert Member is characterized by significant erosional truncation (Van Wagoner *et al.*, 1990). These observations are important for understanding the stratal architecture and speculating about the large-scale sequence stacking pattern of the Blackhawk Formation.

The entire Blackhawk succession in the Book Cliffs of east-central Utah is postulated to represent a highstand sequence set. A sequence set is a large-scale, complex sedimentary succession consisting of numerous high-frequency sequences (Mitchum and Van Wagoner, 1991). The stacking pattern of sequences within sequence sets should be similar to the stacking pattern of parasequences in normal sequences. Therefore, in the early highstand sequence set, an aggradational sequence stacking pattern should occur. Also, because the rate of accommodation is relatively high at this time, little or no truncation associated with sequence boundaries is predicted. This is the case in the Spring Canyon and Aberdeen Members where thick, aggradational to progradational parasequence sets are common and sequence boundaries with good evidence of erosional truncation have not been documented. Changes in parasequence stacking patterns were noted near the top of the Aberdeen Member in this paper (Fig. 10), suggesting the possible existence of a sequence boundary in this interval. The Kenilworth is the first Blackhawk member to have a significant amount of erosion associated with a sequence boundary and suggests that the onset of late highstand sequence set deposition occurred at this time.

The uppermost Desert Member of the Blackhawk appears to have about twice as much truncation along its sequence boundary as the Kenilworth (Van Wagoner *et al.*, 1990), suggesting an increase in the amount and depth of erosion in the latter part of late highstand sequence set deposition. This increase in the amount of erosion can be attributed to reduced accommodation during the late highstand portion of the eustatic sea-level curve. It is speculated that high-frequency sequences will be present in the Sunnyside and Grassy Members of the interpreted late highstand sequence set and that these will be characterized by well developed sequence boundaries. Research currently underway in the Sunnyside and Grassy Members indicates that high-frequency sequences are present in these units (C. O'Byrne, pers. comm., 1990).

The Blackhawk Formation is overlain by a major erosional surface at the base of the Castlegate Member which terminates highstand sequence set deposition and may represent a large-scale sequence boundary at the base of the Price River Formation.

CONCLUSIONS

The Kenilworth Member has been interpreted in this paper as comprising parts of two high-frequency sequences. The lowermost part of the Kenilworth consists of a progradational parasequence set characterized by wave-dominated shorelines and represents the highstand systems tract of one sequence. A sequence boundary overlies the highstand and can be traced basinwards where it is overlain by tidally reworked lags and deltaic deposits of the lowstand systems tract. A backstepping, wave-dominated shoreline parasequence of the transgressive systems occurs at the top of the Kenilworth. The lowstand and transgressive systems tract deposits comprise part of a younger high-frequency sequence.

Observations in this study and from studies by other workers indicate that high-frequency sequences exist in all members of the Blackhawk Formation (Van Wagoner *et al.*, 1990; Kamola *et al.*,1985; C. O'Byrne, pers. comm., 1990). Interpretation of the sequence stacking patterns suggests that these members form a highstand sequence set of high-frequency sequences within the Blackhawk.

The critical factor necessary for interpreting high-frequency cyclicity in the stratigraphic record is the ability to recognize the different building blocks of depositional sequences- sedimentary facies, flooding surfaces and parasequences, parasequence stacking patterns and regional erosional surfaces. The recognition of these elements at any particular position within a sedimentary succession will allow for more accurate predictions of facies distributions, and sandstone thickness, geometry and continuity.

REFERENCES

Balsley J.K. and Horne J.C., 1980, Cretaceous wave-dominated delta systems: Book Cliffs, East Central Utah. unpublished Amoco Production Company Field Guide, Denver Colorado, 161p.

Fouche T.D., Lawton T.F., Nichols D.J., Cashion W.B. and Cobban, W.A., 1983, Patterns and timing of synorogenic sedimentation in Upper Cretaceous rocks of central and northeast Utah. *In:* M.W. Reynolds and E.D. Dolly (eds.), Mesozoic Paleogeography of West-central United States, p. 305-336.

Kamola D.L., Pfaff B.J. and Newman S.L., 1985, Depositional environments in the Book Cliffs: an introduction. *In:* SEPM Mid-year Meeting Field Guide 10, p. 1-6.

Leckie D.A. and Krystinik L.F., 1989, Is there evidence for geostrophic currents preserved in the sedimentary record of inner to middle-shelf deposits? Journal of Sedimentary Petrology, v.59, p. 862-870.

McCabe P., 1988, Depositional environments of coal and coal-bearing strata. Canadian Society of Petroleum Geologists Short Course Notes, 66p.

Mitchum R.M., Jr., and Van Wagoner J.C., 1991, High-frequency sequences and their stacking patterns: sequence-stratigraphic evidence of high-frequency eustatic cycles. *In:* K.T. Biddle and W. Schlager (eds.), The Record of Sea-Level Fluctuations, Sedimentary Geology, v. 70, p. 131-160.

Newman S.L., 1985, Facies interpretations and lateral relationships of the Blackhawk Formation and Mancos Shale, East-central Utah. *In:* SEPM Mid-year Meeting Field Guide 10, p. 69-114.

Speiker E.M., 1931, The Wasatch Plateau coal fields, Utah. USGS Bulletin 819, 206p.

Swift D.J.P., Hudelson P.M., Brenner R.L. and Thompson P., 1987, Shelf construction in a foreland basin: storm beds, shelf sandbodies, and shelf-slope depositional sequences in the Upper Cretaceous Mesaverde Group, Book Cliffs, Utah. Sedimentology, v.34, p. 423-457.

Van Wagoner J.C., Mitchum R.M., Campion K.M. and Rahmanian V.D., 1990, Siliciclastic sequence stratigraphy in well logs, cores and outcrops: concepts for high-resolution correlation of time and facies. AAPG Methods in Exploration Series, No. 7, 55p.

Walker R.G., 1984, Shelf and shallow marine sands. in: R.G. Walker (Ed.), Facies Models, Second Edition, Geoscience Canada Reprint Series 1, p. 141-170.

Young R.G., 1955, Sedimentary facies and intertonguing in the Upper Cretaceous of the Book Cliffs, Utah-Colorado. GSA Bulletin, v.66, p. 177-202.

Young R.G., 1957, Late Cretaceous cyclic deposits, Book Cliffs, Eastern Utah. AAPG Bulletin, v.41, p. 1760-1774.

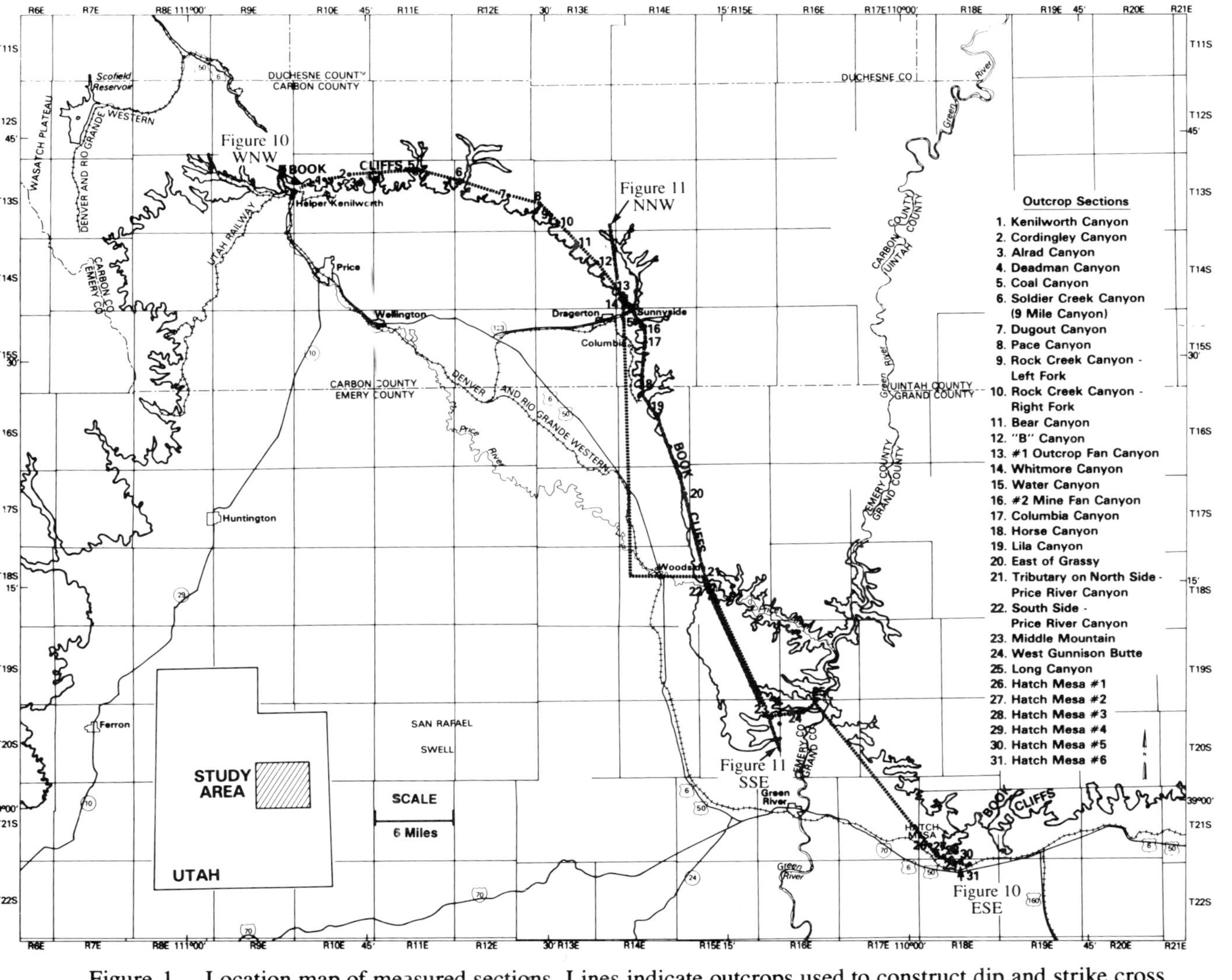

Figure 1 Location map of measured sections. Lines indicate outcrops used to construct dip and strike cross sections.

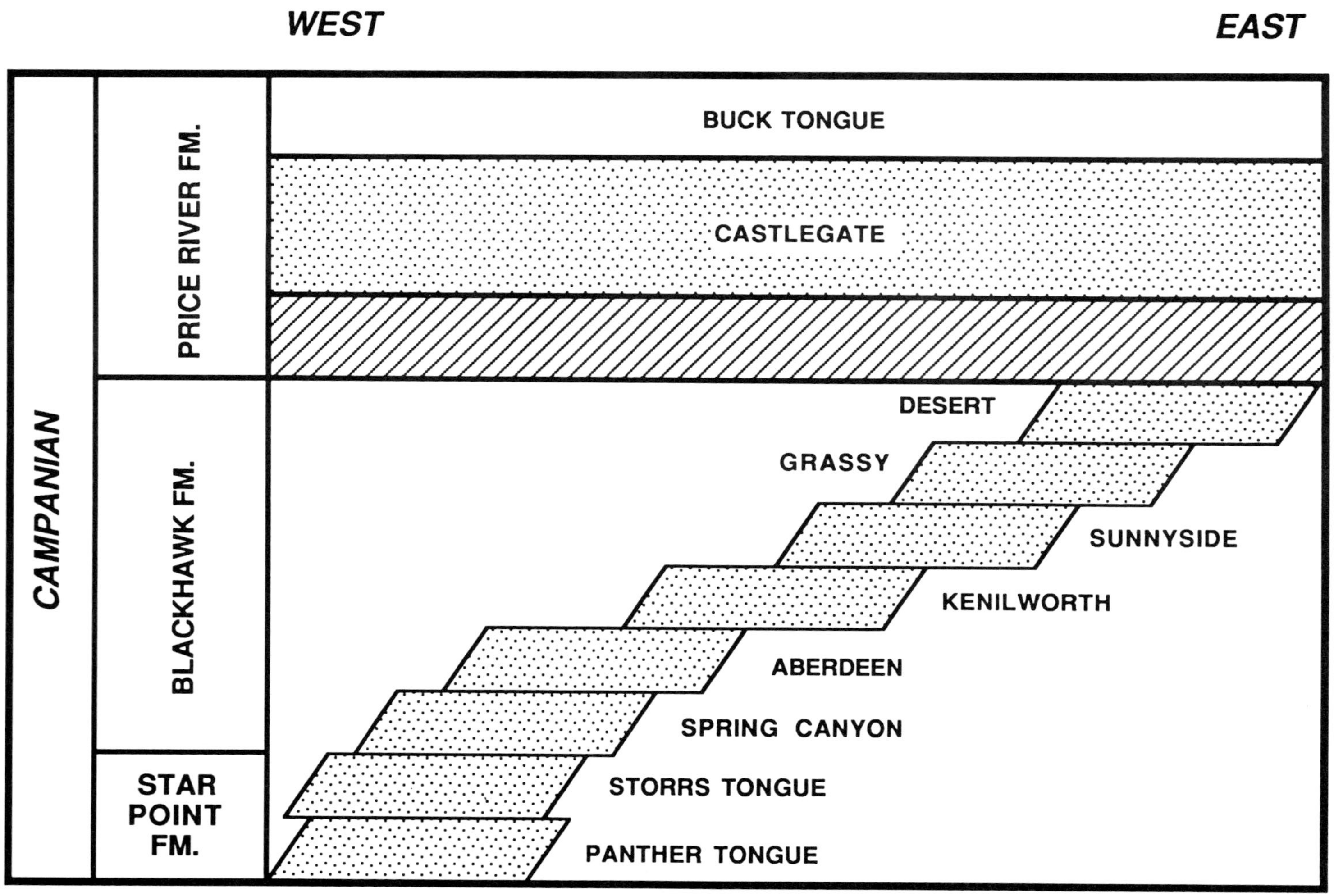

WEST

EAST

CAMPANIAN

PRICE RIVER FM.

BLACKHAWK FM.

STAR POINT FM.

BUCK TONGUE

CASTLEGATE

DESERT

GRASSY

SUNNYSIDE

KENILWORTH

ABERDEEN

SPRING CANYON

STORRS TONGUE

PANTHER TONGUE

Figure 2 Stratigraphic chart of Upper Cretaceous rocks in the Book Cliffs area (modified from Young, 1955).

Figure 3 a) Wave rippled sandstones (WRS) and bioturbated shales in the offshore marine/offshore transition zone at North Price River Canyon.
b) Flute casts (FC) and tool marks (TM) at the base of a current rippled sandstone bed in the offshore transition zone at North Price River Canyon.
c) Hummocky cross stratified sandstones and interbedded bioturbated silty shales in the lower shoreface at North Price River Canyon.
d) Amalgamated hummocky cross stratified sandstone of the lower shoreface at North Price River Canyon. Note the swale with onlapping laminae (arrow).

Figure 4 a) Plan view of trough cross beds in the upper shoreface at North Price River Canyon.
b) Parallel laminated foreshore sandstone at Rock Creek Canyon Right Fork. The upper surface is rooted and is overlain by a coal.
c) Cross bedded fluvial sandstone cutting into coal of the coastal plain at Rock Creek Canyon Right Fork.
d) Vertical succession of deltaic facies at Hatch Mesa.

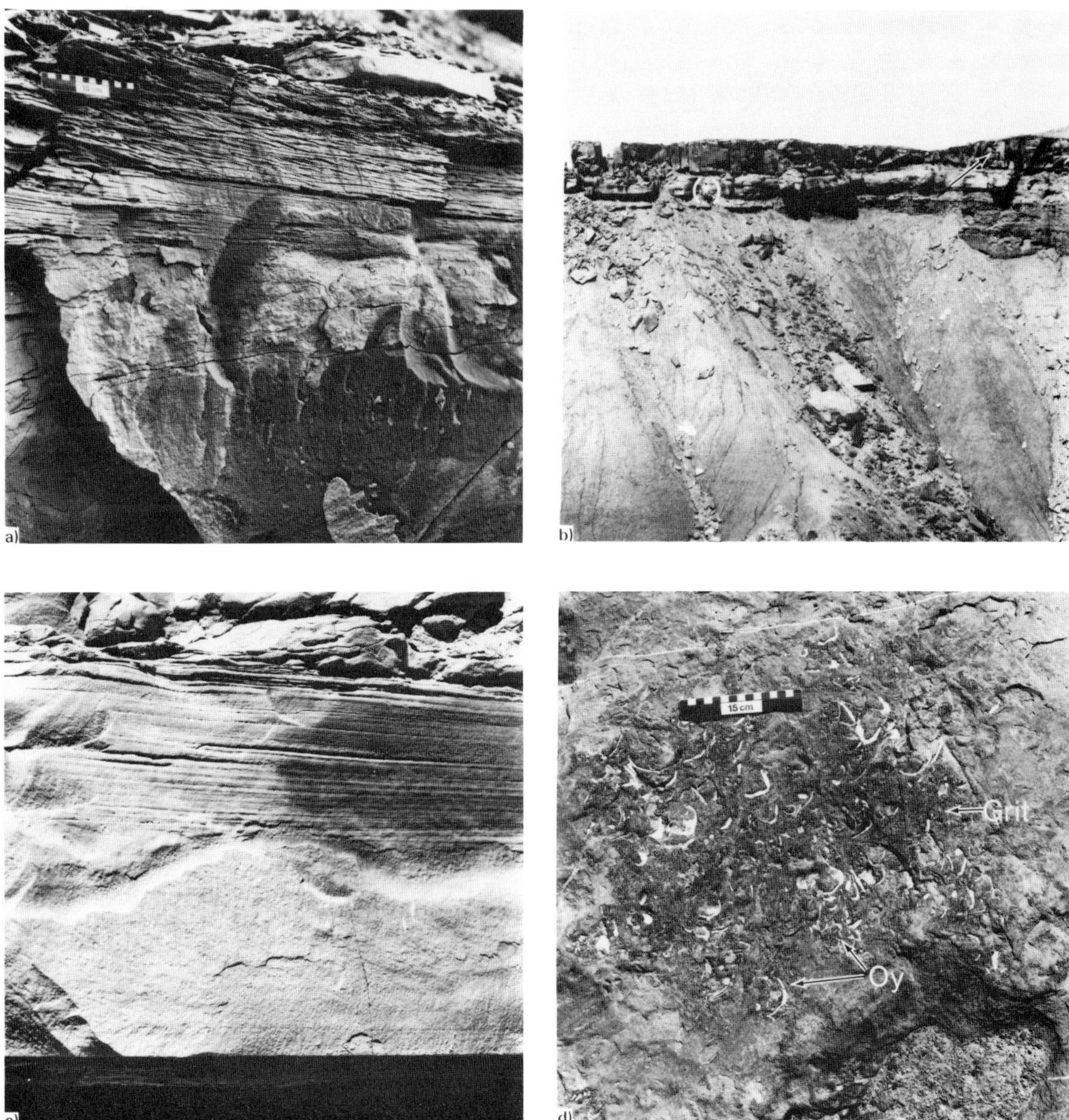

Figure 5 a) Current rippled and parallel laminated sandstone from delta front deposits at Hatch Mesa.
b) Lenticular, "pinch and swell" features (arrow) in distributary mouth bar sandstones at Hatch Mesa.
c) Bouma sequences displaying "a-b" divisions from distributary mouth bar sandstones at Hatch Mesa.
d) Gritty sandstone lag with oyster shell debris (Oy) and bone fragments at North Price River Canyon.

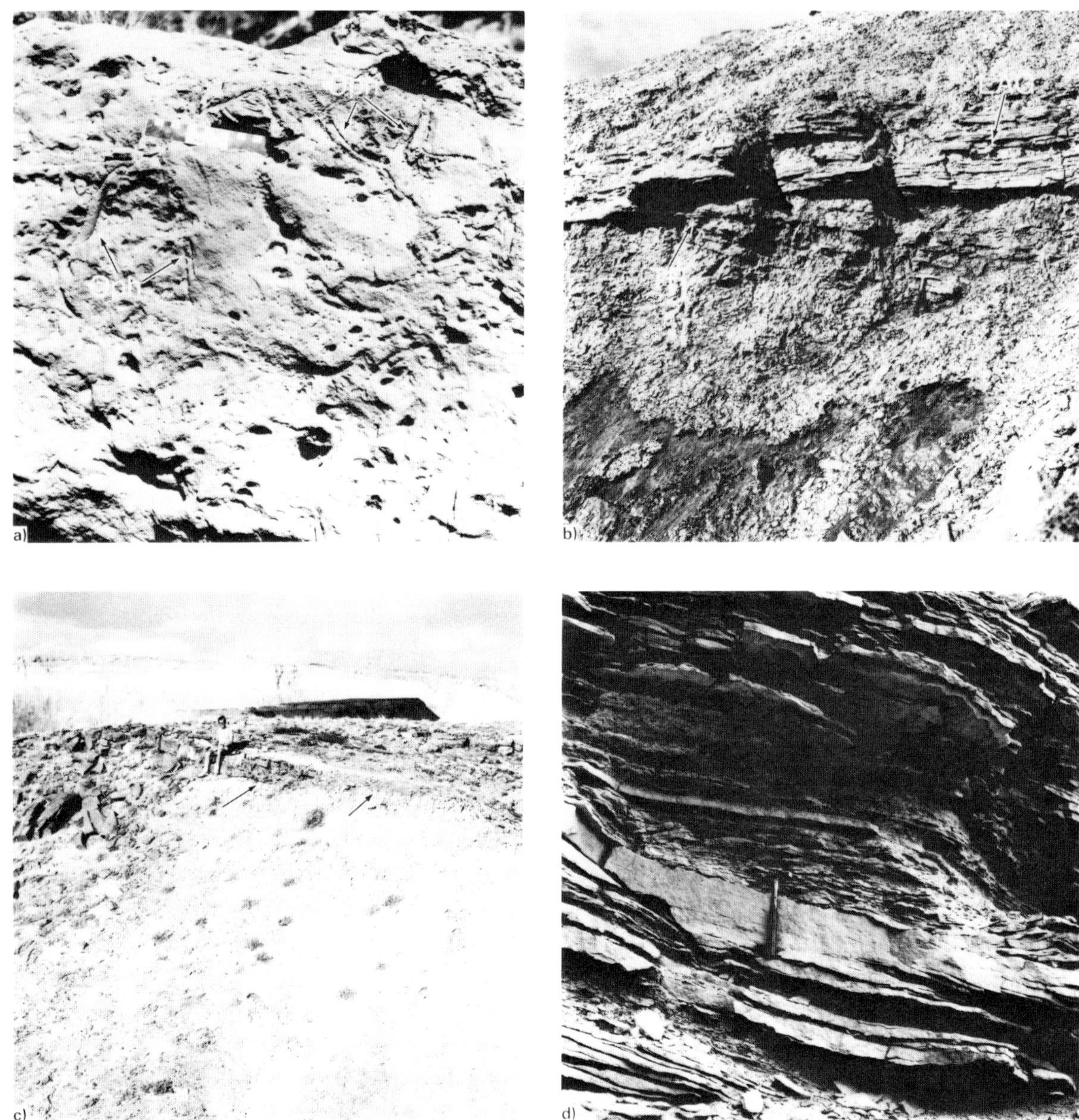

Figure 6 a) Ophiomorpha burrows (Oph) within a grit lag and underlying sandstone at the top of the upper shoreface at North Price River Canyon.
b) Truncation (TR) at the base of the lag deposit encased in marine shales at Hatch Mesa.
c) Trough cross bedded, channelized lag at Hatch Mesa. The base of the lag is marked by an arrow.
d) Sigmoidal bedding and clay drapes from the channelized lag near Hatch Mesa.

Figure 7 a) Oolite deposit at Hatch Mesa (dark area). Note the vehicle in the background for scale and the small syncline marked by the arrows.
b) Laterally accreting fluvial channel above the sequence boundary at the top of Parasequence 4. The channel cuts into offshore transition/distal delta front deposits.

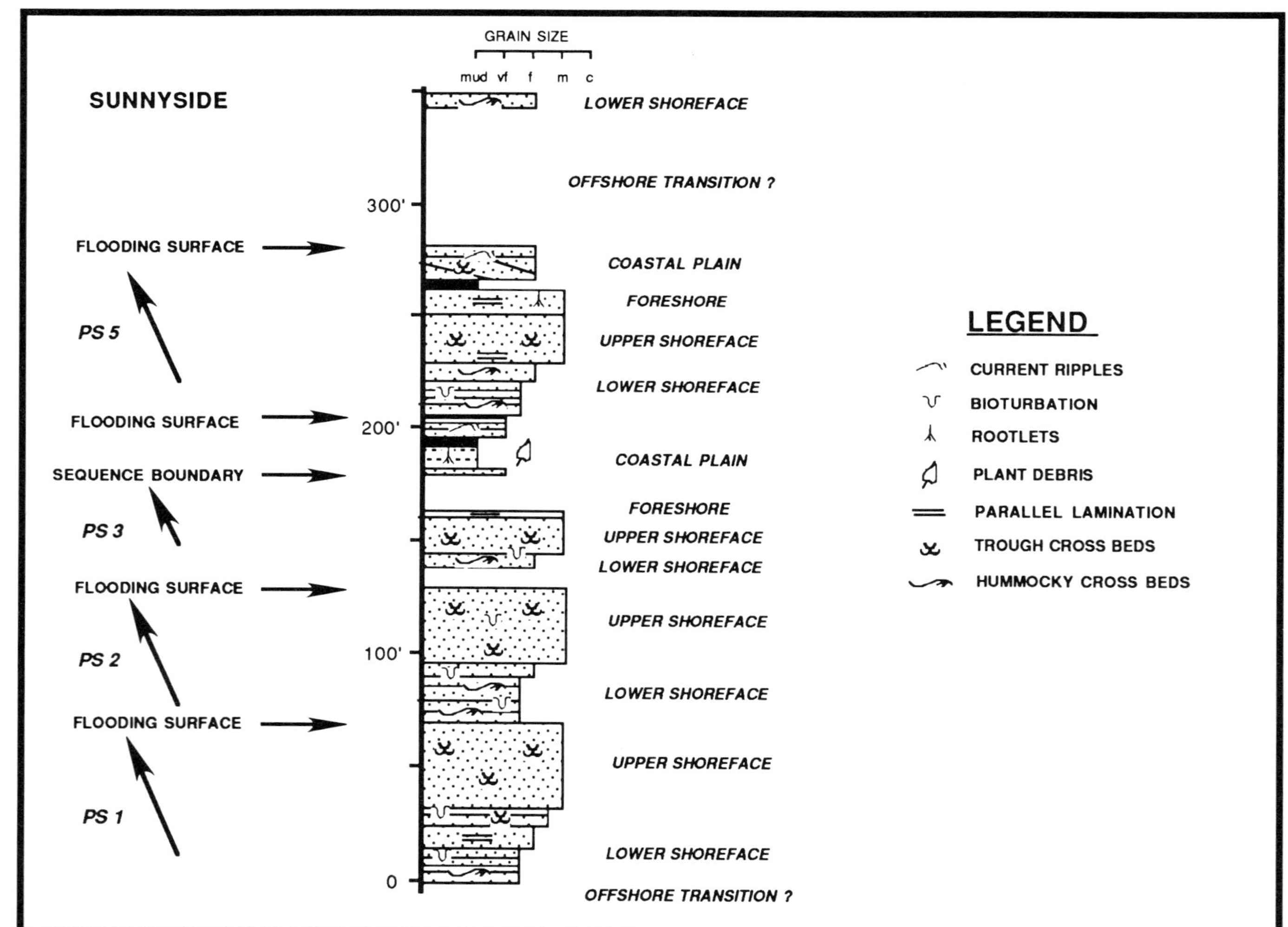

Figure 8. Measured section of Kenilworth parasequences characterized by wave-generated sedimentary features from Rock Creek Canyon Right Fork.

NNW

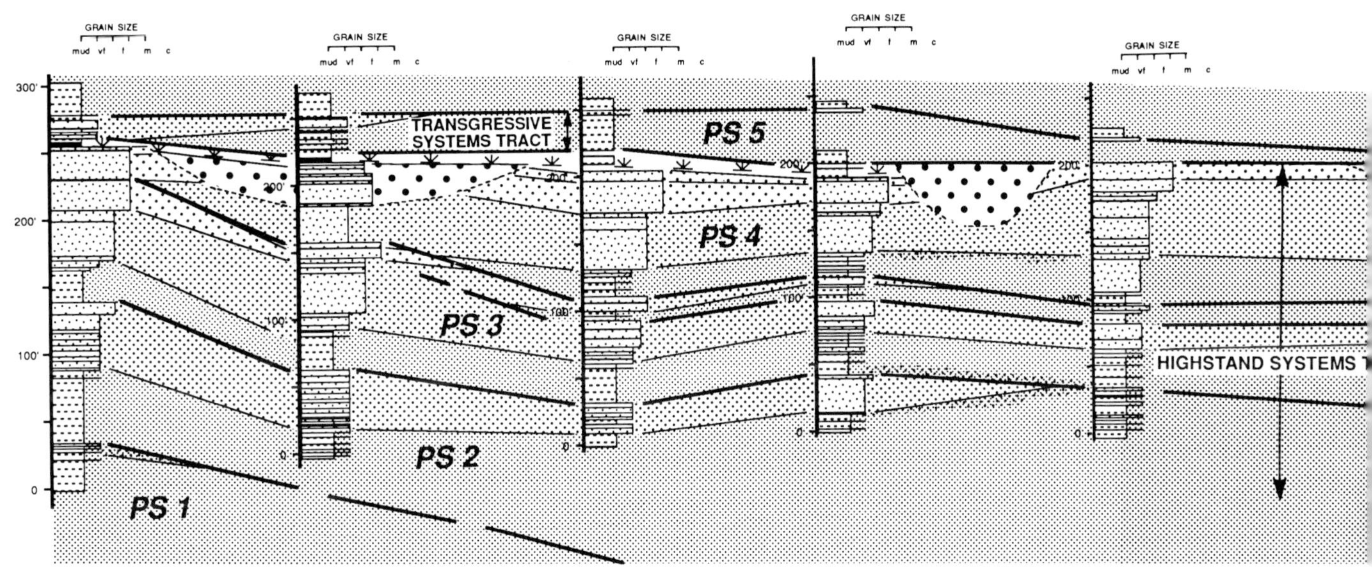

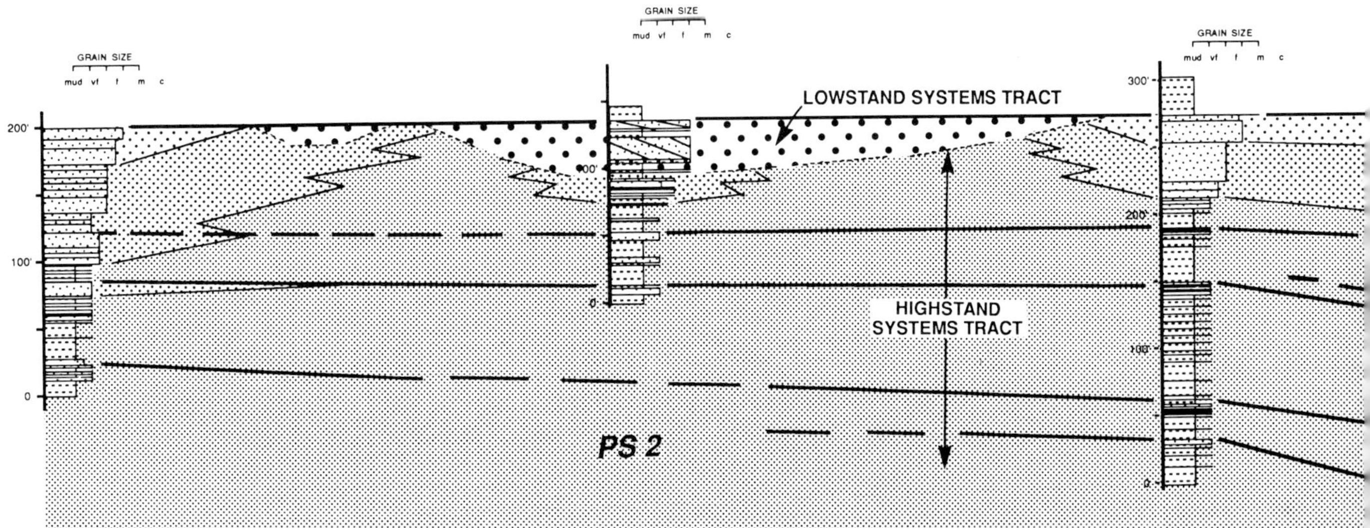

Figure 11 Depositional strike-oriented stratigraphic cross section of the Kenilworth Me observed in the cliffs which originates from the sequence boundary above P PS3 and PS4 indicate areas where the Book Cliffs have changed orientation these deposits are most likely shoreface-attached in the third demension.

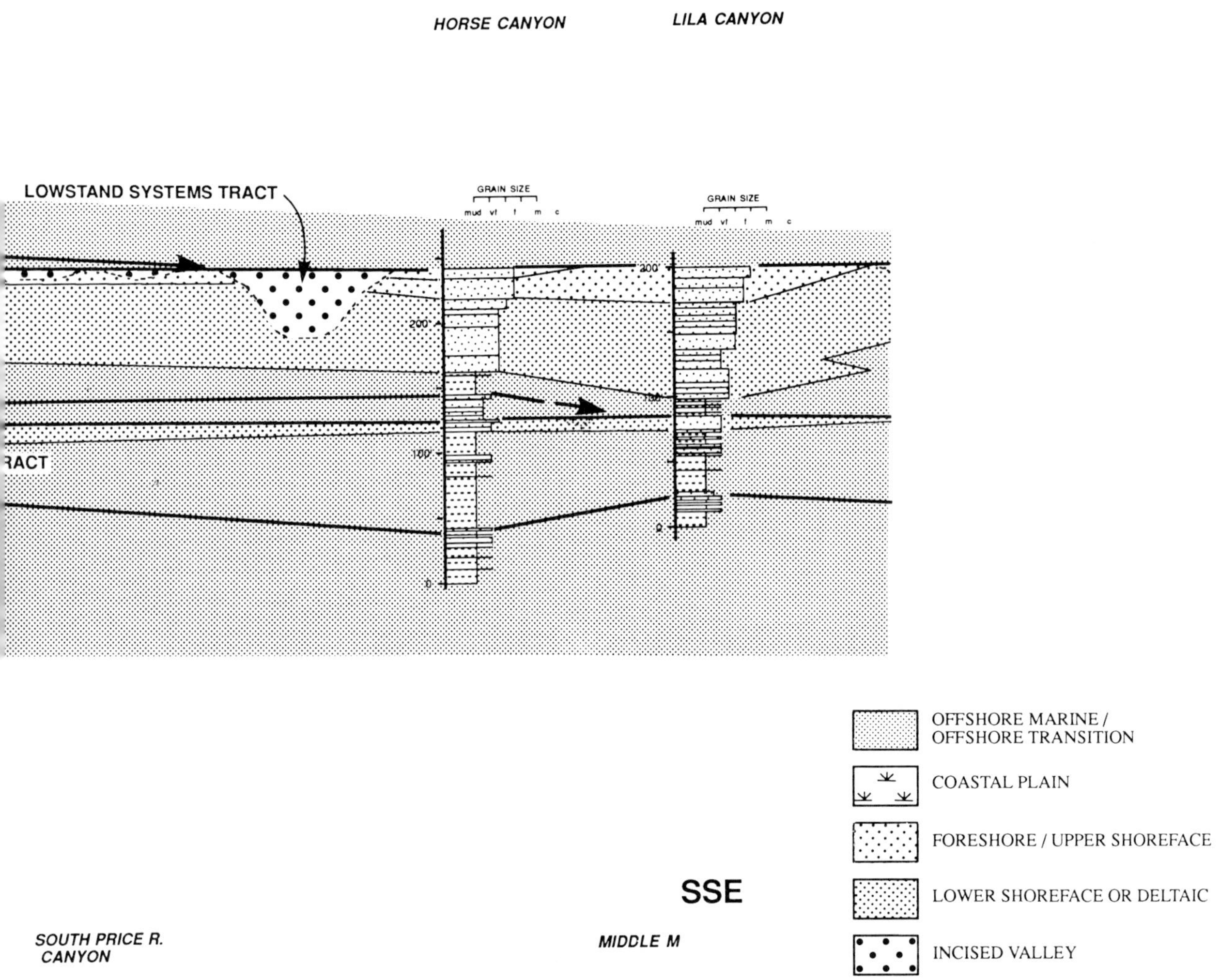

HORSE CANYON *LILA CANYON*

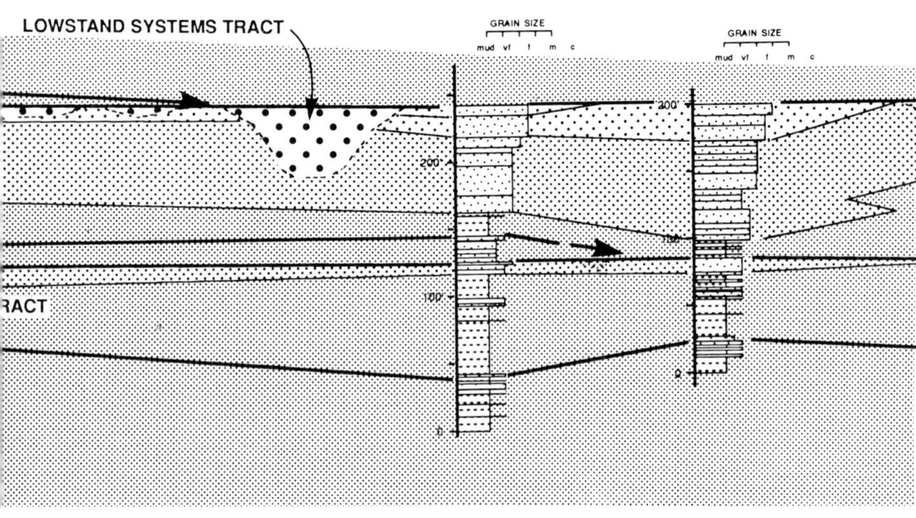

OFFSHORE MARINE / OFFSHORE TRANSITION

COASTAL PLAIN

FORESHORE / UPPER SHOREFACE

LOWER SHOREFACE OR DELTAIC

INCISED VALLEY

SEQUENCE BOUNDARY

FLOODING SURFACE

FACIES CHANGE

SSE

SOUTH PRICE R. CANYON

MIDDLE M

GRAIN SIZE

mud vf f m c

PS 4

PS 3

300'

200'

100'

0

nber. This section illustrates the channeling
4. The thin, discontinuous sandstones in
ith respect to the paleogeograohy and

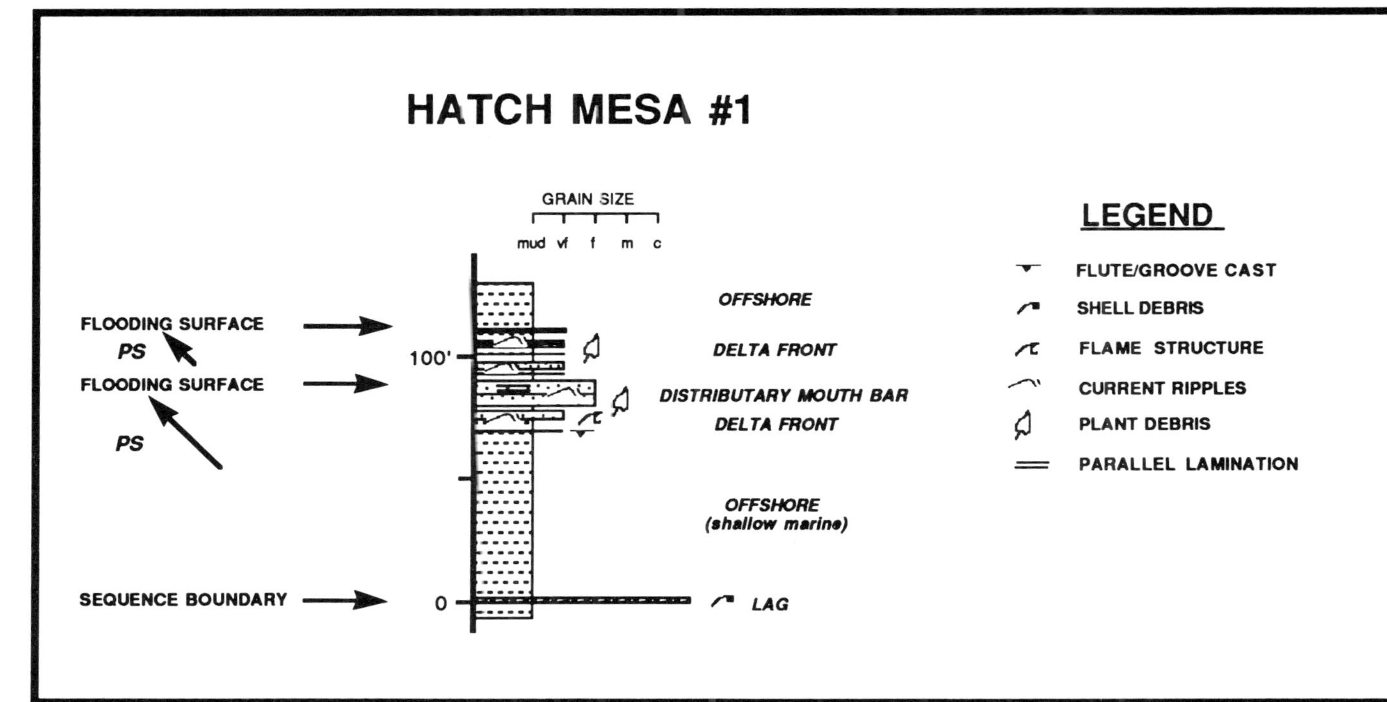

Figure 9 Measured section of parasequences characterized by deltaic deposits from the Kenilworth Member at Hatch Mesa.

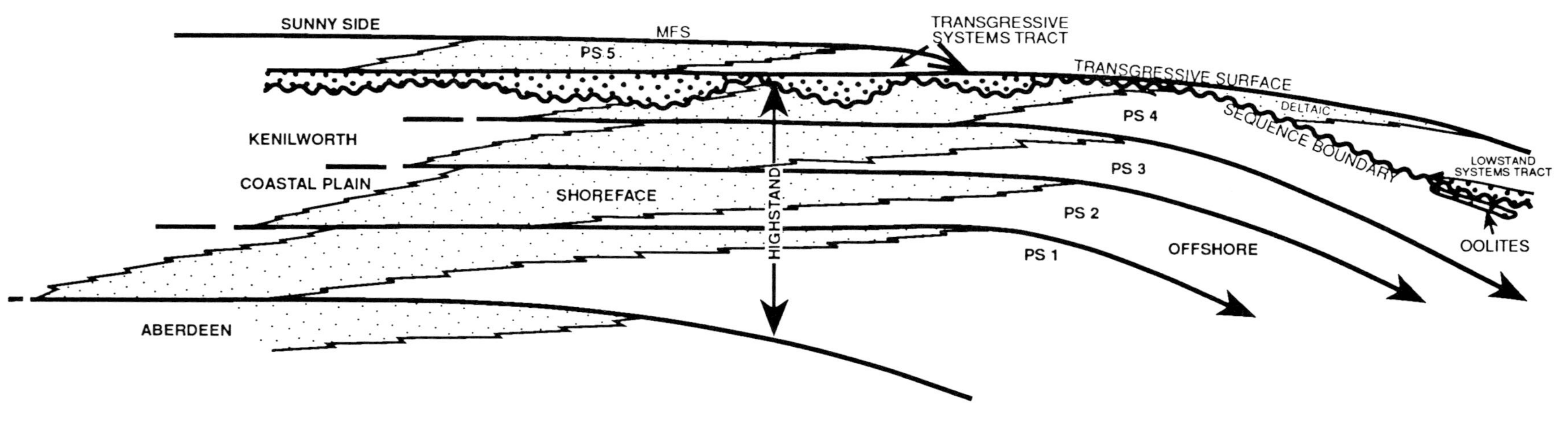

Figure 12. Schematic sequence stratigraphic model for the Kenilworth Member.

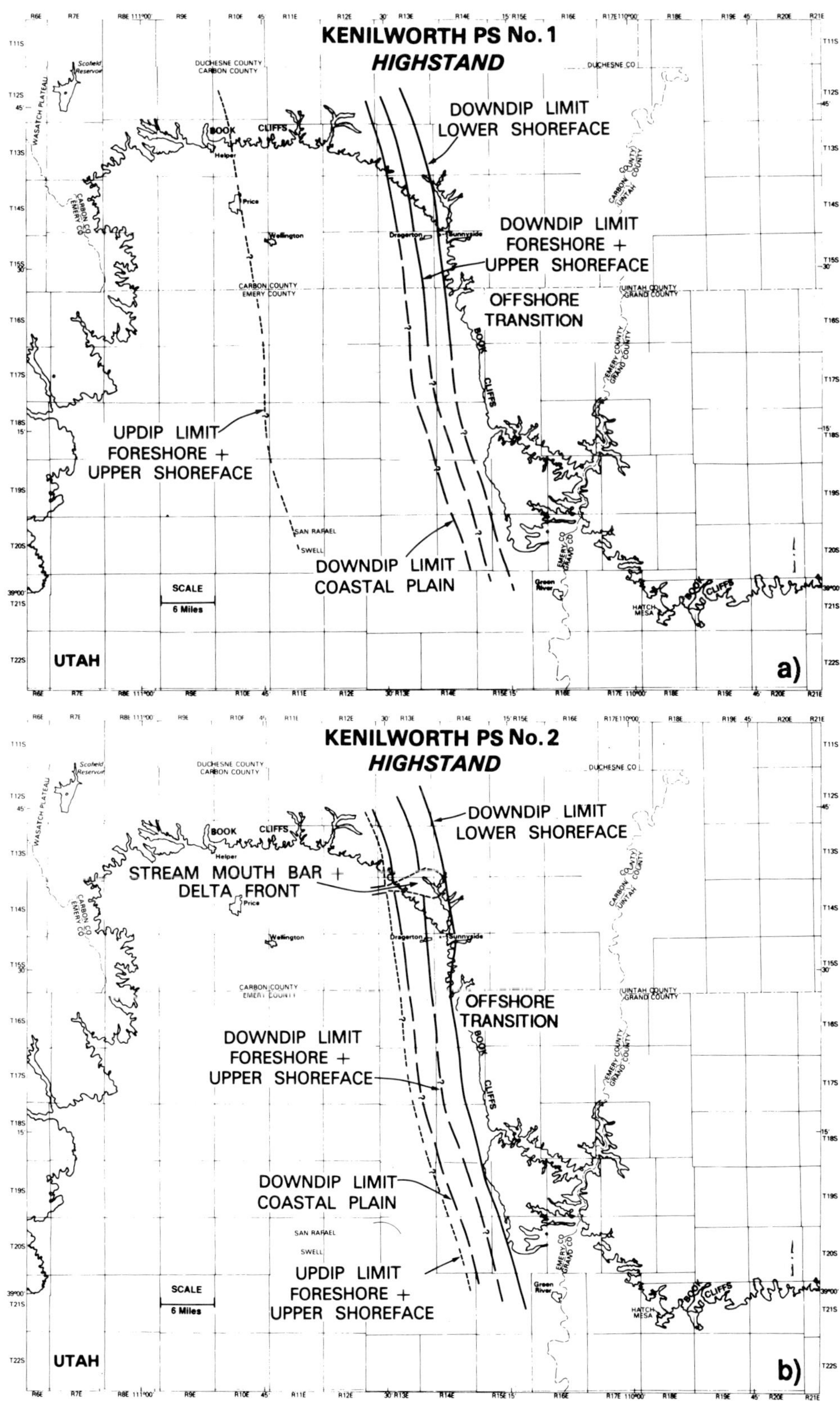

Figure 13 a) Paleogeography of PS1. The interpreted trends and facies distributions have been extrapolated on all maps away from the observations made in the cliff faces for each parasequence. PS1 through PS4 paleogeographic maps show the distribution of environments during the highstand systems tract. b) Paleogeography of PS2.

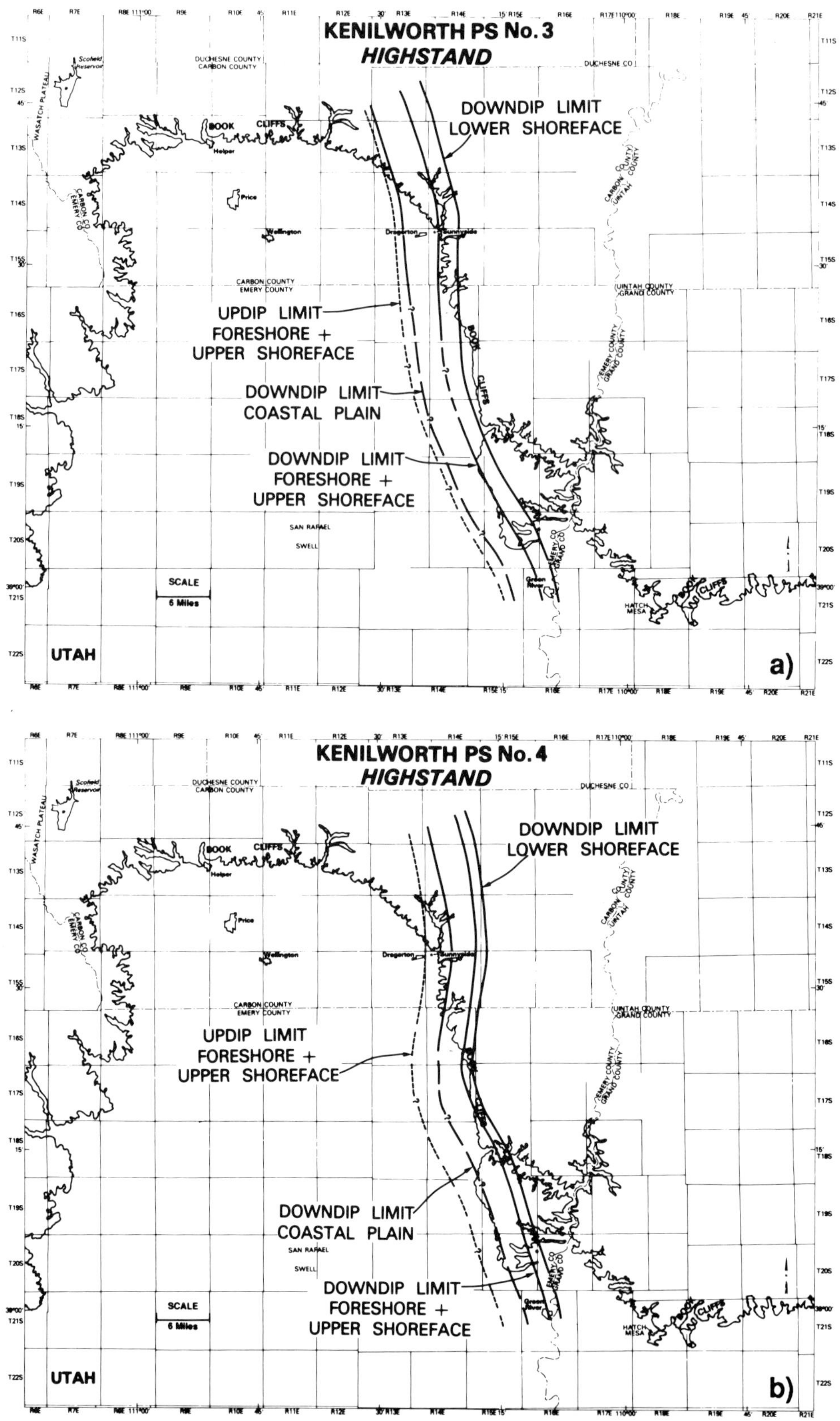

Figure 14 a) Paleogeography of PS3.
b) Paleogeography of PS4.

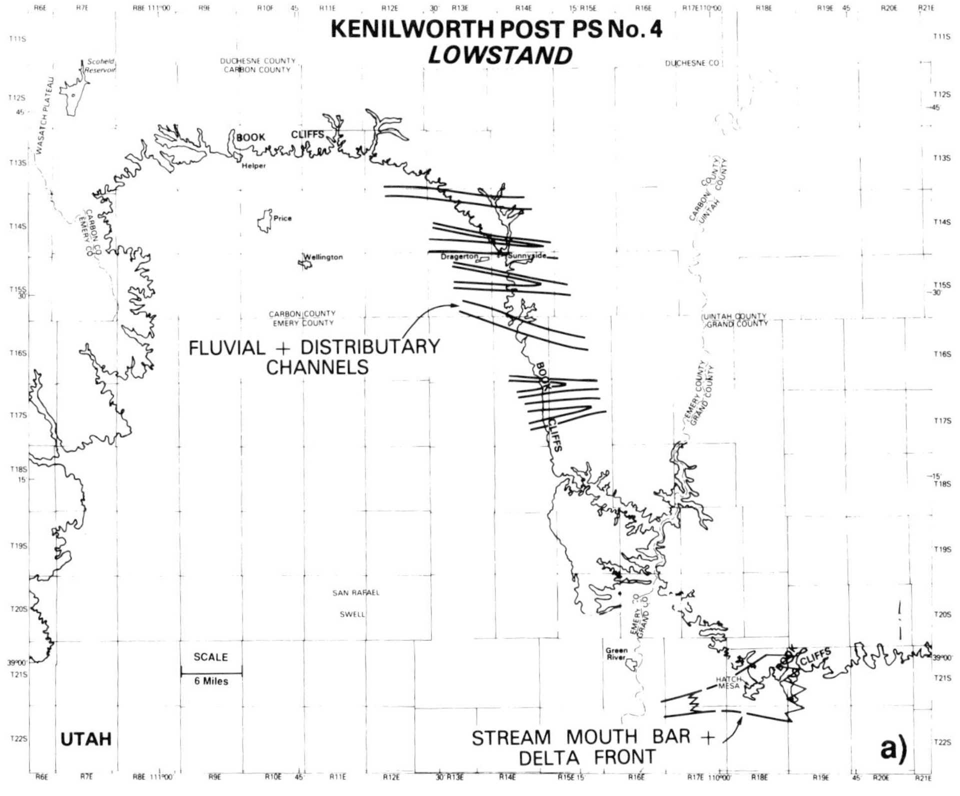

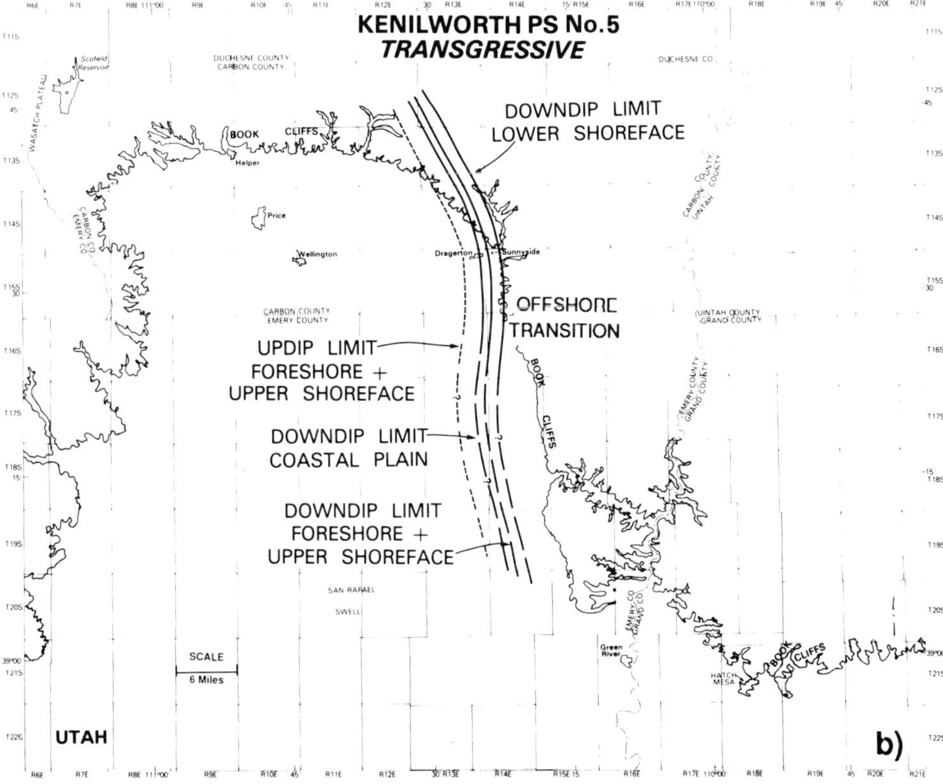

Figure 15 a) Lowstand paleogeography and distribution of incised valleys.
b) Paleogeography of PS5 in the transgressive systems tract.

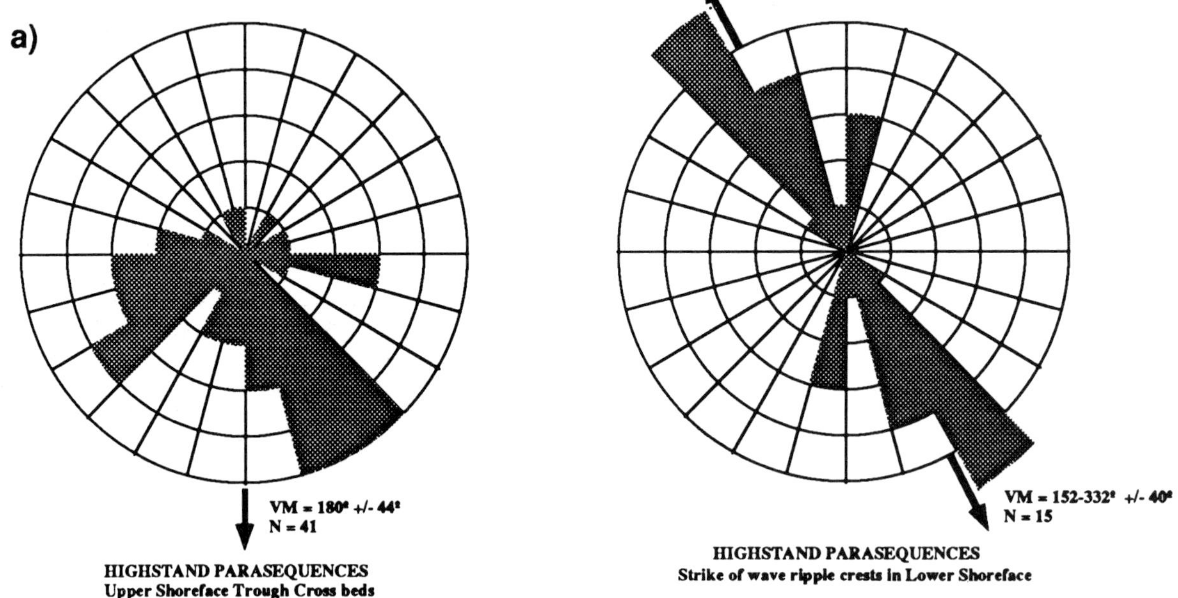

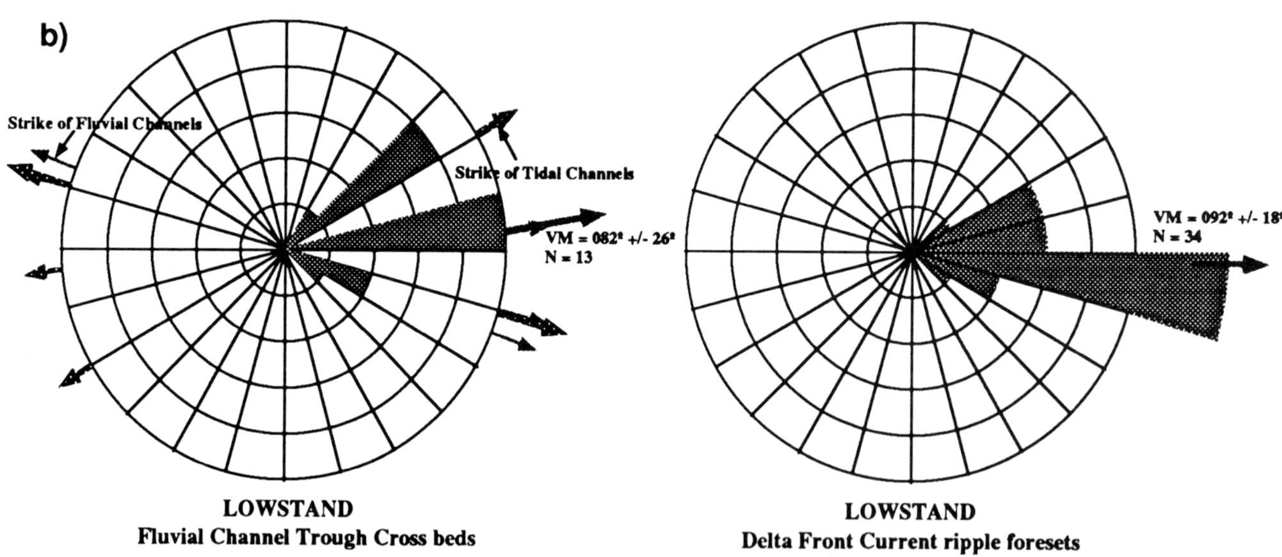

Figure 16 a) Paleocurrent measurements from highstand parasequences (PS1 to PS4).
b) Paleocurrent measurements from lowstand deposits.

DAY THREE

Green River to Moab, Utah

ROAD LOG, DAY THREE: NONMARINE SEQUENCE STRATIGRAPHY AND FACIES ARCHITECTURE OF THE UPDIP DESERT AND CASTLEGATE SANDSTONES IN THE BOOK CLIFFS OF WESTERN COLORADO AND EASTERN UTAH

by
John C. Van Wagoner
Exxon Production Research Company
Houston, Texas

OBJECTIVES: The Desert and updip Castlegate provide an opportunity to study well-exposed sequences and systems tracts in nonmarine to marine settings. Because of the excellent quality of the outcrops, sequence boundaries can be traced continuously from the nonmarine, where the identification of sequence boundaries has historically been less certain, to the marine, where sequence boundaries can be well defined based on basinward shifts in facies. These outcrops also allow regional correlation of systems tracts and identification of exploration-scale facies changes within systems tracts. These stops demonstrate that sequence boundaries are regionally extensive, single physical surfaces that can be used to correlate in outcrop and the subsurface. These stops also show that strata between sequence boundaries are arranged in predictable packages and that facies within these packages commonly, but not always, have predictable characteristics.

0.0 Leave the parking lot of the Best Western River Terrace Motel in Green River, Utah. Turn right on the main road, Highway 6.
0.2 Turn left on the paved road. Drive north toward the Book Cliffs. This road parallels the Green River and is built on the Mancos Shale.
5.3 After ascending a hill which climbs on top of the Mancos out of the Green River alluvial valley, turn off of the paved road onto a gravel road that heads northeast, directly toward Tuscher Canyon.

6.0 STOP ONE. Pull over to the side of the gravel road for an overview of the stratigraphy of the Desert Member of the Blackhawk Formation and the Castlegate Sandstone. This vantage point, looking east and northeast into Tuscher Canyon, affords spectacular views of the Campanian strata in the Book Cliffs. Unfortunately, if you are here in the morning much of what you want to see requires you to look directly into the sun. The stratigraphy at the mouth of Tuscher Canyon is summarized in Figures 3-1 and 3-2a. The Mancos Shale forms the base of the cliffs at the mouth of Tuscher Canyon. Three thin, distal parasequences of the Kenilworth Member of the Blackhawk Formation crop out just above the Mancos. If traced from this vantage point back to the west, these parasequences can be seen to change facies into thick, stacked beach parasequences of the Kenilworth below the Beckwith Plateau and Battleship Butte. Above the Kenilworth, the Sunnyside Member of the Blackhawk Formation is poorly developed in this area. The thick sandstone unit above the Sunnyside is the Grassy Member of the Blackhawk Formation. Well-developed, lower-shoreface hummocky beds compose the bulk of the Grassy. The upper part of the Grassy is locally incised. The incised channels, filled with fluvial deposits, mark a sequence boundary that can be traced regionally (O'Bryne, pers. comm., 1991). The Grassy is overlain by the Desert Member of the Blackhawk Formation. A sequence boundary occurs within the middle of the Desert. Lower-shoreface parasequences composed of hummocky strata compose the lower and middle portions of the Desert. A regionally extensive sequence boundary, marked by truncation and a basinward shift in facies, separates these hummocky strata from overlying coal-bearing coastal-plain rocks filling incised valleys. The coastal-plain strata overlying the sequence boundary in the Desert are in turn overlain by a sequence boundary at the base of the Castlegate. At the mouth of Tuscher Canyon the Castlegate is the thick, prominent sandstone at the top of the cliffs. In this area, the Castlegate is a braided-stream deposit, up to 120-feet thick. The Castlegate is sharply overlain by a flooding surface marking the parasequence set boundary at the base of the Buck Tongue of the Mancos Shale.

At the end of the overview, return to the vehicles and continue east into Tuscher Canyon.

9.1 Cross a small arroyo. Sunnyside parasequences are in the lower part of the cliff.

10.7 Climb a small hill out of the wash. You are driving past hummocky strata of the lower shoreface in the Grassy.
11.1 You are passing the top of the Grassy on the right of the road. In this area, the top of the Grassy is a sharp flooding surface overlain by marine mudstone.

11.5 STOP TWO. Pull over to the side of the road to examine the base of the lower-shoreface sandstone in the Desert. At this stop, shown in Figure 3-2b, the base of the Desert sandstone varies from a sharp contact with the underlying mudstone along some parts of the cliff, to a transitional contact along other parts of the cliff. This variation reflects the storm-driven depositional processes associated with lower-shoreface progradation. The juxtaposition of thick, lower-shoreface hummocky sandstones over marine mudstones in the Desert does not mark a sequence boundary. These thick marine sandstones, often containing well-developed gutter casts and channels, form parasequences which stack in a progradational pattern reflecting a gradual relative rise in sea-level during this time and progradation to the east. This progradational parasequence set is 110-feet thick and extends 40 miles to the east before completely changing facies into marine mudstones. (See Figs. 3 and 4 in the accompanying paper entitled *Sequence Stratigraphy and Facies Architecture of the Desert Member of the Blackhawk Formation and the Castlegate Formation in the Book Cliffs of eastern Utah and western Colorado.* In the remainder of this road log the paper will be referred to as the "Castlegate paper".) Because of the stacking pattern of the parasequences, the parasequence set is interpreted to form in the late highstand systems tract.

These late highstand deposits are truncated by a surface interpreted as a sequence boundary. Fluvial strata rest on the sequence boundary. The sequence boundary is indicated by a white line in Figure 3-2b. In the remainder of the guidebook for this day, this surface will be referred to as the Desert sequence boundary.

At the end of this stop, return to the vehicles and drive a short distance up the canyon, turn left around a corner and park the vehicles opposite a large channel cutting down into the late highstand deposits.

11.7 STOP THREE. This is an excellent outcrop showing the sedimentology and sequence stratigraphy of the Desert and Castlegate. Refer to figures 3-2c and 3-3 for the interpretations of the rocks at this stop. Begin this stop by examining the Desert on the west side of the road illustrated in Figure 3-2c. The most obvious feature in the outcrop is the large fluvial channel, with possible tidal influence, eroding deeply into the lower-shoreface sandstones at the base of the exposure. If a sequence boundary exists at this stop, this erosional surface seems to be the best candidate. However, closer examination reveals that there is an apparent basinward shift in facies at the base of an older sandstone highlighted by the lowest white line and indicated by the arrow in Figure 3-2c. This surface is interpreted as the Desert sequence boundary. The rocks below the sequence boundary are lower very fine-grained, hummocky-bedded sandstones. The hummocky beds are up to 4-feet thick and extensively intercut forming thick, amalgamated bedsets, up to 15-feet thick, of internally truncated and incompletely preserved hummocks. The amalgamated bedsets have sharp, nongradational bases. These bases commonly contain gutter casts and gutter channels. The latter are up to 8-feet wide. As mentioned in the discussion of the previous stop, the hummocky bedsets are interpreted to form in the lower shoreface during a late highstand, based on the regional stacking patterns of the parasequences that contain these bedsets. During the late highstand, accommodation on the shelf would be increasing slowly and might even be slowly decreasing. Under these conditions, lower-shoreface deposits would be subjected to extensive storm reworking over a considerable period of time. This would result in amalgamated bedsets associated with gutter and channel casts forming relatively thin deposits spread over a large area. This facies association, parasequence-stacking pattern, and regional expression of lower-shoreface strata occurring just below erosional surfaces interpreted as sequence boundaries, are a common occurrence in the Grassy, Desert, Castlegate, and Lower and Upper Sego. This common stratal motif is probably controlled most directly by rates of accommodation. If this is the case, this motif would most likely be found in the late highstand systems tract of sequences deposited in epicontinental seas where wave activity predominates.

In Tuscher Canyon, fluvial strata overlie the sequence boundary (Fig. 3-3). The fluvial rocks contain at least 3 distinct channel systems. The youngest system forms the most obvious erosional surface filled with fluvial beds with possible tidal modification. These beds are labelled with the number 3 in Figure 3-2c. Figure 3-4a also shows the channel systems on the east side of the road at STOP THREE. In Figure 3-4a, the youngest channel is labelled with the number 4. Based on this channel's distribution and geometry, it is interpreted as a meandering channel with a point-bar loop exposed at this stop. These fluvial strata are interpreted to form in the lowstand or transgressive systems tract of the Desert sequence.

As mentioned above, the Desert sequence boundary occurs at the base of the oldest channel system, highlighted with a white line and indicated by the arrow and number 2 in Figure 3-4a. However, the sequence boundary is truncated by the youngest channel system (Fig. 3-1c). In this case the sequence boundary is a composite surface locally represented by the base of the downcutting channel. The erosion associated with the youngest channel system is most likely associated with an autocyclic mechanism, although it could also represent a minor relative fall in sea level. Regionally there is evidence for only one sequence boundary in the Desert.

The fluvial sandstones are overlain by coal-bearing finer grained coastal-plain deposits. These are truncated regionally by the base of the Castlegate.

At the end of the stop, return to the vehicles and drive a short way up the canyon.

12.0 STOP FOUR. Pull the vehicles to the side of the road and walk up the south side of the canyon so that you are about 30 feet above the canyon floor. Stacked channels and channel remnants within the Castlegate braided stream can be seen on the cliff walls on the opposite side of the valley. These channels are illustrated in Figures 3-4c, and 3-5a, b. The erosional base of the Castlegate braided-stream complex can also be seen in this area; it is well illustrated in Figure 3-5b. This surface is interpreted as a sequence boundary based on the regional correlation of this surface into the basin. Documentation of the regional basinward extent of the sequence boundary will be examined tomorrow. In this updip position the Castlegate sequence boundary is expressed as a braided-stream lowstand complex, resting erosionally on the underlying coastal-plain facies of the Desert lowstand systems tract. Intervening transgressive and highstand systems tracts are preserved downdip, and will be seen later today and tomorrow. These two systems tracts have been truncated by the Castlegate sequence boundary in this updip position (see Fig. 3 in the accompanying Castlegate paper).

The Desert and Castlegate in Tuscher Canyon illustrate sequence boundaries and systems tracts in nonmarine strata. Late lowstand to early transgressive systems tract deposits commonly are composed of relatively complete fluvial channels similar in geometry and facies to the fluvial channels in the Desert at STOP THREE. These channels are composed of a single upward-fining grain-size succession, and contain a diverse assemblage of bedding types ranging from trough-cross beds to current ripples.

These fluvial deposits with moderate net to gross ratios are transitionally overlain by finer grained flood-plain to coastal-plain deposits interpreted to form in the transgressive to highstand systems tracts. Crevasse splays, bay-fill or bay-head deltas, levees, soil horizons, coals, and point bars, all contained within a thick succession of nonmarine mudstone, are typical components of these systems tracts. The strata in the late highstand systems tract appear to be the most mud-prone strata in the sequence. The rocks in the transgressive to highstand systems tracts have a low net to gross ratio; they are similar to the rocks in the upper portion of the Desert at STOP THREE.

The finer grained deposits of the highstand systems tract are overlain and incised by braided-stream deposits composed of stacked, amalgamated channel remnants, like the Castlegate at this stop; the surface separating the braided deposits from the underlying finer grained rocks is interpreted as a sequence boundary. Bedding in these channel remnants is predominantly trough-cross beds with minor planar and current-ripple laminae. This low diversity of bedding reflects the fact that usually only the lower portions of the channels are preserved. Additionally, each channel remnant commonly starts with a coarser grained accumulation, and fines upward resulting in a sandstone body composed of multiple upward-fining packages. Upward-fining packages can be as thin as several feet. Stacked, amalgamated channel remnants can be seen in Figure 3-5a. Typically, lowstand systems tracts on the shelf have high net to gross ratios and contain the best reservoirs within the sequence.

Stratal units like the Castlegate are the updip parts of regionally extensive incised-valley fills. These braided lowstand systems can be 10's of miles wide as in the Castlegate or in similar Miocene lowstand systems in the subsurface in south Louisiana (Van Wagoner et al., 1990). The incised valleys have different geometries depending where along the depositional profile they are observed. This topic will be addressed throughout the day and tomorrow. Additional discussion of incised valley geometry and variations in the facies of incised-valley fills is presented in the accompanying papers on the Sego and Castlegate in this volume.

Finally, the braided-stream deposits are sharply overlain by either the more complete channel complex or by finer grained flood-plain deposits with complete point-bar deposits. The boundary separating the braided strata from the flood-plain strata is sharp and very areally extensive making an excellent correlation marker. This boundary is interpreted to form in response to base-level rise and is considered a parasequence set boundary.

At the end of this stop, return to the vehicles and continue up the road in the canyon.

12.4 At this branch in the road, take the left fork.

12.6 Drive out of the canyon on a jeep track. As the road climbs out of the canyon it swings to the right. At the top of the curve park on the top of the Castlegate.

STOP FIVE. In this area, the Buck Tongue/Castlegate contact is visible. Within the Buck Tongue in Tuscher Canyon, two backstepping parasequences can be seen on well logs. These parasequences are very silty and so do not weather out well in the outcrop; they represent the transgressive systems tract of the Castlegate sequence. Near the top of the Buck Tongue a progradational parasequence set below the Sego is well exposed. This parasequence set is the highstand systems tract. Return to the vehicles and retrace your steps toward Green River.

18.0 Climb out of the wash on the dirt road heading west.
20.0 Join the paved road. Drive south toward Green River.
26.0 Intersect Highway 6 at the eastern edge of Green River. Turn left and follow Highway 6 west toward I-70.
26.4 Intersect I-70. Turn left and drive east on I-70.
37.2 Exit Ranch Exit 173. At the top of the exit ramp, turn left and cross over the Interstate on the bridge.
37.7 Follow the paved road onto the gravel road.
37.9 Turn right on the gravel road into Floy Canyon. You are driving across the Mancos Shale. Ahead is Hatch Mesa with the Grassy, Desert, and Castlegate cropping out along the top of the Mesa. Parasequences of the Kenilworth can be seen in the lower part of the Mesa. The Kenilworth outcrops in the Hatch Mesa area are discussed in the paper entitled *High-Frequency Sequence Stratigraphy and Facies Architecture of the Kenilworth Member of the Blackhawk Formation* by D.R. Taylor in this guidebook.
38.2 Cross the railroad tracks.
39.7 Cross a cattle guard.
39.9 Turn sharply left onto a jeep track just after driving out of a small wash. Drive on top of the Mancos to view the Desert sequence stratigraphy along the eastern cliff face of Hatch Mesa. Alternatively, continue straight on the main dirt road for another one-half mile, pull over to the side of the road and walk up onto the Mancos on the west side of the road to view the Desert along the eastern face of Hatch Mesa.

STOP SIX. A photo of this face is shown in Figure 3-6a. At this stop the Desert sequence boundary is clearly marked by truncation and a basinward shift in facies. Coal-bearing coastal-plain and fluvial strata of the Desert lowstand erosionally overly lower-shoreface beds of the underlying late highstand systems tract. The sequence boundary defined in this area by the truncation and basinward shift in facies is the same surface that we saw in Tuscher Canyon at STOP THREE. STOP SIX is 11 miles in a basinward direction from STOP THREE . The correlation between STOPS THREE and SIX are shown in Figure 3 in the Castlegate paper.

The Castlegate crops out as a fluvial sandstone unit at the top of the ridge, although recent erosion has stripped most of the Castlegate off the top of the cliff. At the end of the stop continue driving on the Mancos until you reach a small turn-around area if you drove up onto the top of the Mancos to view the stratigraphy. If you parked on the dirt road and climbed up onto the Mancos, return to the vehicles. The road log includes the mileage necessary to turn the vehicles around on top of the Mancos.

40.3 Turn the vehicles around at this point. Retrace the route back to I-70.
40.6 Rejoin the main Floy Canyon dirt road. Turn right.
42.3 Recross the railroad tracks.
42.5 Turn left onto the paved road. Return to I-70 east.
42.9 Cross the highway bridge and turn left onto the I-70 east access ramp driving toward Crescent Junction and Thompson.
54.4 Leave the Interstate at Exit 185 to Thompson.
54.9 Turn left under the Interstate and drive north into Thompson and Thompson Canyon.
55.9 Pass through Thompson and cross the railroad tracks by the Amtrak office. Continue driving north into Thompson Canyon.
59.3 Cross a small wash. Petroglyphs are on the lower-shoreface hummocky beds to the left of the road. A small Forest Service sign at the base of the petroglyph-covered cliff explains the origins of these excellent examples of Native American art.

59.5 STOP SEVEN. Stop the vehicles at the base of the Desert lower-shoreface deposits on the right side of the road. View the petroglyphs at the base of the cliff just east of the parked vehicles. The petroglyphs are painted onto very fine-grained, well-sorted hummocky-bedded sandstones of the late highstand systems tract. Figure 3-6b,c show the Desert outcrop at the petroglyph site. A description of the rocks, and interpretations of depositional environments and sequence stratigraphy at this stop are presented in Figure 3-7. The Desert sequence boundary incises into the lower-shoreface sandstones just above the top of the petroglyphs. This incision is illustrated in Figure 3-6c. This incised surface is continuous with the sequence boundary examined at the last STOP SIX on the northeast side of Hatch Mesa (Fig. 3-6a,, Castlegate paper). Fluvial sandstones, probably deposited in a braided stream, rest on the sequence boundary at this stop. These sandstones are medium-grained, with trough-cross beds up to 1-foot thick. Some of the cross beds are contorted and overturned. Current ripples are common. The base of the fluvial sandstones contains abundant clay rip-up clasts. The sequence boundary, underlain by lower very fine-grained, lower-shoreface, hummocky-bedded sandstones and overlain by medium-grained fluvial sandstones is documented in Figure 3-8a-d.

Walk up through the section above the petroglyphs on a path that follows a small fault through the Desert and Castlegate. The fluvial sandstones resting on the sequence boundary are overlain by thin bedded sandstones and mudstones interpreted to have been deposited in a coastal plain as abandoned channel fill (Figure 3-9b). These deposits are overlain by a coal (Fig. 3-9a,b). The coal represents the top of the Desert lowstand systems tract. A transgressive surface caps the coal. This sharp, regionally extensive surface is overlain by three bedsets of hummocky strata, totalling 10-feet thick, interpreted to have been deposited in the lower shoreface. These deposits represent the transgressive systems tract. In a basinward direction, these deposits form backstepping parasequences (Fig. 3, Castlegate paper).

At the coal bed, follow the path along the cliff face a short distance to the south and then climb over the hummocky strata above the coal. A carbonaceous mudstone rests sharply on the underlying hummocky bedsets. This mudstone with interbedded current-rippled sandstones and coals is interpreted to have been deposited in a nonmarine to marginal marine part of a coastal plain. The thick fluvial sandstone capping the mudstone contains the trace fossil *Teredolites* indicating a nearby marine source. The base of the carbonaceous mudstone represents a basinward shift in facies. This fact, plus regional correlation, indicates that the contact between the lower-shoreface sandstones and the carbonaceous mudstone is the Castlegate sequence boundary.

Finally, climb over the large fluvial sandstone at the top of the outcrop by following the path to the north. This sandstone and the underlying mudstone composes the Castlegate lowstand systems tract. From the top of this sandstone, the Buck Tongue can be seen to the north. Figure 3-10 shows the well log, stratification, depositional environments, and sequence stratigraphy of the Exxon Production Research Co. Sego Canyon no. 2, drilled 1 mile north of this outcrop in Sego Canyon. The figure shows only the lower half of the well. Note the backstepping parasequences in the lower part of the Buck Tongue. These represent the transgressive system tract in the Castlegate sequence. The transgressive surface marking the top of the lowstand is sharp without any lag or transgressive deposits. At the end of this stop return to the vehicles and continue driving north up the Thompson Canyon road.

59.7 Pass the right fork into Sego Canyon. Continue straight in Thompson Canyon. About fifty yards past the right fork, turn left onto a jeep track heading northwest toward a small house with a tin roof. Cross a wash and follow the track to the left, as it climbs up a hill on to the top of the Castlegate.
62.1 Passing a branch of Blaze Canyon. There are excellent exposures of the Desert and Castlegate in this canyon.
62.3 Passing a "saucer-shaped" water catchment structure on the right.
63.9 Blaze Canyon can be seen to the southeast.
64.9 Passing the head of the west branch of Blaze Canyon.
65.1 Climbing a steep slope onto the top of the Castlegate. The LaSals are visible in the distance to the south.

65.6 STOP EIGHT. Park the vehicles and walk about fifty yards to the east. At the canyon edge, stop to study the opposite canyon wall. All of the sequence stratigraphy in the Desert and Castlegate is exposed along this wall. Figure 3-11 is a photo of this cliff face and provides a summary of the geology at this stop. Additionally, the stratification, depositional environments, and sequence stratigraphy of the Exxon Production Research Co. Blaze Canyon No.1, drilled 200 yards behind this cliff face, are shown on Figure 3-11.

At the end of the stop retrace the route back to Thompson Canyon.

72.2 Cross Thompson Canyon wash. Take the left fork.
72.3 Intersect Thompson Canyon road. Turn right and head south toward Thompson.
77.2 Intersect I-70. Head west on I-70 toward Crescent Junction.
82.2 Leave I-70 at Exit 180. After exiting the Interstate, turn left, pass under the bridge and drive south toward Moab on Highway 191.
103.2 Pass the turnoff to Dead Horse Point.
109.1 Pass the turnoff to Arches National Park. On the east side of the road you can see the thick wall of the Moab Tongue of the Jurassic Entrada Formation. Most of the arches in the park are in the Moab Tongue. The Entrada Formation rests on the Triassic to Jurassic Navajo Formation. The Moab fault is on the west side of the road. This is a major fault with up to 3000 feet of displacement.
114.7 Crossing the Colorado River and entering Moab.

REFERENCES

Van Wagoner, J. C., R. M. Mitchum, K. M. Campion, V. D. Rahmanian, 1990, Siliciclastic sequence stratigraphy in well logs, cores, and outcrops; concepts for high-resolution correlation of time and facies, AAPG Methods in Exploration Series, No.7, 55 p.

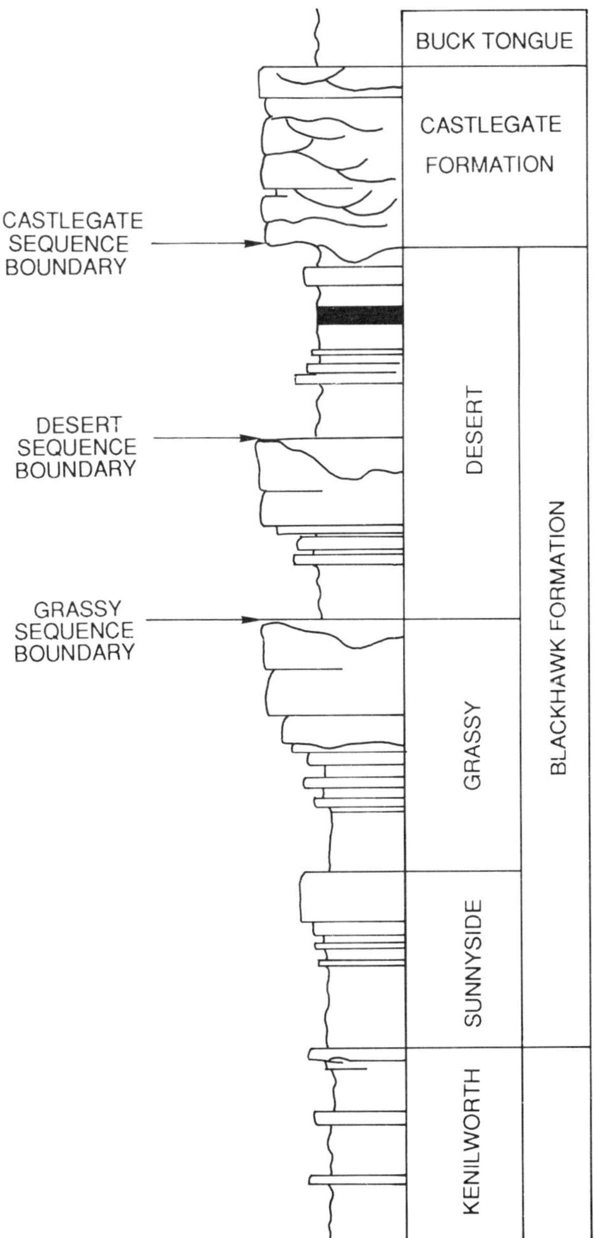

Figure 3-1: Generalized stratigraphic section for the Upper Cretaceous strata at the mouth of Tuscher Canyon located in the north half of T20S R16E, in eastern Utah. Sequence boundary positions are indicated on the Figure. A sequence boundary may exist at or near the top of the Sunnyside. Additional work needs to be done on the Sunnyside in this area. The thickness of this stratigraphic interval is approximately 650 feet.

Figure 3-2: **A.** Panorama of the Book Cliffs at the mouth of Tuscher Canyon located in the north half of T20S R16E. Three Kenilworth parasequences change facies into the basin to the east or right on the photo. The Kenilworth is indicated by K, Sunnyside by S, Grassy by G, Desert by D, and the Castlegate by C. The Kenilworth, Sunnyside, Grassy, and Desert are members within the Blackhawk Formation. **B.** The Desert and Castlegate within Tuscher Canyon at Stop Two. Number 1 indicates the thick, amalgamated hummocky beds within the upper part of the Desert. The hummocky-bedded sandstones have sharp bases, commonly associated with gutter casts. The sequence boundary near the top of the Desert is indicated by a white line and the lower arrow. Number 2 rests on the nonmarine strata overlying the sequence boundary. The upper arrow pointing to the black line indicates a channel within this fluvial lowstand fill. Number 3 rests on the braided-stream deposits of the Castlegate. The sharp base of the braided-stream deposits in the Castlegate is a sequence boundary. The vertical-scale bar is 25-feet high. **C.** Desert and Castlegate stratigraphy in Tuscher Canyon at Stop Three, 100 yards north of photo B. Number 1 rests on the thick, amalgamated hummocky beds of the Desert. These beds are interpreted as late highstand deposits. The lower arrow points to the sequence boundary, referred to as the Desert sequence boundary. Number 2 rests on fluvial strata composed of intercutting channels. The upper arrow points to one of the downcutting channels in this alluvial section. Notice that the sequence boundary indicated by the lower white line is truncated by the downcutting channel. Below number 3, the sequence boundary is a composite surface locally represented by the channel base. Number 3 rests on the fill of the downcutting channel. Number 4 rests on the Castlegate.

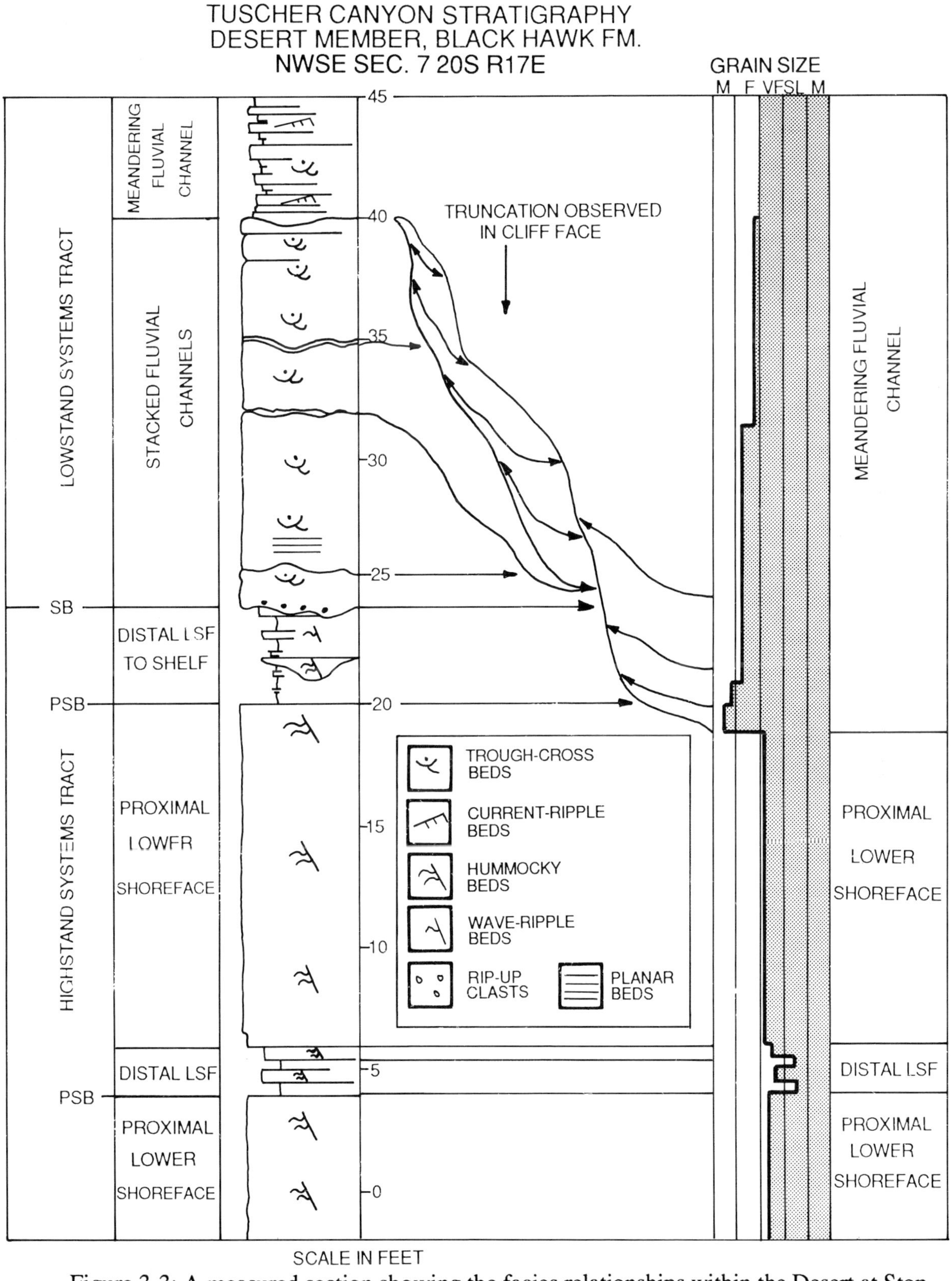

TUSCHER CANYON STRATIGRAPHY
DESERT MEMBER, BLACK HAWK FM.
NWSE SEC. 7 20S R17E

GRAIN SIZE
M F VF SL M

LOWSTAND SYSTEMS TRACT

MEANDERING FLUVIAL CHANNEL

STACKED FLUVIAL CHANNELS

TRUNCATION OBSERVED IN CLIFF FACE

MEANDERING FLUVIAL CHANNEL

SB

DISTAL LSF TO SHELF

PSB

HIGHSTAND SYSTEMS TRACT

PROXIMAL LOWER SHOREFACE

TROUGH-CROSS BEDS

CURRENT-RIPPLE BEDS

HUMMOCKY BEDS

WAVE-RIPPLE BEDS

RIP-UP CLASTS

PLANAR BEDS

PROXIMAL LOWER SHOREFACE

DISTAL LSF

PSB

PROXIMAL LOWER SHOREFACE

DISTAL LSF

PROXIMAL LOWER SHOREFACE

45
40
35
30
25
20
15
10
5
0

SCALE IN FEET

Figure 3-3: A measured section showing the facies relationships within the Desert at Stop THREE. This section was measured on the vertical face shown in Fig. 3-2b at the location of the number 2.

Figure 3-4: **A.** Photograph showing the Desert on the east face of Tuscher Canyon at Stop Three. Number 1 rests on the thick lower-shoreface deposits in the Desert. Number 2 points to a channel cast within the lower shoreface. Number 3 points to the sequence boundary marked by the lower white line at the base of the fluvial strata. The upper white line is at the base of a meandering channel cutting in this area through the previously deposited fluvial unit. Number 4 rests on the fluvial fill of the meandering channel. **B, C.** Stratification within the Castlegate in Tuscher Canyon at Stop Four. In this area, the Castlegate is a braided-stream sandstone. The outcrop consists of a complex of incomplete channel fills stacked in an aggradational pattern. The Castlegate is 95-feet thick in Tuscher Canyon.

A

SB

B

SB

1

2

Figure 3-5: **A.** Castlegate Formation cropping out in Tuscher Canyon at Stop Four. This photo illustrates some aspects of the architecture of the braided-stream complex. In this position within the Castlegate sequence, the sequence boundary is the erosional surface separating the thick, amalgamated braided-stream complex from the underlying finer grained coastal-plain, coal-bearing deposits of the Desert. The sequence boundary is indicated by the white line and arrow labelled SB. Within the Castlegate, relatively continuous, through-going surfaces define the major channelized units. These surfaces are indicated with white lines without arrows. Smaller channels and channel remnants make up the fill between these surfaces.These channels are not laterally continuous and cannot be easily correlated around small areas of the outcrop. **B.** The surface interpreted as the sequence boundary at the base of the Castlegate is marked with white tape and labelled SB. The striking difference between the fluvial strata within the Castlegate (1) and the fluvial strata below (2) the sequence boundary is clearly shown in this photo. The Castlegate is 95-feet thick in these photos.

Figure 3-6: **A.** A panorama of the Desert along the northeast side of Hatch Mesa located in NW NW Sec. 27 T22S R18E at Stop Six. Number 1 rests on the lower-shoreface sandstones of the highstand systems tract below the Desert sequence boundary. These sharp-based, lower-shoreface sandstones are part of a complex of lower-shoreface parasequences that are observed to stack regionally in a progradational parasequence set; they clearly precede the fall in sea level. Number 2 points to the white line highlighting the Desert sequence boundary. In this area, the sequence boundary is strongly erosional and incises the underlying lower shoreface. Coal-bearing nonmarine to slightly marginal marine strata rest on the sequence boundary in the lowstand systems tract. Number 3 indicates the probable Castlegate Sandstone on the east side of Hatch Mesa. Much of the Castlegate has been eroded away in this area. **B.** The Desert at Stop Seven. Number 1 rests on thick, lower-shoreface hummocky strata of the late highstand systems tract. Number 2 points to Indian petroglyphs drawn on the lower-shoreface strata. Number 3 points to the Desert sequence boundary truncating the lower-shoreface deposits. This sequence boundary is the same surface identified at the previous stop and illustrated in photo A of this figure. Number 4 rests on the braided-stream deposits in the Desert lowstand systems tract. Number 5 points to the small point on which the Desert sequence boundary is accessible for detailed inspection. Number 6 points to three sharp-based, hummocky-sandstone beds that rest directly on top of a coal. These marine strata represent the transgressive systems tract on top of the Desert lowstand. The transgressive systems tract is not present to the west at Stop 6 because the overlying Castlegate sequence boundary truncates progressively more deeply westward and erodes the transgressive systems tract away completely so that 2 miles west of Stop 7, the Castlegate lowstand rests unconformably on the Desert lowstand. **C.** This photo shows a close up of the Desert sequence boundary truncating the underlying lower-shoreface deposits. These vertical facies relationships constitute a basinward shift in facies. The outcrop in this photo is 60-feet thick.

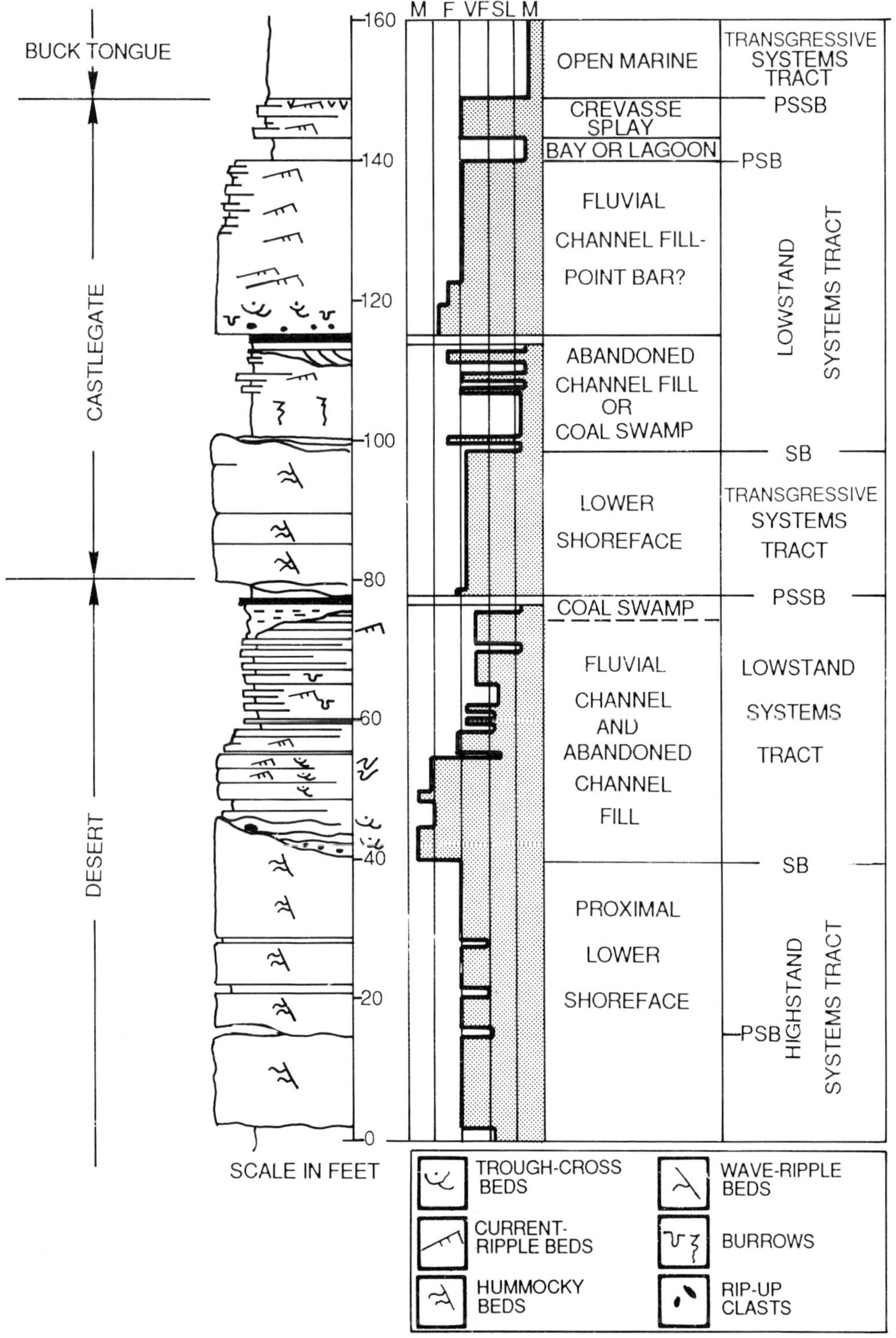

Figure 3-7: Measured section through the Desert and Castlegate in Thompson Canyon at Stop Seven.

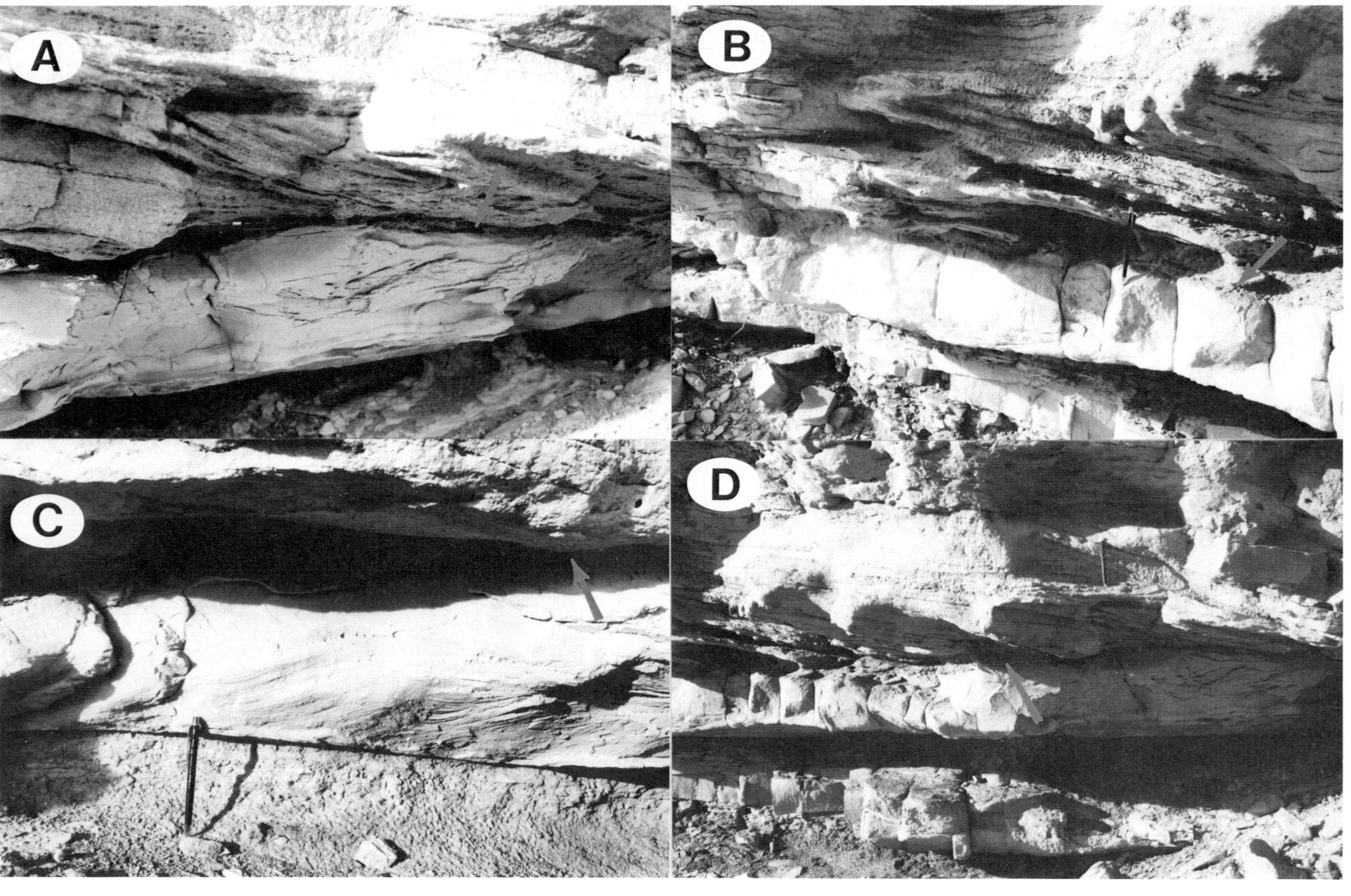

Figure 3-8: The Desert sequence boundary at Stop Seven indicated by the arrow in each photo. Each of the four photos was taken on the small point indicated by number 5 in Fig. 3-6B. **A.** The sequence boundary is overlain by medium-grained sandstone containing abundant clay rip-up clasts. In this area, the sandstone contains trough-cross beds. The sequence boundary is underlain by lower very fine-grained hummocky beds and thin mudstones. This vertical juxtaposition of facies is an excellent example of a basinward shift in facies. The pencil resting on the sequence boundary is for scale. **B.** Another view of the sequence boundary. Weathered clasts can be seen just above the unconformity. **C.** The hummocky bedding can be clearly seen in the bed just below the sequence boundary. **D.** Minor erosion on the sequence boundary can be seen in this photo.

Figure 3-9: **A.** The top of the Desert lowstand systems tract and the transgressive systems tract in Thompson Canyon at Stop Seven. The arrow is pointing to a 1-foot thick coal bed capping the lowstand fluvial deposits. This is a widespread coal in the Thompson area and may record the onset of transgression. Petrographic analysis needs to be done to determine if the coal is fresh water. The coal is overlain by three beds containing hummocky strata interpreted to have been deposited in the lower shoreface. Therefore, the sharp base of these beds is a parasequence set boundary. The vertical-scale bar is 8-feet high. **B.** Another view of the upper lowstand systems tract at Stop Seven. The upper two people in the photo are examining the coal bed that caps the top of the lowstand. The other people are examining thin interbedded sandstones and mudstones within the lowstand. These strata onlap to the south over a distance of 100 yards onto an erosional surface. The strata are interpreted as abandoned channel fill.

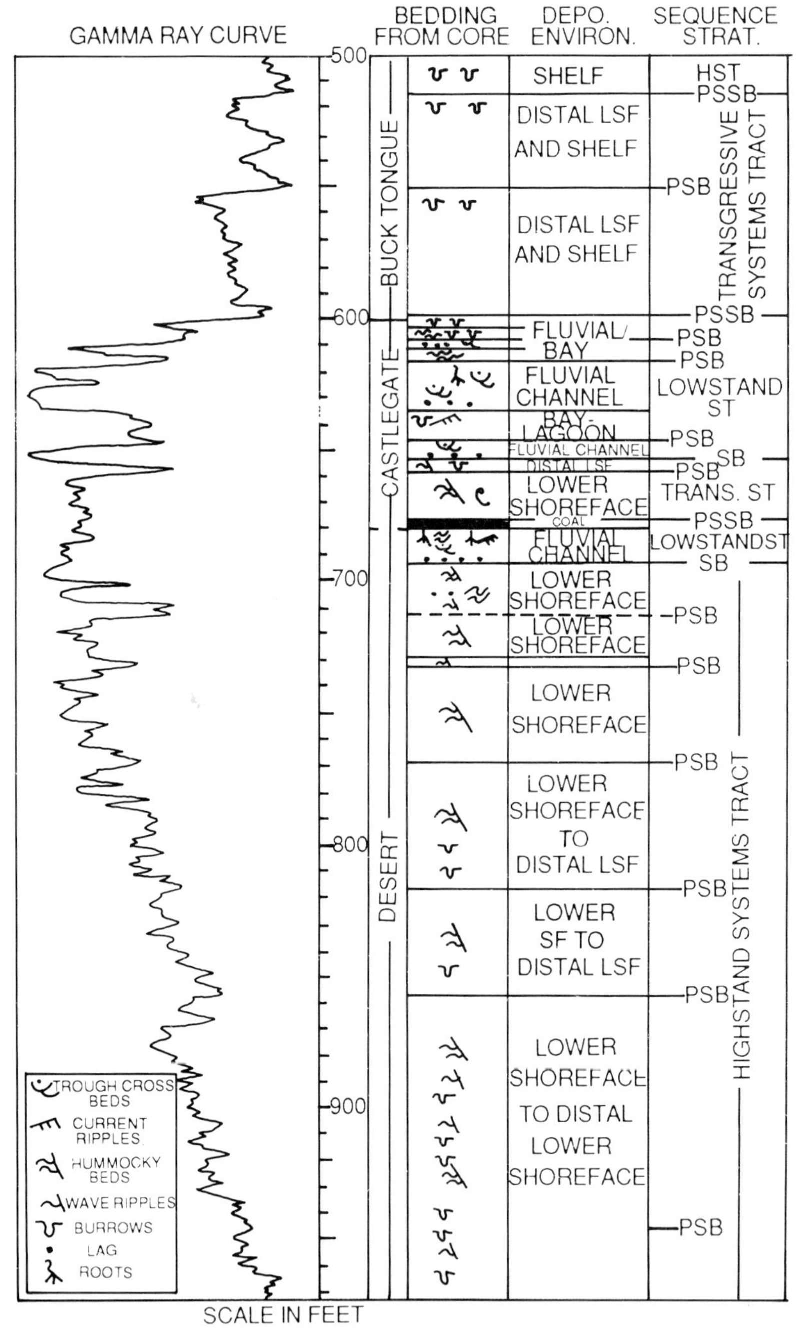

Figure 3-10: Well-log response, facies, and sequence stratigraphy of the Desert, Castlegate, and lower part of the Buck Tongue in the Exxon Production Research Co. Sego No. 1, drilled and cored in Sego Canyon 1.75 miles to the north of Stop Seven.

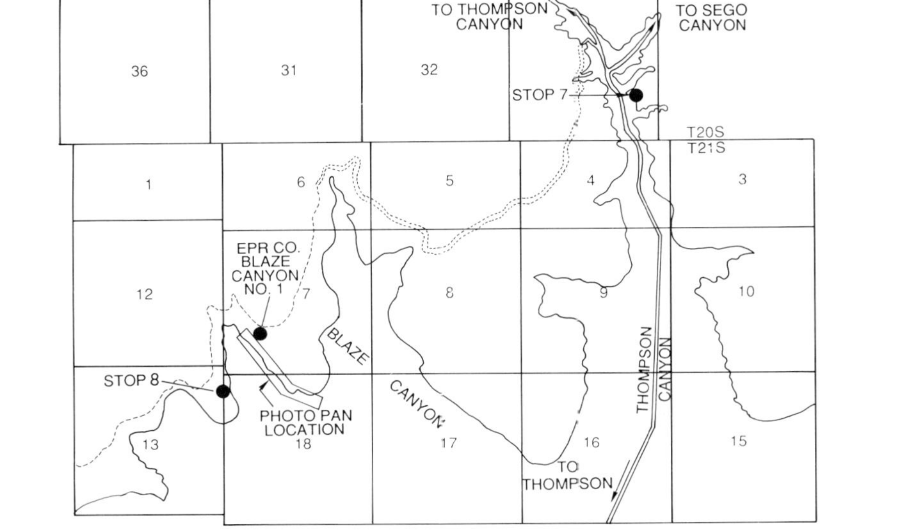

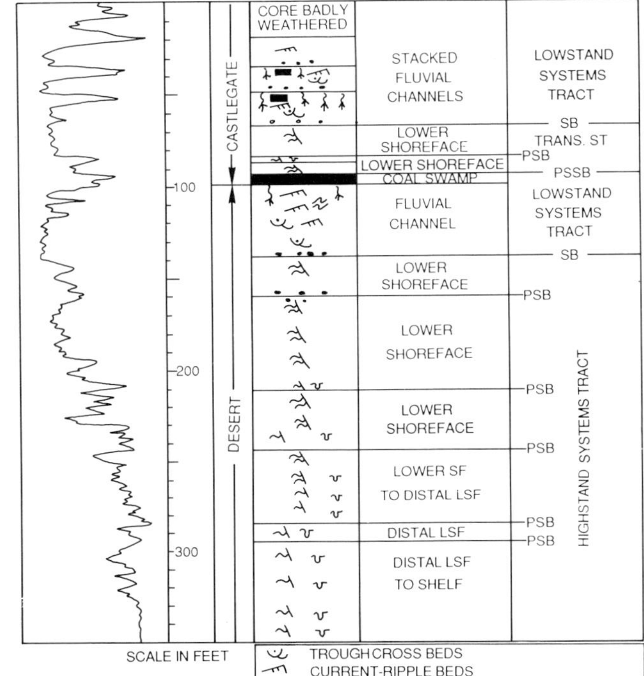

Figure 3-11: **A.** Photo showing the stratigraphic relationships of the Desert, Castlegate, Buck Tongure, and Sego at Stop Eight exposed along a spectacular cliff face one canyon west of Blaze Canyon in location NE SE Sec.7 T21S R20E. The Desert and Castlegate sequence boundaries can both be seen in this photo as well as all of the systems tracts that have been studied during the day. Number 1 rests on the lower-shoreface, hummocky-bedded deposits of the highstand systems tract. Number 2 rests on the Desert sequence boundary. Number 3 points to the fluvial fill of the Desert lowstand systems tract. Number 4 points to hummocky-bedded sandstones that sharply overlie a coal. This is the same coal observed in the previous stop in Thompson Canyon. The sandstones, also seen in Thompson Canyon, are part of the transgressvie systems tract within the Desert sequence. Number 5 points to the Castlegate fluvial deposits within the Castlegate lowstand systems tract. The Castlegate sequence boundary is marked by the white line at the base of these fluvial deposits. In this canyon, the Castlegate sequence boundary truncates the lower-shoreface sandstones, locally removing them. The fluvial deposits are marked by well-developed accretion surfaces oriented in a down-stream direction. Number 6 rests on the Buck Tongue and number 7 rests on the Lower Sego. **B.** The well-log response of the Exxon Production Blaze Canyon No. 1 with associated facies interpretations from continuous cores. The log is annotated with the sequence-stratigraphic interpretations based on the integration of the subsurface data with the outcrop. **C.** A map showing the location of the Blaze Canyon well and the photo panorama in A. This map also shows the relationship of Stop Eight to Stop Seven.

Day Four

Moab to Grand Junction, Colorado

ROAD LOG, DAY FOUR: NONMARINE SEQUENCE STRATIGRAPHY AND FACIES ARCHITECTURE OF THE DOWNDIP CASTLEGATE SANDSTONE IN THE BOOK CLIFFS OF WESTERN COLORADO AND EASTERN UTAH

by
John C. Van Wagoner
Exxon Production Research Company
Houston, Texas

OBJECTIVES: Today the Castlegate will be traced to the most basinward facies exposed in the Book Cliffs completing an updip to downdip analysis of the Castlegate sequence boundary and systems tracts. The Desert will be seen only at STOP ONE. As the Castlegate sequence boundary is traced basinward, changes in the sequence-boundary expression will be examined. These expressions will be compared with the expression of the sequence boundary seen yesterday. Changes in incised-valley geometry will also be studied, especially in the Horsepasture area of STOP THREE, time and weather permitting. Laramide deformation appeared to influence the Castlegate sequence boundary around the Colorado-Utah border. This influence will be examined later in the day.

Leave the motel parking lot in Moab. Drive north on Highway 191. The road log begins at the bridge crossing the Colorado River outside of Moab.

0.0 Crossing the Colorado River on the north side of Moab.
2.7 Passing the entrance to Arches National Park. This park contains the greatest density of natural arches in the world from a three-foot opening to the largest, Landscape Arch. This 105-foot high ribbon of rock measures 291 feet across. Much of the faulting in the Jurassic and Triassic rocks around Moab is controlled by movement of Pennsylvanian salt, 1000's of feet thick in some places.
8.6 Pass the turnoff to Dead Horse Point.
25.0 The Mancos, Desert, and Castlegate can be seen along the Book Cliffs to the north.
29.1 Turn right onto the access road for I-70. Drive east toward Thompson.
30.8 The three prominent points to the northeast are cliffs consisting of both Desert and Castlegate. They stand guard over the small town of Thompson. Today Thompson is a stop for Amtrak and the center of sulfuric acid production in eastern Utah.
34.2 Passing exit 185 to Thompson.
35.2 If the morning is clear enough, Fisher Towers can be seen through a notch in the Mancos, nestled at the base of the La Sal Mountains to the southeast.
37.6 Ahead, the broad domal outline of the Uncompahgre fills the skyline.
38.7 The red, jagged skyline to the south is the jointed Jurassic sandstones in Arches National Park.
40.3 Leave the Interstate on Ranch exit 190. At the top of the ramp, turn left.
40.8 Drive straight as road crosses bridge over Interstate. Drive toward Book Cliffs.
41.3 Turn right on old Hwy 6, now a frontage road.
41.8 Turn left on dirt road toward Sagers Canyon.
42.9 Drive through a wash and pass under Denver and Rio Grande Western railroad bridge. Climb out of wash. At top of small hill follow main dirt road to left.
43.1 Cross wash again.
43.5 Cross wash again.
43.8 Cross wash again.
44.1 Pass turn off to left. Keep straight on main dirt road.
44.3 Cross wash.
45.1 Climb hill in Mancos.
46.0 Climb small hill in Mancos capped with blue sign stating "U.S. Hwy 6 & 50 four miles".
46.1 Descend hill, turn left on dirt road heading into Sagers Canyon.
46.7 Road bends sharply right and descends steep hill
47.1 Cross wash.
47.15 Pull off road, climb to Stop 1.

continued.....

STOP ONE. See Figure 4-1 for a measured section of this stop. At this stop the Desert sequence boundary is expressed as a correlative conformity coincident with a flooding surface bounding a parasequence set (Fig. 4-1, Figs 3a and 4, Castlegate paper). Above the Desert sequence boundary, deeper water marine mudstones and thin, hummocky-bedded sandstones grade vertically into distal lower-shoreface hummocky strata. These deposits form several thin parasequences, which stack into a progradational parasequence set. The Castlegate sequence boundary is well developed at this stop. The sequence boundary is expressed as an erosional surface marked by up to thirty feet of truncation into the distal lower-shoreface strata (Fig. 4-2a,c). Fluvial channel fills overlie the sequence boundary (Fig. 4-2c). In some places, rip-up clasts up to 18 inches in length rest on the bases of the channel fills (Fig. 4-2b). The rip-up clasts are derived from dolomite concretions forming in the marine strata below the sequence boundary at the time of sea-level fall. As incised channels migrated across the shelf during the drop in sea level at Castlegate time, sediment containing the concretions collapsed into the channels. The concretions were broken up, leaving large, angular, slightly deformed, finely laminated clasts as a lag.

To reach the Castlegate outcrop at this stop, climb a long, steep ridge up from the arroyo directly opposite from where the vehicles are parked. A small path has been worn into the ridge and can be easily followed up to the top of the Desert. From there, it is easy to walk along the base of the Castlegate and to find a way up to the top of the cliff. From the top of the Castlegate, the Buck tongue can be seen to the west. At the end of the stop, retrace your steps back to the vehicles.

Be very cautious at this stop. There are loose rocks at the top of the ridge. It is important to walk carefully to prevent rocks from sliding or rolling down hill onto members of the trip who have not yet reached the top.

When everyone reaches the vehicles, turn the vehicles around and begin to retrace the route back to old Highway 6 and 50.

56.3 Intersect the dirt road. Turn left and descend the hill into the arroyo.
56.7 Reach a Y-intersection. At this intersection, turn right.
63.1 Passing under the railroad bridge. Climb out of the wash, heading southeast toward Highway 6 and 50.
63.9 Intersection of Highway 6 and 50. Turn left and travel to the east-southeast.
69.5 Join the newly paved access road for I-70. Cross over I-70 on the Interstate overpass. Drive east on Utah 128 parallel to I-70.
70.7 Crossing a wash on a small bridge.
71.5 Climbing a hill, passing a T-intersection sign.
71.7 Turn left on the gravel road at the T intersection. Drive north toward Nash Canyon and the Horse Pasture.
72.0 Cross the Denver and Rio Grande Western Railroad tracks.
72.9 Pass under I-70. For the next 25 miles you will be riding on the Mancos Shale.
92.5 Crossing a gas pipeline.
92.6 Passing the turn off to Horse Pasture on the right. Proceed toward Nash Canyon on the main dirt road.
93.0 Passing through a fence line. This land was the Lazy Y Cross ranch owned by the Cunningham Cattle Company. In the summer of 1991 it was bought by the Nature Conservancy and the State of Utah. The Cattle Company owned 8000 acres and had the grazing rights to an additional 250,000 acres. The land will now be used as a wildlife preserve. In the higher elevations on the ranch, elk, bear, and mountain lions are present today.
93.7 Passing a small corral on the left. Stay on the main dirt road.
94.2 Passing an intersection with a road to the right. Stay on the main dirt road and proceed straight ahead.
94.7 Reach a Y-intersection. Turn right and cross the bridge.
94.9 Crossing through a fence line.
95.5 Crossing another fence line.
95.7 Reach the turn into the ranch house of the Lazy Y Cross Ranch. Turn right and follow the dirt road around a small hill.
97.4 A dirt road leading from the ranch house joins the main dirt road. Drive straight ahead.
97.7 You are at the base of a Castlegate and Desert cliff. Turn left on a jeep track that leads to a corral shaded beneath a stand of cottonwood trees.
98.1 Pass through the gate at the corral.

98.2 STOP TWO. Pull the vehicles over to the edge of the road in a grove of trees. Use the measured section and short cross section for this stop in Figure 4-3 as a guide to the stratigraphy and facies interpretations. Climb up into the Castlegate on the east side of the road. As you climb through the scrub to reach the Castlegate, you pass over two parasequences of the Desert. The thickest parasequence in the Desert is well exposed on a narrow ridge cropping out just above the corral. The sequence boundary, expressed as a correlative conformity, is coincident with the flooding surface on top of this parasequence set. One thin parasequence lies above the sequence boundary. Above the thin parasequence, four parasequences are arranged in a progradational parasequence set. Each successively oldest parasequence in the set is thicker and contains a higher percentage of more proximal facies than the parasequence below. The oldest three parasequences are dominated by lower-shoreface hummocky bedding. The youngest parasequence in the set is a complete beach parasequence with distal lower-shoreface up to foreshore facies. The lower part of this beach parasequence contains wave-rippled and hummocky-bedded sandstones and interbedded mudstones. Gutter casts and channels are numerous in this succession. Seven feet of very fine-grained, hummocky-bedded sandstone rest with a sharp contact on the underlying mudstone and interbedded sandstone. Although this is a sharp lithologic contact, it does not represent a basinward shift in facies marking a sequence boundary for these reasons: 1) the interbedded mudstones and wave-rippled sandstones with gutter casts and the overlying hummocky beds resting sharply on the mudstones and thin sandstones are both lower-shoreface deposits. 2) The sharp-based hummocks reflect rapid deposition during high-energy storms. Hummocky beds are nearly always sharp based. The rapid introduction of sand from the nearshore into a more distal lower shoreface resulting in a sharp lithologic contrast reflects changes in energy rather than significant changes in water depth. 3) Seaward of this beach parasequence there are additional parasequences within this parasequence set. This indicates that sea-level was gradually rising during deposition of these parasequences, not falling. Perhaps these deposits herald the approaching fall in sea level. However, they are not the product of that fall.

The top of the beach parasequence is truncated by two separate incising events. The oldest incision can be seen on the south or southwestern end of this outcrop directly above the corral (Fig. 4-2d, 4-4). The youngest and most significant incision can be seen along the cliff face southeast of the corral (4-3, 4-4). Both of these incised valleys are filled with fluvial sandstones with paleocurrents to the east and southeast. At the locality of the measured section Bull Canyon A, the beach has been completely removed by truncation. Both of these surfaces of truncation are interpreted as sequence boundaries, based partly on observations made in the rocks updip but largely on observations made in the rocks downdip. These interpretations are discussed at more length in the text and appropriate figures for the next stop in the guidebook.

At the conclusion of this stop, return to the vehicles, turn them around and retrace the route back through the corral toward the main road.

98.7 Drive back through the corral and return to the main dirt road. Turn right and retrace the route back to the ranch house turnoff.
99.0 At the Y-intersection, take the left fork.
99.7 Reach the intersection with the road to the ranch house. Turn left and follow the main dirt road to the southeast.
102.9 Turn left on the jeep trail that goes to the Horse Pasture. (The stop at the end of this jeep trail is important to get the full picture of the Castlegate. However, the road is bad and the drive takes about 30 minutes. If the weather is poor or if there is insufficient time, go on to STOP FOUR).
103.1 Passing through the fence line. If the fence is closed, close the fence behind the last vehicle.
103.7 Turn right on the dirt road. This is the third dirt road on the right after passing through the fence.
104.5 Passing a side road into a well pad. A tin shack is on the pad. Continue straight.
104.6 The road curves right and descends a hill. At the bottom of the hill take the left fork in the road.

106.8 STOP THREE. The road ends at this mileage. This stop is informally called the Horsepasture. Circle the vehicles and park. Climb the slope east of the vehicles to the base of the Castlegate bench. This is the first bench above the Mancos Shale. The stratigraphy in the Horsepastures area is summarized on a cross section in Fig. 4-5. Measured section CG-1 on this cross section is on the edge of the bench just above the vehicles. As in the last stop, there are two episodes of incision in the Castlegate in the Horsepastures area. The oldest incision, labelled sequence-boundary 1 on Figure 4-5 is overlain by cross-bedded sandstones with uni- and bidirectional reactivation surfaces, a sigmoidal geometry, wave-rippled toes, and paleocurrent transport directions to the west, north, and southwest (Fig. 4-5, measured section CG-6).

These sandstones are interpreted as tidal or estuarine deposits. They are preserved in this area as remnants; most of this sandstone has been eroded by the younger sequence boundary. Still, careful tracing of beds between the Horsepasture and Bull Canyon confirms that these remnant sandstones are temporally equivalent to the oldest fluvial channels at Bull Canyon. In the Horsepasture, the sandstones on sequence-boundary 1 are interpreted to have been deposited in the distal ends of an incised valley complex where the fluvial deposits can be reworked by tidal processes. 1.5 miles to the east of CG-1, the estuarine sandstones are overlain by a 3-foot thick hummocky bed which is overlain by the younger sequence boundary. This suggests that there was a period of sea-level rise following deposition of the estuarine sandstones but before the major incision associated with sequence-boundary 2.

The youngest sequence boundary in the Castlegate is sequence-boundary 2; it is the major unconformity in the Castlegate. This surface is correlative with sequence-boundary 2 in Bull Canyon (Fig. 4, Castlegate paper). In Horsepastures, sequence-boundary 2 is marked by significant incision (Fig. 4-5) and it is overlain by a stack of fluvial channels up to 25-feet thick forming the lowstand systems tract of the Castlegate. Figure 4-6a,b shows the sequence boundary at the base of the channel complex and the fill within some of the channels. Figure 4-6c shows the incision of the fluvial channels into the underlying hummocky-bedded sandstones of the late highstand systems tract of the Desert. CG-3 and CG-2 were measured along the cliff shown in Figure 4-6c. This figure also shows the transgressive surface over the top of the Castlegate lowstand marked by the sharp contact with the Buck Tongue.

As Figure 4-5 shows, the fluvial channels fill incised valleys. These valleys range in width from 50- to 200-yards and in thickness from 25 to 5 feet. The valleys trend in a southeast direction which is the consistent downdip paleocurrent direction for the Castlegate. The incised-valley geometry in this part of the Castlegate is very different from the geometries observed updip. In Tuscher Canyon (STOPS ONE-FOUR yesterday) the incised valley formed a broad sheet almost 100-feet thick. Interfluves were absent. In the Thompson Canyon area (STOPS SEVEN and EIGHT yesterday) the valley also formed a broad sheet but was only about 40- to 50-feet thick. Interfluves were uncommon. In Sagers Canyon (STOP ONE today) the valley system began to break up. Discrete, relatively narrow valleys were up to 0.25 miles wide and ranged in thickness from 10- to 50-feet thick. Interfluves were present there and in Bull Canyon. This pattern of downdip thinning and narrowing of the valleys with an increase in the area of interfluves is consistent with the incised-valley geometry observed in the Horsepasture. Southeast from CG-1 the valley thicknesses continue to decrease. In section 8 (Fig. 4-5), the southernmost valley is only 3-to 5-feet thick. The easternmost Castlegate valley system observed in the Book Cliffs is found in section 10. It is 120-feet wide and 10-feet thick (Fig. 4-7a). In between these two separate valleys is a broad interfluve marked by a *Glossifungites* ichnofacies. Evidence that the top of this interfluve was subaerially exposed or submerged under very shallow, brackish water is discussed at the next stop.

At this point, the reason for the downstream decrease in valley thickness and width is not clear. However, these changes appear to be regional and not restricted to one part of the Castlegate. Perhaps the gradient decreased from the west to the east and the streams gradually lost their ability to incise as they flowed across the broad, flat surface of this portion of the shelf. There may be a tectonic control on the geometry of the sequence boundary. Although the structural control on valley geometry is uncertain, there does appear to be a tectonic control on the Castlegate incised valley orientation. In this area of the Book Cliffs there are a series of northwest trending anticlines and synclines formed by Laramide deformation (Young, 1983) starting about 70ma (Tweto, 1977). Ammonites from this area (Gill and Hail, 1975) indicate that the age of the Castlegate is about 78 ma (absolute age dates provided by W. J. Devlin, pers. comm.). Quite likely the incised-valley drainage patterns were influenced by the incipient Laramide folds. Examine the cross section in Figure 4 of the accompanying Castlegate paper for more data about the interaction between tectonics and sea-level change.

At the end of the stop, return to the vehicles and retrace the route back to I-70.

121.4 Passing under the I-70 bridge.
122.2 Cross the railroad tracks.
122.5 Intersect the paved road, Utah 128, at the stop sign. Turn right and return to I-70.
124.0 Turn right onto the I-70 east bound entrance ramp. Proceed east on I-70 for about 15 miles.
133.5 Passing Exit 212 to Cottonwood Canyon. Continue east on I-70.
140.9 Leave I-70 at Exit 220 for Westwater and Sulphur Canyons. At the top of the exit ramp, turn left across the bridge and drive toward the Book Cliffs.
141.7 Intersect a T junction. The right turn goes to Westwater Canyon. Drive straight to Sulphur Canyon. In 0.1 miles the road bends sharply to the left.
143.7 At this Y intersection go right. The road drives on the Mancos Shale for the next 9 miles.

146.2 Climbing a Mancos Shale hill. During May and June it is not uncommon to see golden eagles perched on this hill.
150.3 Descend a small hill into a wash.
150.5 Cross the wash and pass a road curving to the right. Continue straight toward the Castlegate cropping out as cliffs on top of the Mancos on either side of the road. This is the mouth of Sulphur Canyon.
152.1 Pull over to the side of the road at the mouth of a small amphitheater in the Castlegate east of the road.

STOP FOUR. Walk up the deer trail that diagonally ascends the Castlegate bench . At this location, the Castlegate is composed of 8 parasequences arranged in a progradational parasequence set. These parasequences and their stacking pattern are illustrated in Figure 4-7b and in the measured section from this locality shown in Figure 4-8. Each parasequence is wholly composed of lower-shoreface hummocky beds. These beds are all very fine-grained or finer. The Buck Tongue sharply overlies the Castlegate. Between the last hummocky bed and the Buck Tongue, there is locally developed along this segment of the Castlegate an upper medium- to lower coarse-grained carbonaceous sandstone ranging in thickness from 1 to 3 feet. Biostratigraphic analysis indicates that this sandstone contains abundant spores, pollen, and large, well-preserved plant cuticles, as well as occasional dinoflagellates. The plant cuticles are particularly fragile. Their excellent preservation indicates no reworking and deposition at or very close to their origin. For this reason and because the grain size of the sandstone containing these cuticles is much coarser than the hummocky beds below, this coarser sandstone is interpreted to have been deposited in brackish to nonmarine conditions with a fluvial source. Additionally, the ichnofossil *Scoyenia* has been found associated with this coarser sandstone. Examination of the Castlegate cross section (Fig. 4) presented in the Castlegate paper, incorporating a measured section described here in Sulphur Canyon, indicates significant regional angular truncation of the underlying hummocky-bedded parasequences below this coarser sandstone. The Castlegate sequence boundary is placed below the coarser sandstone. This portion of the sequence boundary was an interfluve subjected to fluvial bevelling of the underlying strata but not incision. The angular incision attests to structural movement in this area during the time of the Castlegate sea-level fall.

At the conclusion of the stop return to the vehicles and retrace the route back to I-70.

162.6 Join the paved road. Cross over I-70 and turn left onto the access ramp. Proceed east on I-70 for about 21 miles.
167.7 Passing Exit 225.
169.3 Passing the coals, shales, and fluvial sandstones of the Dakota.
171.9 Passing from Utah into Colorado.
173.4 Passing Exit 2 to Rabbit Valley. variegated red and green mudstones to the north of the Interstate are in the Morrison Formation.
175.3 Variegated red and green mudstones to the southeast along the low ridge are part of the Brushy Basin Member of the Upper Jurassic Morrison Formation. In the Uncompahgre area the Morrison is subdivided into the lower Salt Wash Member and the upper Brushy Basin Member. The red and green shales of the Morrison are overlain by the tan and light brown beds of the Lower Cretaceous Dakota Sandstone.
177.6 The rocks are dipping steeply to the north off the nose of the Uncompahgre Uplift. Cretaceous marine mudstones crop out to the north of I-70 and the Dakota Sandstone crops out to the southwest. The folding in this area of the Uncompahgre is caused by Laramide deformation occurring in the Late Cretaceous. The monoclinal fold on the east and northeast side of the uplift is a drape over steep basement-involved reverse faults (Stearns and Jamison, 1977). The Uncompahgre was uplifted earlier in the Late Pennsylvanian as part of the Ancestral Rockies (Stone, 1977).
181.2 The coal seams to the west are part of the Lower Cretaceous Dakota Formation.
182.7 Leave the Interstate at Exit 11 to Mack.
183.0 Turn left under the I-70 bridge. Pass under both lanes and continue straight toward Highway 6 and 50.
183.7 Crossing the Denver and Rio Grande Western Railroad tracks.
183.8 Turn left on Highway 6 and 50. Pass through Mack and continue west.
184.7 Crossing Mack Wash.
186.3 Turn right on 8 road to Baxter Pass.
188.2 Crossing R road. Continue straight.
189.2 The paved road bends sharply left. Follow this bend onto S road.

192.3 The paved road ends. Continue on the dirt road as it bends sharply to the right. As you can tell, you are driving across the Mancos Shale.
194.7 Crossing a bridge over a wash.
195.3 Passing the left turn to Prairie Canyon. Continue straight toward West Salt Creek Canyon.
200.4 Passing the turn on the right for the Mitchell Road. This road passes some excellent oolite deposits on top of the Castlegate. If time permits, we may investigate these outcrops. For now continue north toward West Salt Creek Canyon.
201.3 Passing the entrance to the West Gas Baxter Compression Station.
202.4 Turn right onto a jeep trail. The reddish brown beds about 100 yards to the east are oolite beds on top of thin distal sandstone and mudstone beds of the Castlegate.
202.6 Crossing a wash.
202.7 Turn right at the Y intersection. Drive to the base of the oolite bed. Park the vehicles and climb the Mancos to get on top of the Castlegate.

STOP FIVE. Figure 4-9 is a measured section at this locality. Figure 4-10a shows the outcrop and summarizes the stratigraphy. The reddish brown oolites are made up of small concentrically layered ooids cored with medium-grained quartz grains (Fig. 4-10b,c) in an accumulation ranging in thickness from 7 to 9 feet (Fig. 4-10a). Some of the ooids have been replaced with phosphate. In the outcrops along Mitchell road, the oolites contain abundant broken ammonites, especially *Baculites*. The oolite occurs as a continuous bed from the east side of Prairie Canyon to approximately 3 miles east of West Salt Creek Canyon for a total distance of 4.5 to 5.0 miles. The oolite bed rests on the distal lower-shoreface and shelf equivalents of the beds exposed in the Castlegate at STOP FOUR. At STOP FIVE the Castlegate strata are thin, lower very fine-grained hummocky-bedded sandstones and thicker interbedded mudstones. Two or three distal parasequences can be identified which represent the distal late highstand systems tract deposited in paleowater depths of 10's to perhaps 100 feet. The trace fossils suggest a Cruziana Ichnofacies.

The oolite unit is noticeably red everywhere it outcrops. The oolite bed rests on a 2-foot thick bed of upper fine- to lower medium-grained sandstone with abundant laminae of sulphur and wood fragments. The lower contact of this interval rests sharply on underlying marine mudstones and siltstones with a low-angle cross-cutting relationship. The oolite bed and associated sandstones are interpreted to sit on the Castlegate sequence boundary. The iron cementation of the oolite may be due to subaerial exposure and soil development. Additional work needs to be done to confirm this hypothesis.

When the stop is completed, return to the vehicles and retrace the route back to I-70.

222.8 Turn right onto the access ramp, enter I-70 east and drive south toward Grand Junction.
220.8 Leave I-70 at the Horizon Drive Exit. Turn right onto Horizon Drive and right again into the Hilton parking lot.

REFERENCES

Gill, J. R., W. J. Hail, Jr., 1975, Stratigraphic sections across Upper Cretaceous Mancos Shale-Mesaverde Group boundary, eastern Utah and western Colorado, U.S.Geological Survey Oil and Gas Investigations Chart OC-68, 1 sheet.

Stone, D. S., 1977, Tectonic history of the Uncompahgre uplift *in* H. K. Veal, ed., Exploration Frontiers of the Central and Southern Rockies, Rocky Mountain Association of Geologists, p. 23-30.

Stearns, D. W., and W. R. Jamison, 1977, Deformation of sandstones over basement uplifts, Colorado National Monument, *in* H. K. Veal, ed., Exploration Frontiers of the Central and Southern Rockies, Rocky Mountain Association of Geologists, p. 31-39.

Tweto, O., 1977, Tectonic history of west-central Colorado, *in* H. K. Veal, ed., Exploration Frontiers of the Central and Southern Rockies, Rocky Mountain Association of Geologists, p. 11-22.

Young, R. G., 1983, Petroleum in northeastern Grand Co., Utah, *in* W. R. Averett, Northern Paradox Basin - Uncompahgre Uplift, Grand Junction Geological Society Field Trip Guidebook, October 1-2, 1983, p. 1-7.

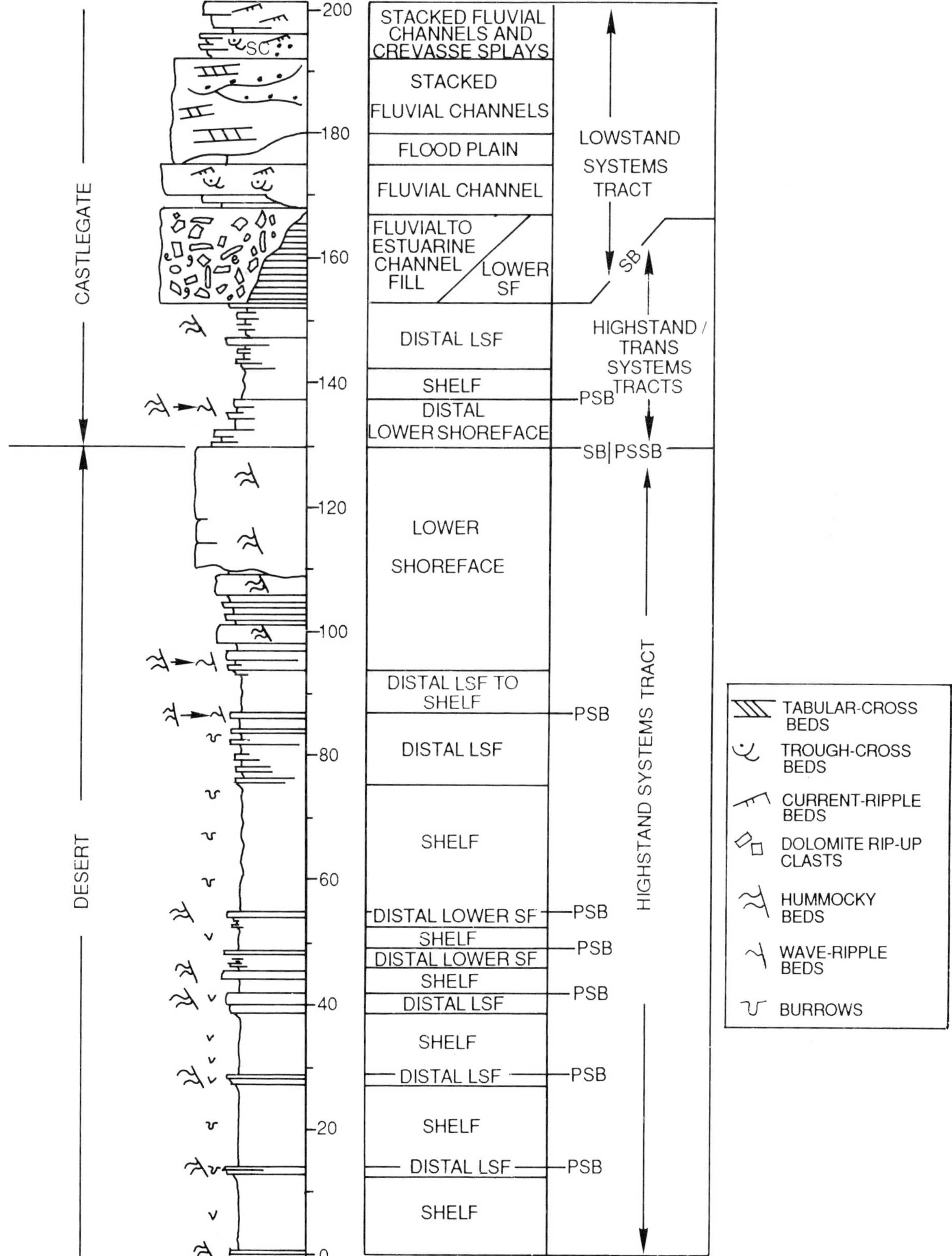

Figure 4-1: Measured section through the Desert and Castlegate at the mouth of Sagers Canyon, Stop One located in NE Sec.5 T21S R21E. In this area, the Desert sequence boundary is coincident with a parasequence set boundary; the fluvial facies of the lowstand systems tract is missing. The Castlegate sequence boundary is a significant surface of truncation marked by a basinward shift in facies. The lower part of the Castlegate fluvial-channel complex overlying the unconformity contains angular, laminated dolomite clasts up to 18 inches long. These clasts are broken pieces of dolomitic concretions that grew in the distal marine, late highstand deposits beneath the sequence boundary. During the sea-level fall that produced the Castlegate sequence boundary, incised valleys cutting across the shelf eroded these concretions and incorporated their broken fragments as clasts in the valley channels.

Figure 4-2: **A.** A portion of the Castlegate sequence boundary in Sagers Canyon at Stop One. Number 1 rests on distal lower-shoreface, hummocky-bedded sandstones of the Desert highstand systems tract. The Castlegate sequence boundary is shown by the white line. Number 2 rests on the medium- to upper fine-grained fluvial sandstone of the Castlegate Formation. **B.** Dolomitic clasts, pelecepod and gastropod shells in the lower part of the fluvial channel complex filling the incised valley at Stop One. The field notebook is 7 inches long. **C.** Incision associated with the Castlegate sequence boundary at Stop One shown by the white line. Number 1 rests on fluvial sandstones, number 2 on distal lower-shoreface deposits, number 3 on the youngest lower-shoreface parasequence in the Desert. The position of the portion of the sequence boundary shown in A is indicated by the small box in C. **D.** Fluvial incision into a complete beach parasequence at Stop Two in Bull Canyon. Number 1 rests on the fluvial channel fill, number 2 rests on the beach which consists of lower shoreface, upper shoreface, and foreshore. This beach is part of the Desert late highstand.

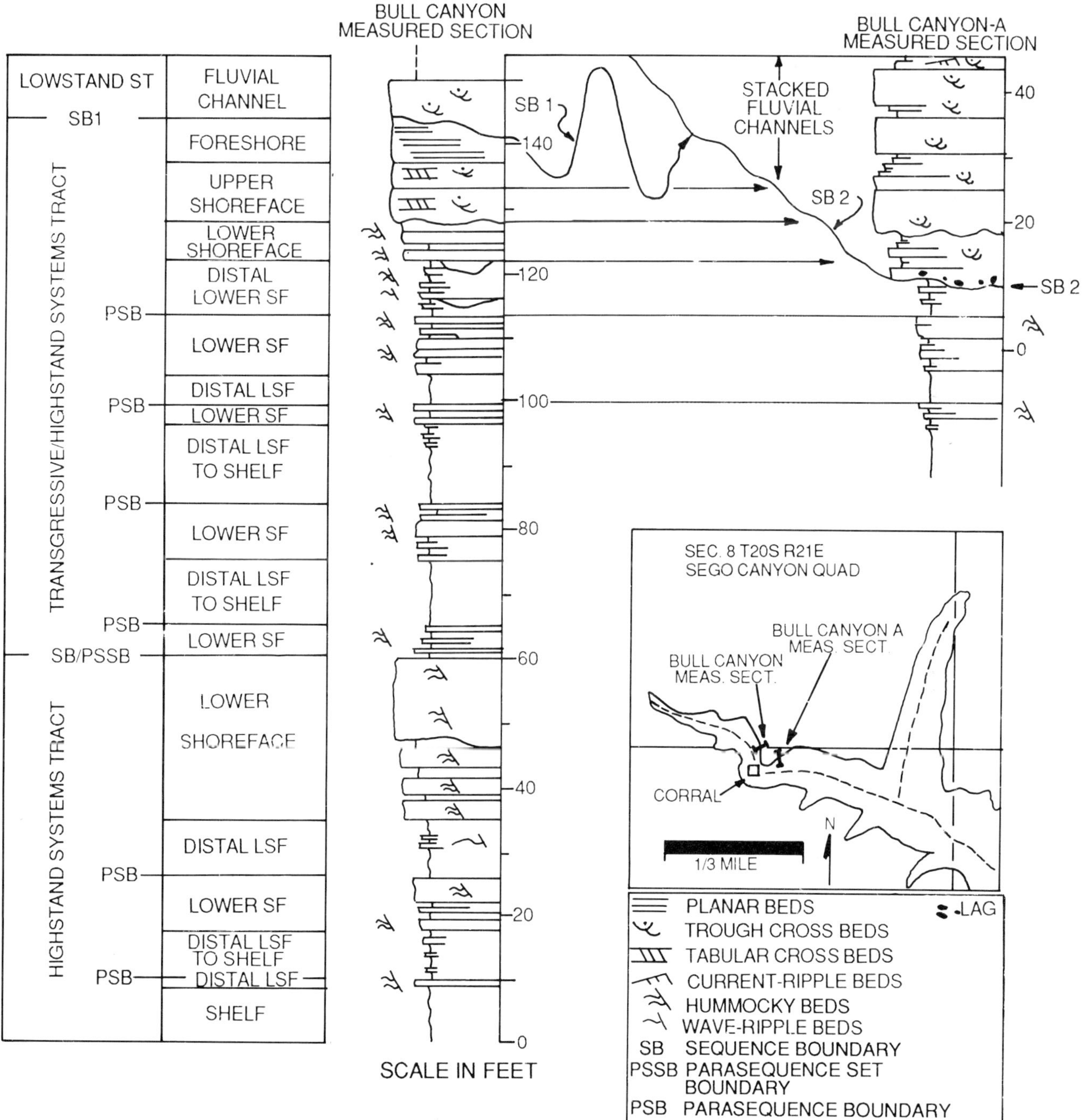

Figure 4-3: Two measured sections and a small cross section showing the stratigraphic and facies relationships between two locations in Bull Canyon at Stop Two. The spatial relationship of the two locations is shown on the inset map. Two generations of channels have been observed at this location. Both of these erosional surfaces are interpreted as sequence boundaries based on regional stratigraphic relationships.

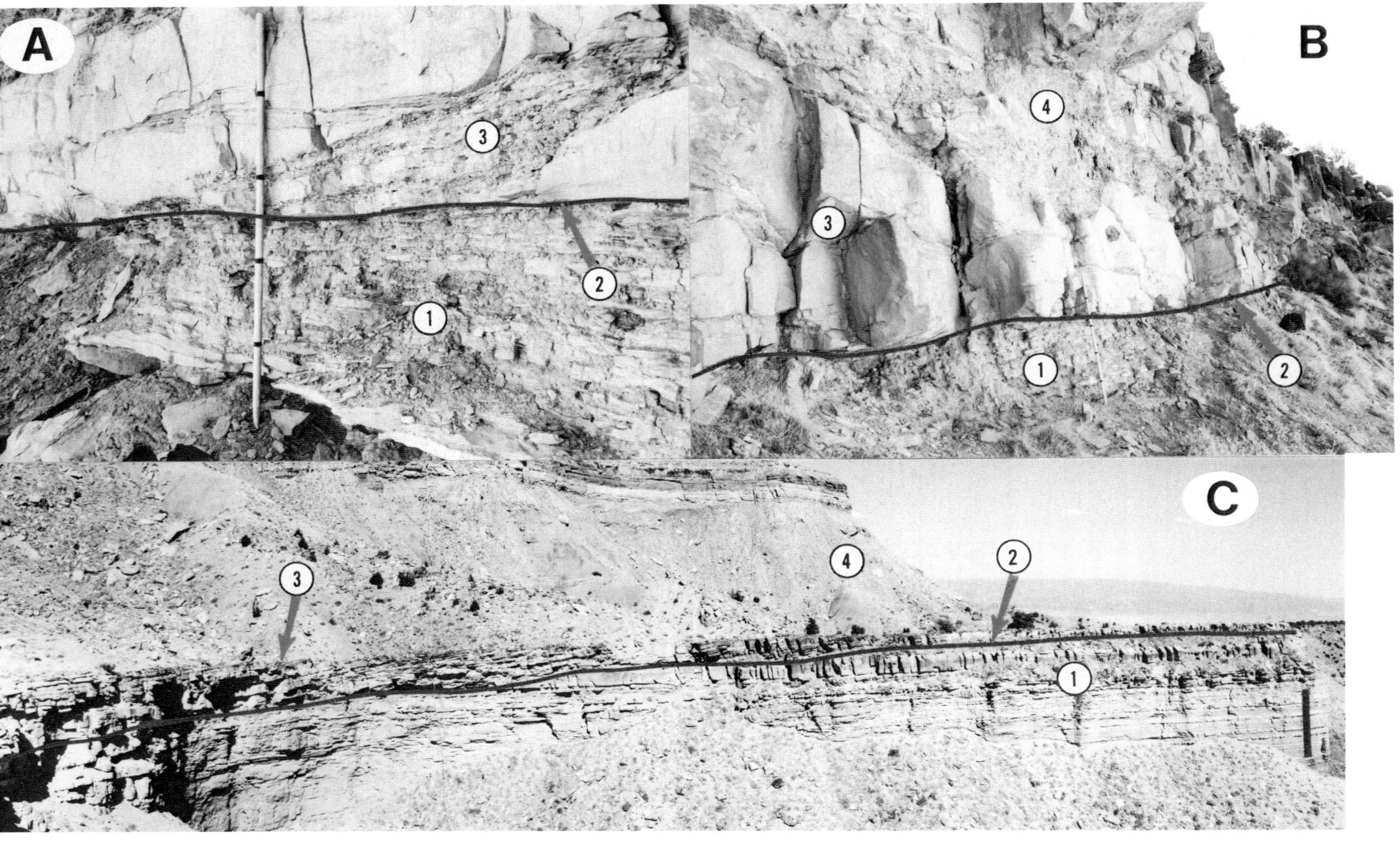

Figure 4-6: **A.** Castlegate sequence-boundary 2 in Horsepastures at Stop Three. Number 1 rests on very thin, discontinuous wave-rippled sandstones and interbedded mudstones interpreted as distal lower-shoreface deposits. Number 2 points to the sequence boundary. Number 3 rests on the lower part of the fluvial-channel complex described in CG-1 in Fig. 4-5. The staff is 5-feet long. **B.** Another view of sequence-boundary 2 in Horsepastures. Number 1 is on the distal lower-shoreface strata, number 2 points to the sequence boundary, number 3 is on fluvial-channel sandstones, and number 4 rests on a channel base filled with clay rip-up clasts. The staff is 5-feet long. **C.** Truncation and a basinward shift in facies at Stop Three. Number 1 rests on lower-shoreface sandstones, number 2 points to sequence-boundary 2. The sequence boundary truncates the lower shoreface. The sandstone above the sequence boundary is interpreted as fluvial. Sections CG-2 and CG-3 (Fig. 4-5) were measured along this face. See Fig. 4-5 for the paleocurrent directions and map view of this fluvial sandstone. Number 3 points to the flooding surface on top of the Castlegate lowstand fluvial deposit. Number 4 rests on the Buck Tongue which is especially thick because this area is an interfluve for the Lower Sego. The scale bar is 40-feet high.

CASTLEGATE SE
BC

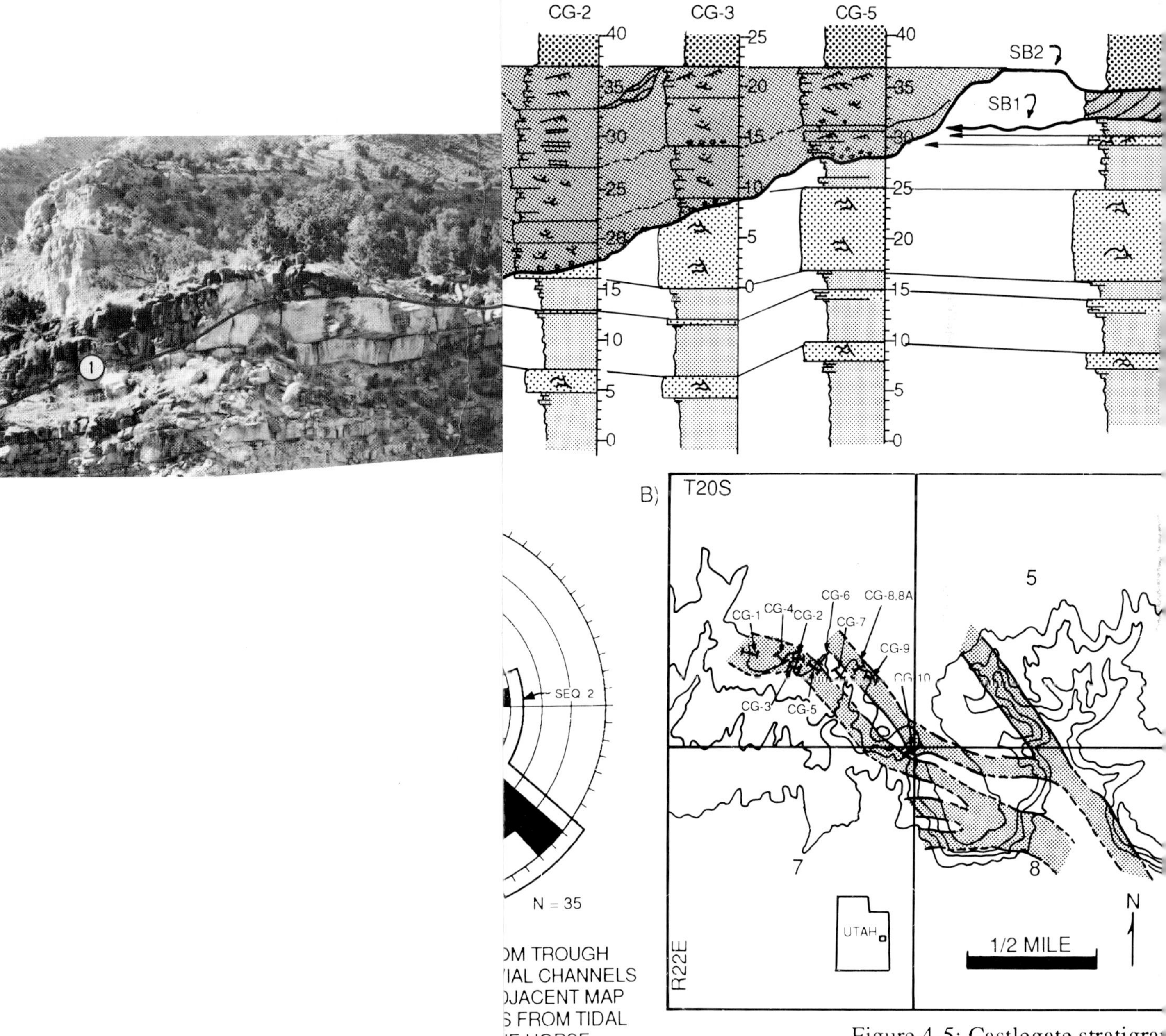

Figure 4-5: Castlegate stratigra
A. Cross section showing the st
sections CG-1 to CG-10. **B.** Ma
valley distribution within seque
diagram showing the paleotrans
filling the incised valleys of seq
sequence 1.

QUENCE STRATIGRAPHY, HORSE PASTURE AREA
OK CLIFFS, WESTERN UTAH

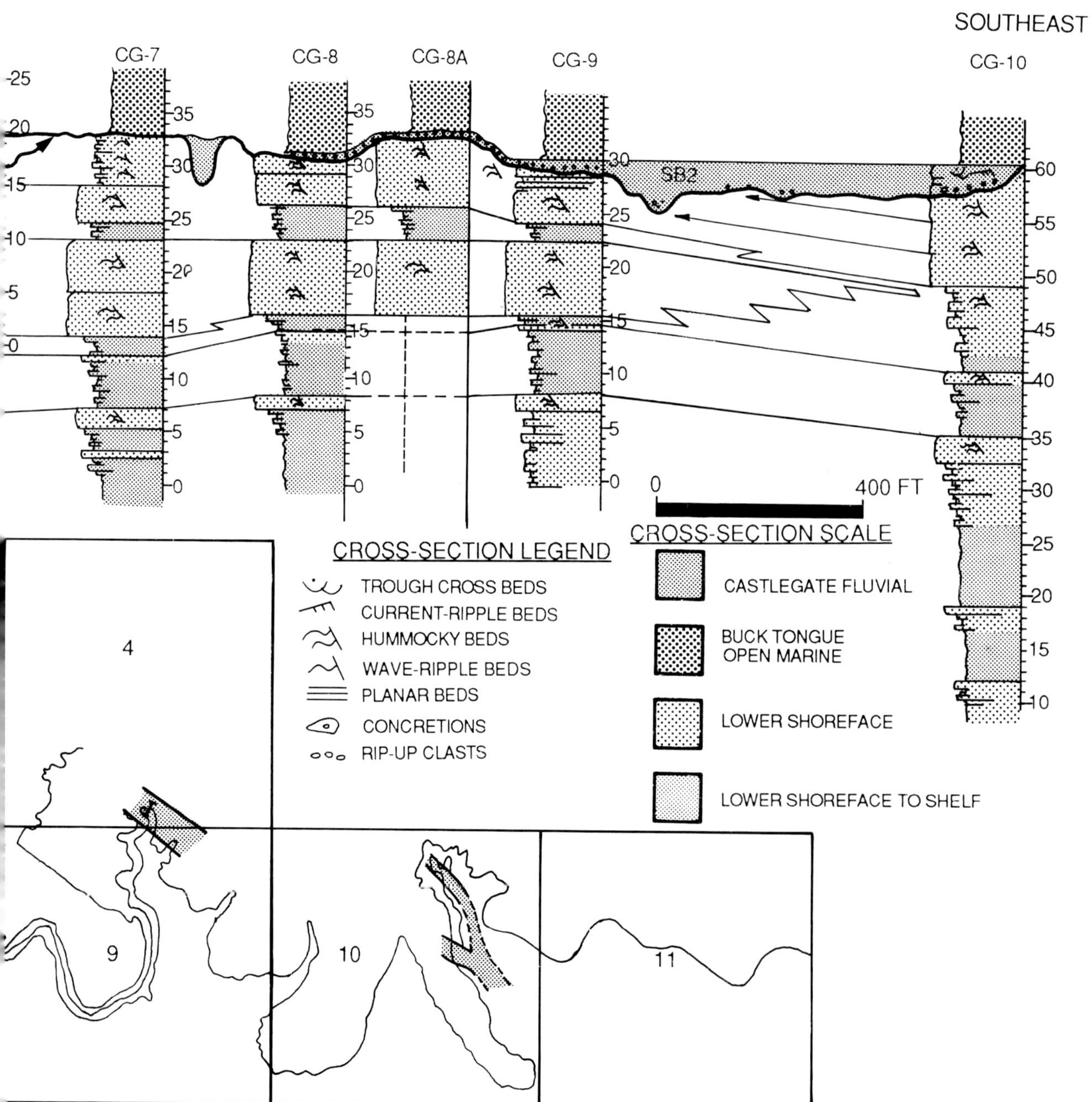

hy in the Horsepasture area, Book Cliffs of western Utah.
atigraphy of the Castlegate in closely spaced measured
p showing the location of CG-1 to CG-10 and the incised-
ıce 2 of the downdip part of the Castlegate. **C.** Rose
port directions of trough-cross beds within the fluvial strata
ence 2 and tidal-cross beds in the estuarine channels in

Figure 4-7: **A.** The easternmost Castlegate fluvial channel that has been found along the Book Cliffs incising underlying lower-shoreface sandstones. This channel is located in section 10 on the location map, Fig. 4-5. The channel is 13-feet thick and 150-feet wide. It is filled with upper fine-grained sandstone containing small-scale trough-cross beds and abundant current ripples. The channel is oriented to the southeast. **B.** The Castlegate at the mouth of Sulphur Canyon at Stop Four. This section is composed of hummocky and wave-rippled sandstones and interbedded mudstones arranged in parasequences stacked in a progradational parasequence set. Detailed correlation indicates that these parasequences are the basinward equivalents of the parasequences seen beneath the Castlegate sequence boundary in Bull Canyon and Horsepastures. Therefore, the parasequences in Sulphur Canyon represent the Desert Highstand. The Castlegate sequence boundary is coincident with the parasequence set boundary and is a surface of subaerial exposure. These rocks in Sulphur Canyon are older than all of the Castlegate Formation updip from the Horsepastures.

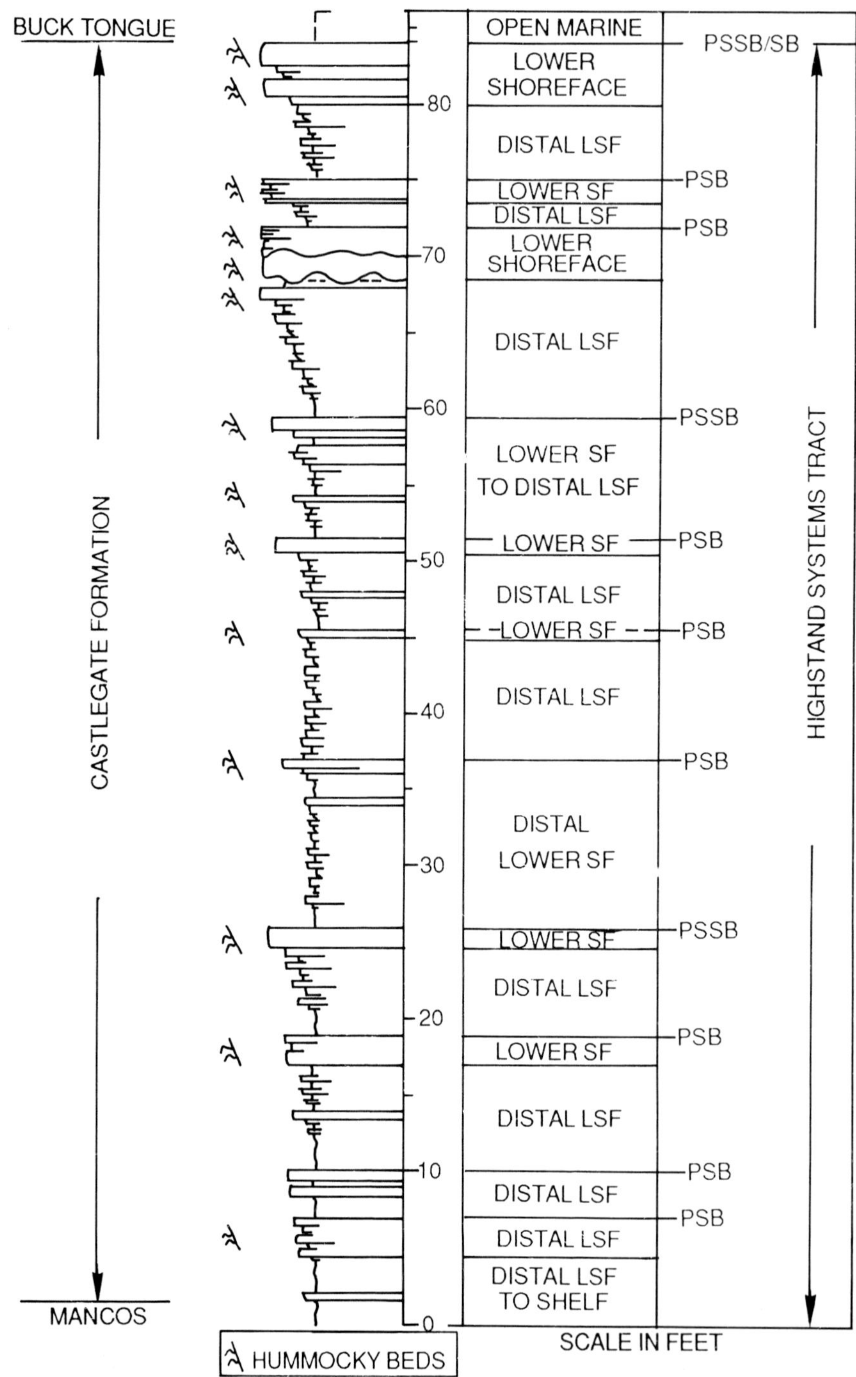

Figure 4-8: Measured section through the Castlegate Formation at the mouth of Sulphur Canyon.

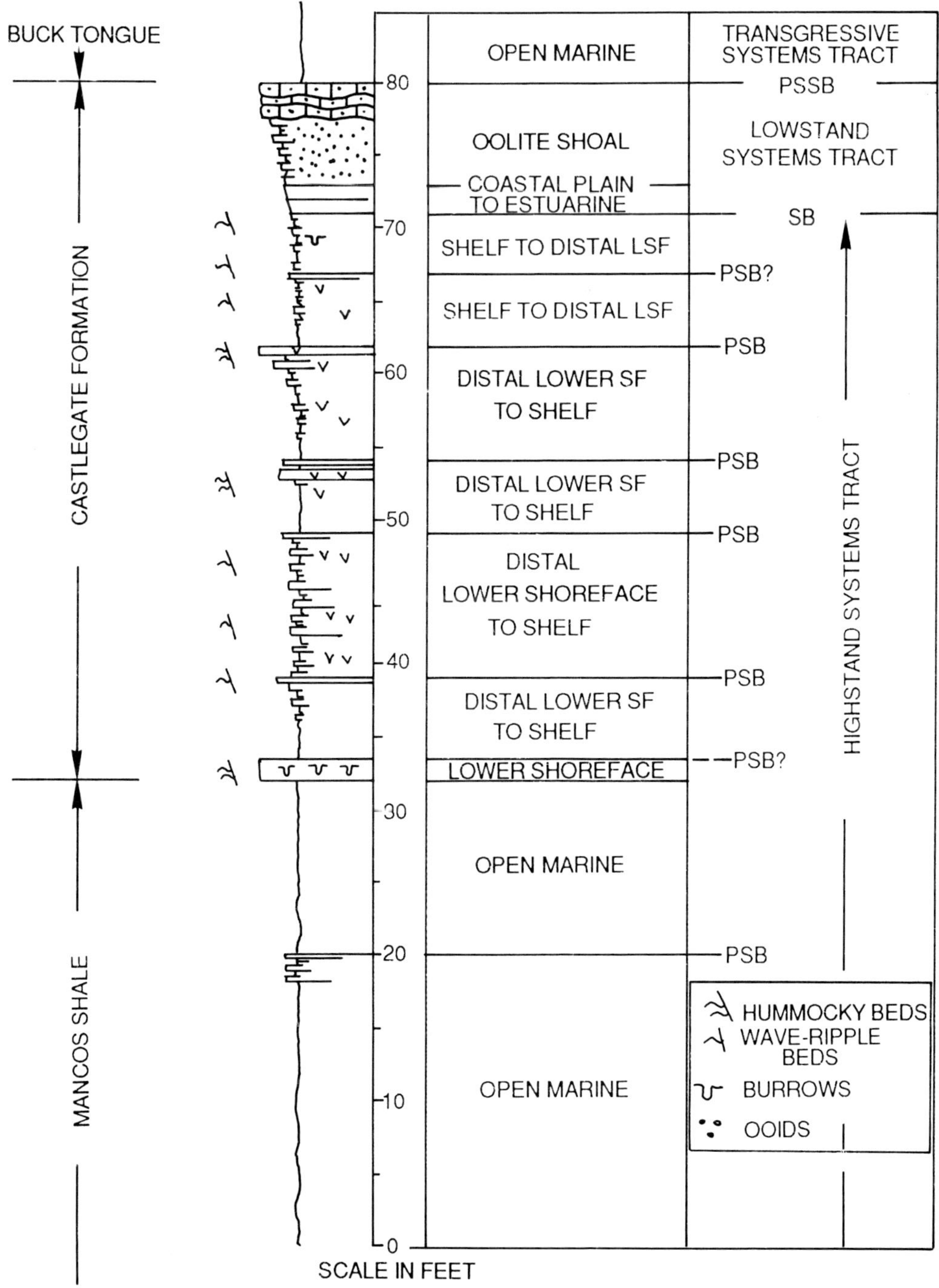

Figure 4-9: Measured section through the distal lower-shoreface deposits of the Castlegate Formation along the West Salt Creek road. These shelf strata are sharply overlain by a calcareous oolite bed which has been extensively cemented with iron-rich clay. The oolite contains fragments of ammonites. The Castelgate sequence boundary is interpreted to occur at the base of the oolite deposit. The oolite bed is overlain by the deeper water deposits of the Buck Tongue.

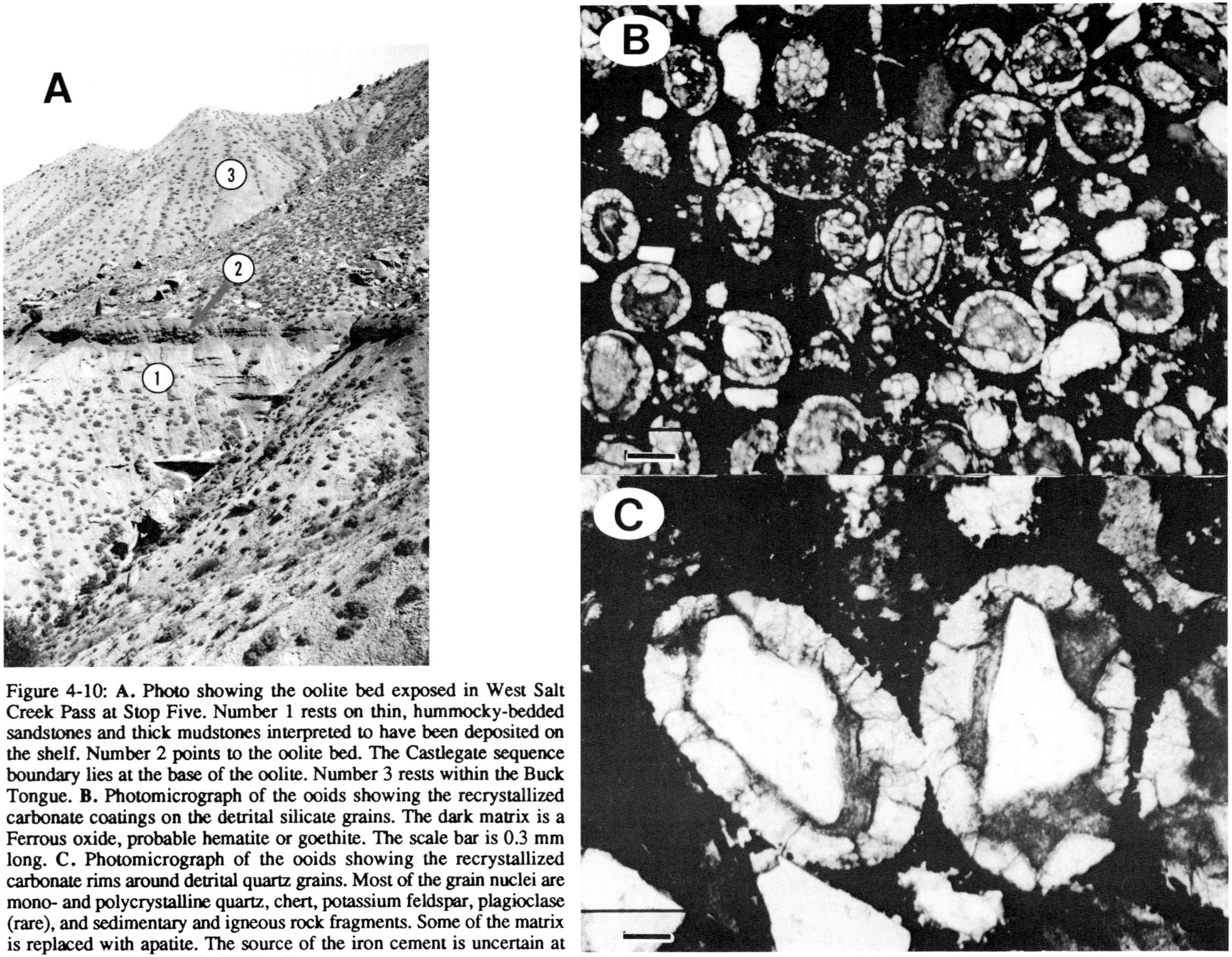

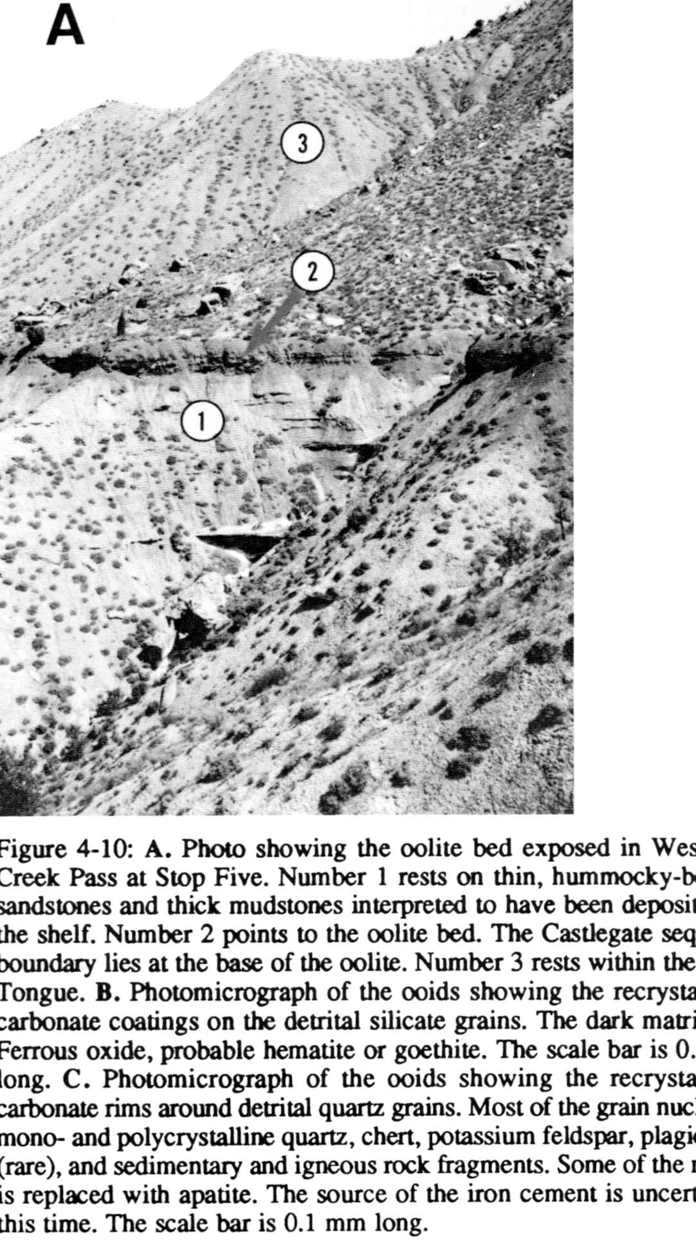

Figure 4-10: **A.** Photo showing the oolite bed exposed in West Salt Creek Pass at Stop Five. Number 1 rests on thin, hummocky-bedded sandstones and thick mudstones interpreted to have been deposited on the shelf. Number 2 points to the oolite bed. The Castlegate sequence boundary lies at the base of the oolite. Number 3 rests within the Buck Tongue. **B.** Photomicrograph of the ooids showing the recrystallized carbonate coatings on the detrital silicate grains. The dark matrix is a Ferrous oxide, probable hematite or goethite. The scale bar is 0.3 mm long. **C.** Photomicrograph of the ooids showing the recrystallized carbonate rims around detrital quartz grains. Most of the grain nuclei are mono- and polycrystalline quartz, chert, potassium feldspar, plagioclase (rare), and sedimentary and igneous rock fragments. Some of the matrix is replaced with apatite. The source of the iron cement is uncertain at this time. The scale bar is 0.1 mm long.

Sequence Stratigraphy and Facies Architecture of the Desert Member of the Blackhawk Formation and the Castlegate Formation in the Book Cliffs of Eastern Utah and Western Colorado

by
John C. Van Wagoner
Exxon Production Research Co.
Houston, Texas

INTRODUCTION

The Desert Member of the Blackhawk Formation and the Castlegate Formation are well exposed along the Book Cliffs in eastern Utah and western Colorado. The cliffs are oriented relative to the southeasterly paleotransort direction within the Desert and Castlegate so that outcrops expose both strike and dip views of these stratal units. Starting in Tuscher Canyon outside of Green River, Utah and ending in West Salt Creek Canyon, one can traverse the Castlegate and Desert from a totally nonmarine depositional environment updip to shelf mudstones, in the case of the Desert, and red, ferruginous oolites in the case of the Castlegate downdip. Three sequence boundaries are developed in the interval, one within the Desert and and two within the Castlegate. Because of the high-quality of the exposures, the superb access to the rocks, and the orientation of the cliffs, these two units provide an unsurpassed natural laboratory to study nonmarine to marine sequence stratigraphy. Variations in sequence-boundary expression from updip to downdip, systems tract variations across a basin, changes in incised-valley geometries, and fluvial architecture in nonmarine lowstand deposits can all be analyzed in the Castlegate and Desert.

LOCATION AND LITHOSTRATIGRAPHY

Figure 1 shows the Campanian stratigraphic succession at the mouth of Tuscher Canyon near the town of Green River, Utah. In this area, four members of the Blackhawk Formation crop out. They are, from oldest to youngest, the Kenilworth, Sunnyside, Grassy, and Desert. With the exception of the Sunnyside, they have each been interpreted to have sequence boundaries at or near their tops (see Taylor, this guidebook; Van Wagoner, this guidebook, O'Bryne, pers. comm., 1991). Additional work needs to be done on the Sunnyside to evaluate the sequence stratigraphy of that unit. The Castlegate rests erosionally on the Desert (Young, 1955). In this area, the base of the Castlegate is the base of the massive cliff capping the strata at the mouth of Tuscher Canyon. Fine-grained, coal-bearing sandstones and mudstones separate the Desert from the Castlegate in Tuscher Canyon. To the west, the Desert is completely truncated beneath the Castlegate erosional surface. To the east, the Desert changes facies into the Mancos Shale between Sagers Canyon and to a position about 3 miles east of Bull Canyon. The Castlegate persists as a correlatable unit to Dry Canyon several miles east of the Colorado-Utah border, a distance of about 65 miles. The Castlegate is sharply overlain by the Buck Tongue of the Mancos Shale. This dark-gray marine mudstone can be traced from Tuscher Canyon to East Salt Creek Canyon in western Colorado. The transgressive surface at the base of the Buck Tongue is planar, regionally extensive, and makes an excellent datum to hang outcrop and subsurface cross sections.

Figure 2 shows the location of the Castlegate and Desert outcrops along the Book Cliffs. The locations indicated with letters are the sections described on the cross section in Figure 3. The other line of cross scection marked on the map east of Sagers Canyon refers to the measured-section locations on Figure 4. A detailed location map is included on Figure 4.

AGE

Fouch (1983) places the age of the Castlegate between *Baculites asperiformis*, based on the work of Gill and Hail (1975) and *Exiteloceras jennyi* or between about 79 and 74 ma in the mid to late Campanian. New dates (W. J. Devlin) suggest an age of about 79 to 77 ma using the same ammonite data. Fouch (1983) interprets the age of the Desert to be about 79 to 80 ma.

PREVIOUS WORK

Young (1955) correlated the base of the Castlegate from Price River Canyon in the west to the Colorado-Utah border in the east,where the Castlegate is represented by thin distal marine deposits. Van De Graff (1972) recognized 5 depositional environments within the Castlegate from piedmont updip through braided stream, delta plain, shoreline, and delta front. He showed that the Castlegate shoreline was oriented northeast-southwest with a paleotransport direction to the southeast. Pfaff (1985) studied the Castlegate in the Price River Canyon where the Castlegate crops out as thick braided-stream deposits. Van Wagoner (Van Wagoner et al., 1990) summarized the sequence stratigraphy of the Castlegate and Desert. He showed that a significant sequence boundary separated these two units and that there was an additional sequence boundary within the Desert.

Young's (1955) classic work in the Book Cliffs included the stratigraphic correlation of the Desert from about Woodside Canyon to just beyond Sagers Canyon in eastern Utah. Balsley (1983) also correlated the Desert along the Book Cliffs from price to about Green River, Utah.

SEQUENCE STRATIGRAPHY

The sequence stratigraphy of the Desert Member of the Blackhawk Formation (called Desert in the remainder of the paper) and the Castlegate Formation (called Castlegate in the remainder of the paper) is summarized on two cross sections: Figure 3 and Figure 4. Figure 3 covers the stratigraphy of the Desert and Castlegate from the mouth of Tuscher Canyon to Sagers Canyon (Fig. 2). Figure 4 covers the stratigraphy of the Desert and Castlegate from Sagers Canyon to one canyon east of West Salt Creek Canyon, about 6 miles east of the Colorado-Utah border (Fig. 2). Most of Figure 4 illustrates Castlegate stratigraphy, the Desert changes facies into the Mancos Shale to the east of Sagers Canyon.

One sequence boundary is interpeted within the Desert. Two sequence boundaries are interpreted within the Castlegate downdip; only one sequence boundary appears to carry updip beyond Bull Canyon (Fig. 2). For this reason, Figure 3 shows one Castlegate sequence boundary while Figure 4 shows 2. The sequence boundaries in the Desert and Castlegate subdivide the mid-to late-Campanian rocks along the Book Cliffs into three sequences within which facies associations can be grouped into systems tracts. The following discussion describes the systems tracts and sequence boundaries in stratigraphic order from the oldest system tract, the highstand systems tract below the Desert sequence boundary, to the youngest systems tract, the highstand systems tract within the Castlegate sequence.

Grassy Highstand Systems Tract: This systems tract lies above the Grassy sequence boundary (O'Bryne, pers. comm.) and below the Desert sequence boundary (Fig. 3). It is composed of at least 7 upward-thickening beach parasequences stacked in a progradational parasequence set. Each parasequence ranges in thickness from 10 to 30 feet. The set extends from the mouth of Tuscher Canyon (Fig. 3) to measured section 13 on Figure 4 where the youngest parasequences in the set change facies into the Mancos Shale, a distance of about 40 miles.

The parasequences in this systems tract were deposited in a beach environment and consist of foreshore, upper shoreface, lower shoreface and shelf strata. The shallow-water portion of the beach makes up a small part of each parasequence. Most of the parasequences are composed of hummocky strata deposited in the lower shoreface. The change from mudstones and thin, wave-rippled sandstones deposited on the shelf into hummocky strata of the lower shoreface is gradational in many parasequences. In some parasequences, the base of the lower shoreface rests sharply on the underlying finer grained marine rocks. In all of the lower-shoreface strata in these parasequences, the hummocky beds are amalgamated forming hummocky bedsets up to 2- to 3-feet thick without any intervening marine mudstones within the bedsets. Gutter casts and gutter channels are common, especially at the sharp bases of the lower shoreface. These types of deposits have been interpreted as the products of "forced regresssions" (Posamentier and Vail, 1988) with sequence boundaries placed at the bases of the sharp-based, lower-shoreface deposits. This interpretation is incorrect for these deposits in the Grassy highstand. The fact that these parasequences are part of a progradational parasequence set indicates that sea level was rising while they were deposited, not falling. Examination of the sharp-based hummocky deposits shows that the sharp bases represent normal progradation, probably due to high-energy storms. As the sharp-based, lower-shoreface hummocks are traced seaward they quickly become transitionally based indicating that there is no regional extent to the minor "basinward shift in facies" suggested by the "forced regression".

Based on the parasequence stacking patterns, the amalgamation of the hummocky beds within this highstand, the presence of the gutter and channel casts, the presence of the Desert sequence boundary just above the hummocks, and the extensive basinward distribution of the hummocks, these deposits are interpreted to form in the late highstand at a time when accommodation is decreasing rapidly on the shelf but before a relative fall in sea level occurs. Decreasing accommodation would create the conditions allowing extensive progradation across the shelf. It would result in lower-shoreface beds being subjected to more storm reworking before being moved below storm wave base producing the amalgamation, sharp bases and gutter and channel casts..

The associaton of the lower-shoreface strata described above and an overlying sequence boundary is repeated in the Desert highstand below the Castlegate sequence boundary and in the Castlegate highstand below the Lower Sego sequence boundary. In none of these examples does there appear to be any evidence to support deposition of these lower-shoreface strata during a sea-level fall.

Desert Sequence Boundary: This is a regional surface that can be correlated, as a single surface, from the mouth of Tuscher Canyon to Sagers Canyon, a distance of about 25 miles into the basin. The surface is marked by truncation and a basinward shift in facies. The truncation is so extensive along the cliffs between Coal Canyon (section J, Fig. 3) and Hatch Mesa Southwest (section I, Fig. 3) that coals of the Desert lowstand systems tract rest on the Mancos Shale. In Thompson Canyon (STOP SEVEN, Day Three), point-bar deposits erode into lower-shoreface hummocky strata. At Sagers Canyon, the sequence boundary lacks incision and does not exhibit a basinward shift in facies (Fig. 3, also see Fig. 4-1). The reason for this change in expression is not presently understood. Perhaps the sequence boundary is on an interfluve in this area. Additional work is planned this fall to resolve this question.

Desert Lowstand Systems Tract: This systems tract is fluvial and marginal marine throughout the Desert. In many places coal seams up to 1-foot thick are present within the lowstand. Many of these coals are indicated on Figure 3. A coal seam is visible at the top of the lowstand in Thompson Canyon (STOP SEVEN, Fig. 3-7 and 3-9, Day Three). In Tuscher Canyon, the Desert lowstand systems tract is a complex of fluvial channels, possibly containing some tidal influence. One of the channels exhibits meander loops in the outcrop and is interpreted to be the deposit of a point bar. In Thompson Canyon, the Desert lowstand fluvial units are fluvial channels interpreted to be point bars. Some of the channels in this canyon might be braided. Abandoned channel-fill deposits are also common in the lowstand systems tract. Overall, this fluvial succession is interpreted as a low-lying coastal plain with discontinuous coal swamps, meandering streams, and possibly some higher energy fluvial systems that in places might be braided.

Depending upon where the lowstand is observed, it is overlain either by the Castlegate sequence boundary or by hummocky strata of the Desert transgressive systems tract. From the Long Face (section G, Fig. 3) to the west, the Castlegate sequence boundary rests directly on the lowstand of the Desert. This means that from the Long Face, west to Tuscher Canyon (section K, Fig. 3) the fluvial deposits of the Castlegate rest on the fluvial deposits of the Desert. Seaward of the Long Face, the Desert lowstand is overlain by a continuous coal up to 2-feet thick. The coal is sharply overlain by hummocky beds. The hummocky beds are about 10-feet thick everywhere between sections G and C (Fig. 3). These beds are interpreted to have been deposited in the lower shoreface and represent the transgressive systems tract of the Desert.

Desert Transgressive Systems Tract: The transgressive systems tract consists of two parasequences composed of hummocky beds interpreted to have been deposited in the lower shoreface. No equivalent updip shoreline deposits have been identified. In a landward direction the hummocky beds are truncated by the Castlegate sequence boundary which erodes progressively deeper into the Desert toward the west. Any shoreline sandstones laterally equivalent to the hummocky strata were probably truncated by the sequence boundary.

The hummocky-bedded parasequences rest on a continuous coal in the Thompson Canyon area. This coal might be the first deposit of the transgression if the coastal plain was initially flooded to create broad bay and lagoons. Coal petrography will indicate if the coal was deposited in fresh or brackish water. This work has not been done to date and no references with this information have been discovered.

The parasequences in the transgressive systems tract are observed to backstep in section D (Fig. 3).

Desert Highstand Systems Tract: The highstand systems tract is locally preserved in areas that are interfluves for the incised valleys defining the Castlegate sequence boundary. An upward-coarsening and thickening parasequence lies above the youngest parasequence in the transgressive systems tract in the Jeep Trail section (D, fig. 3). This upward-coarsing parasequence passes upward from hummocky beds, into trough-cross beds, and up into planar laminae. These laminae are capped with roots and a very thin coal (Fig. 3). This succession of strata is interpreted to be a beach within the late highstand systems tract. An identical vertical association of facies in exactly the same stratigraphic position is seen in Bull Canyon at STOP TWO on Day Four. In Bull Canyon, the beach is clearly preserved as isolated remnants below the Castlegate sequence boundary (measured section 26, Fig. 4).

The late highstand systems tract in Bull Canyon has the same stratal characteristics as the Grassy late highstand deposits. In Bull Canyon, the lower shoreface at the base of the beach parasequence described in the previous paragraph, has a sharp base and is associated with gutter casts. The Desert highstand parasequences below the Castlegate sequence boundary in Bull Canyon are arranged in a strongly progradational parasequence set that extends all the way from Bull Canyon to measured section 1 on Figure 4, a distance of 45 miles. At measured-section 1, only the distal ends of the highstand parasequences remain. The highstand deposits are composed of hummocky bedsets up to 5-feet thick. In some places, three bedsets stack at the top of the systems tract to form accumulations 15-feet thick. Hummocky beds are interbedded with mudstones and thinner beds of wave ripples and small-scale hummocks. The hummocky beds are uniformally well sorted with a lower very fine-grain size. No unidirectional deposits were observed in the highstand deposits seaward of the beach in Bull Canyon.

The parasequences in the highstand thin toward the northeast along the cliffs. This direction is not directly downdip, which appears to be to the southeast, but the direction must have a component of dip to explain the uniform northeastward thinning and shaling out of the hummocky beds. One of the reasons that the highstand systems tract extends so far to the east might be because it is being transected at an angle due to the orientation of the cliffs. Figure 4 illustrates another important fact about this systems tract: the hummocky beds at the top of the highstand systems tract are bevelled beneath a surface that lies on top of the hummocks and below the flooding surface of the Buck tongue. This surface is interpreted to be the Castlegate sequence boundary; it will be discussed in the next section.

The areal extent of the beds in the highstand systems tract, individual bed continuity, which is considerable even taking into account the orientation of the cliffs relative to the paleoshoreline, and the strongly progradational character of the parasequence set suggests that these deposits were deposited during the late highstand when accommodation on the shelf was being reduced.

Castlegate Sequence Boundary: This surface is the major unconformity in this part of the Cretaceous in the Book Cliffs. Between Tuscher Canyon and the Colorado-Utah border, the sequence boundary at the base of the Castlegate exhibits more than 150 feet of erosional relief. West of Tuscher Canyon, the Castlegate sequence boundary truncates completely through the Desert. In addition to regional truncation, the sequence boundary is marked by a basinward shift in facies. From Tuscher Canyon (section K, Fig. 3) to Hatch Mesa Southwest (section I, Fig. 3) the sequence boundary is broadly erosional without much dramatic, local incision. But from section I (Fig. 3) to measured section 11 (Fig. 4) the Castlegate sequence boundary exhibits significant local incision into the underlying Desert highstand deposits. The incision at measured section 11 is illustrated in Figure 4-7a. This is the easternmost incision by the Castlegate sequence boundary that could be found along the Book Cliffs. Seaward of measured section 11 (Fig. 4) there is bevelling below the surface interpreted to be the sequence boundary, but no incision.

The Castlegate sequence boundary from northeast of measured section 11 (Fig. 4) to the eastern end of the cross section is interpreted to lie directly on top of the highstand systems tract. Over this area, the sequence boundary is interpreted to be a surface of subaerial exposure. A cursory look at Figure 4 suggests that the top of the highstand systems tract should be just a marine-flooding surface. This is how it has been interpreted by workers in the past (Young, 1955). Four observations, taken together, suggest that the surface is a sequence boundary marked by subaerial exposure which is nearly coincident with the parasequence set boundary marked by the base of the Buck Tongue. First, the surface significantly bevels the marine strata beneath. 50 feet of bevelling has taken place below the surface. Second, an oolite shoal lies at the eastern end of the cross section in Figure 4 resting on the surface interpreted to be the Castlegate sequence boundary. This shoal was deposited along a shoreline suggesting that the surface updip of the shoal was exposed. Third, the surface is marked by a *Glossifungites* substrate (MacEachern and Pemberton, 1991). This substrate is intermittently exposed along the cliffs between measured sections 9 and 12 (Fig. 4). The trace fossil *Scoyenia* has been found on this surface, suggesting deposition in a nonmarine environment. (Frey and Pemberton, 1984). Finally, there is a 1-to 3-foot carbonaceous, medium- to coarse-grained sandstone bed with local, thin coal lenses and beds resting upon the hummocky strata of the underlying highstand systems tract and below the marine mudstones of the Buck Tongue. Biostratigraphic analysis of three samples from this carbonaceous sandstone by Martin B. Farley and P.P. McLaughlin of EPR show that the samples are characterized by a high abundance of spore-pollen and other terrestrial debris. All of the samples had abundant fragments of vascular plant cuticle, and large pieces of vascular cell fragments. These last constiutents are especially significant because plant cuticles and cell fragments are fragile. Even small amounts of transport will destroy these particles. This means that the rocks with the plant cuticles were in place and had not been transported. This fact alone rules out marine erosion as a process for creating the surface on top of the highstand. Marine dinoflagellates were rare in the samples studied. Based on the biostratigraphic assemblage found in the three samples, Farley and McLaughlin concluded that the depositional environment for the carbonaceous sandstone was coastal bays or lagoons to estuaries.

At the far eastern end of the Castlegate outcrop, an oolite shoal rests on the sequence boundary. Siliciclastic shelf strata lie below the oolites. The shoal is made up of carbonate ooids coating medium- to coarse-grained grains of quartz, chert, and sedimentary rock fragments. The oolites are extensivley iron cemented. The source of the iron is unknown. If the iron is not a recent weathering phenomena, the iron may indicate soil development during the Campanian. Some of the ooids have been partially replaced with apatite cement. Broken ammonite fragments occur in the oolite, especially in the Michell Road measured section (measured section 1, Fig. 4). High-energy oolites resting on siliciclastic shelf strata represents a basinward shift in facies. The presence of the oolites indicates a shallow-water, high-energy shoreline without significant influx of siliciclastic mud. No siliciclastic shoreline sandstone bodies are found along this lowstand shoreline. The Castlegate sequence boundary updip of the oolites is interpreted to be entirely subaerially exposed (Fig. 4).

It is not clear that the oolites mark the final lowstand shoreline for the Castlegate. It is possible that the oolites mark an intermediate lowstand position. If so, the unconformable portion of the Castlegate sequence boundary could extend much father basinward beyond this outcrop.

In the area of measured sections 26 to 13 (Fig. 4) there appears to be two sequence boundaries within the Castlegate.The youngest sequence boundary is the major surface. It is the surface that can be correlated up and downdip. The older sequence boundary appears to be only locally developed in the area mentioned. The relationship of these two sequence boundaries to each other is discussed in the guidebook under STOP THREE, Day Four.

An interesting nomenclatural problem occurs in eastern Utah and western Colorado where the Castlegate sequence boundary lies on top of the hummocky-bedded Desert highstand strata. Using sequence stratigraphy, all of the time-equivalent, genetically related units within the Desert and Castlegate are bracketed by the same sequence boundaries and are in the same systems tracts. Time equivalency is established by correlation using bed, parasequence, parasequence set, and sequence boundaries. Using lithostratigraphy, the hummocky beds in measured sections 1 to 13 on Figure 4 are called Castlegate because they are the only sandstone between the Mancos Shale and the Buck Tongue. But these beds are not temporally or physically related to the Castlegate in Tuscher Canyon or to the Castlegate at the type section in Price River Canyon. The Castlegate strata at Tuscher and Price River Canyons sit above the surface interpreted as the sequence boundary. The Castlegate out in the basin sits below that surface.

Castlegate Lowstand Systems Tract: Between Tuscher Canyon (K on Fig.3) and Hatch Mesa Southwest (I on Fig. 3) the Castlegate lowstand is a braided-stream boundary deposit up to 100-feet thick interpreted to be a sheet sandstone within a broad incised-valley complex. The sandstone rests directly on the underlying lowstand fluvial deposits of the Desert sequence. In Thompson Canyon the Castlegate fluvial deposits are point bars resting on proximal lower-shoreface deposits. Farther to the east, individual incised valleys are separated by discrete interfluves. In the Horsepastures (measured sections 14-23, Fig. 4), the incised valleys in the Castlegate lowstand are shallow and narrow, separated by broad interfluves (Fig. 4-5, STOP THREE, Day Four). The valleys flow to the southeast. Down the axis of the incised valleys, incision progressively decreases and the valleys get narrower.

As mentioned above, the Castlegate sequence boundary between measured sections 3 and 10 on Figure 4 is interpreted to be a surface of subaerial exposure and bevelling, but not a surface incised by paleovalleys. The lowstand deposits on this surface are 1-to 3-feet thick. They are interpreted to have been deposited in brackish-water bays and the updip ends of estuaries. Some of the lowstand deposits may have formed in lakes or swamps. The bevelling is interpreted to have been accomplished by fluvial processes during the sea-level fall. The angular discordance below the sequence boundary is interpreted to record the interaction of the sea-level fall creating the sequence boundary and nearly contemporaneous structural movement of the Laramide-aged anticlines and synclines that broadly deform the strata along this part of the Book Cliffs in front of the nose of the Uncompaghre Uplift (Young, 1983).

The following characteristics summarize the Castlegate lowstand systems tract:

1. Lowstand deposits are fluvial with some minor tidal modification.
2. The fluvial deposits fill incised valleys.
3. Updip the valleys are broad, sheet-like with no interfluves.The fill is up to 100-feet thick and rests erosionally on finer grained coastal-plain strata.
4. In a downdip direction the valleys progressively narrow and shallow, interfluves become progressively broader. The thickness of the fill in these valleys decreases basinward along valley axes from 50 feet updip to 2-3 feet downdip. Beyond this point, incised-valley floors are projected to merge with the subaerially exposed surface discussed above.
5. Downdip the incised valleys rest on distal marine strata.
6. The lowstand shoreline is marked by carbonate deposition. Younger lowstand-shoreline deposits might exist farther basinward if the ooid shoal marks "temporary" shoreline position.
7. Incised-valley fills and oolite shoals are the only lowstand systems tract deposits seen within the Castlegate. No shallow-marine lowstand deposits have been found.

The characteristics of the Castlegate lowstand fluvial channels summarized above suggest a fluvial system flowing to the southeast across a surface whose gradient is gradually decreasing. Ultimately, the gradient becomes so low that the rivers no longer can maintain their channels and overland flow occurs. The flow has some erosive capability at times, bevelling off uplifting strata to create the angular unconformity seen Figure 4.

Castlegate Transgressive Systems Tract: The Buck Tongue of the Mancos Shale overlies the Castlegate lowstand everywhere along the Book Cliffs. The base of the Buck Tongue is a regionally extensive transgressive surface. This transgressive surface is well exposed. Detailed examination reveals that the surface is extremely planar with no detectable relief, even in the subsurface, nor is it marked by transgressive lag or any other type of transgressive deposit. Two backstepping parasequences occur within the lower part of the Buck Tongue. These can be seen in the well log from the Exxon Production Research Co. Sego Canyon No. 2 (Fig. 3-10 in the guidebook, STOP SEVEN, Day Three).

Castlegate Highstand Systems Tract: The upper portion of the Buck tokngue consists of several distal lower-shoreface parasequences arranged in a progradational parasequence set. This set is the upper part of the Castlegate highstand systems tract. The top parasequence is truncated by the sequence boundary at the base of the Lower Sego.

CONCLUSIONS

1. Regionally extensive surfaces marked by truncation and a basinward shift in facies can be traced in the Desert and Castlegate, one surface near the top of the Desert and one surface at the base of the Castlegate (updip). These surfaces are interpreted to be sequence boundaries.

2. The strata between the sequence boundaries can be subdivided into systems tracts. Systems tracts have relatively predictable associations of facies and stacking patterns.

3. Lowstand systems tracts are fluvial showing little or no tidal modification. These fluvial systems range from braided streams (updip Castlegate), through point bars (Desert, Castlegate) to very thin fluvial channel fills composed entirely of current ripples (downdip Castlegate). Coals are common within the lowstand systems tracts.

Fluvial systems fill incised valleys ranging in geometry from broad valleys miles to 10's of miles wide (updip Castlegate) to narrow, shallow valleys less than 150 feet wide and 5 feet deep (downsip Castlegate). Siliciclastic shallow-marine, lowstand-shoreline deposits were not observed in the Desert or Castlegate.

4. Transgressive systems tracts consist of distal to proximal lower-shoreface parasequences arranged in a retrogradational parasequence set. Transgressive systems tracts rest on a major flooding surface called the transgressive surface.

5. Highstand systems tracts are either truncated and so preserved locally as remnants (Desert) or composed of thin, mud-prone parasequences stacked in a progradational parasequence set (Castlegate).

6. The Castlegate sequence boundary shows a variety of physical expression. These include: 1. Braided-stream deposits resting on finer-grained, coal-bearing coastal plain strata (updip), 2. Point bars resting on proximal lower-shoreface deposits, 3. Narrow and shallow incised channels eroding into distal lower-shoreface deposits, 4. A bevelled surface of bypass nearly coincident with a transgressive surface, and 5. Oolites resting on siliciclastic shelf strata.

7. Distribution of potential reservoir strata and associated traps within the Castlegate and Desert are controlled by the sequence boundary. Distribution of seal facies is controlled, in part, by the transgressive surface.

REFERENCES

Balsley, J. K., 1982, Cretaceous wave-dominated delta systems: Book Cliffs, east-central Utah, AAPG Continuing Education Guidebook, 219 p.

Fouch, T. D., T. F. Lawton, D. J. Nichols, W. D. Cashion, W. A. Cobban, 1983, Patterns and timing of synorogenic sedimentation in Upper Cretaceous rocks of central and northeast Utah, *in* M. W. Reynolds and E. D. Dolly, eds., Mesozoic Paleogeography of West-Central United States, Rocky Mountain Section, Society of Economic Paleontologists and Mineralogists, Rocky Mountain Paleogeography Symposium 2, p. 305-336.

Gill, J. R., and W. J. Hail, 1975, Stratigraphic sections across Upper Cretaceous Mancos Shale-Mesaverde Group boundary, eastern Utah and western Colorado, U. S. Geological Survey Oil and Gas Investigations Chart 0C-68, 1 sheet.

Pfaff, B. J., 1985, Facies sequences and the e volution of fluvial sedimentation in the Castlegate Sandstone, Price Canyon, Utah *in* 1985 SEPM Midyear Meeting Field Guides, Rocky Mountain Section, Society of Economic Paleontologists and Mineralogists, Denver, Colorado, p. 10- 7 to 10-32.

Posamentier, H. W., and P. R. Vail, 1988, Eustatic controls on clastic deposition II- sequence and systems tract models, *in* C. K. Wilgus, B. S. Hastings, C. G. St. C. Kendall, H. W. Posamentier, C. A. Ross, J. C. Van Wagoner, eds., Sea-Level Changes: an Integrated Approach, Society of Economic Paleontologists and Mineralogists Special Publication No. 42, p. 123-154.

Van De Graff, 1972, Fluvial-deltaic facies of the Castlegate Sandstone (Cretaceous), east-central Utah, Journal of Sedimentary Petrology, v. 42, no. 3, p. 538-571.

Young, R. G., 1955. Sedimentary facies and i ntertonguing in the Upper Cretaceous of the Book Cliffs, Utah-Colorado, Geological Society of America Bulletin, v. 66, p. 177-202.

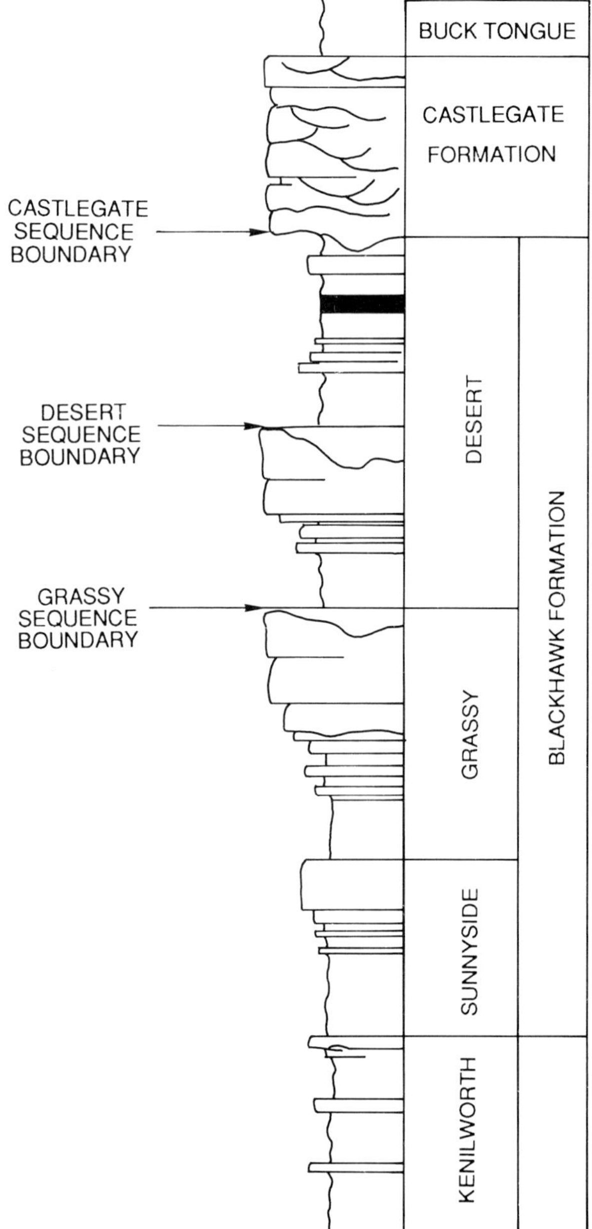

Figure 1: Generalized stratigraphy of the Campanian strata in the book Cliffs at the mouth of Tuscher Canyon in eastern Utah.

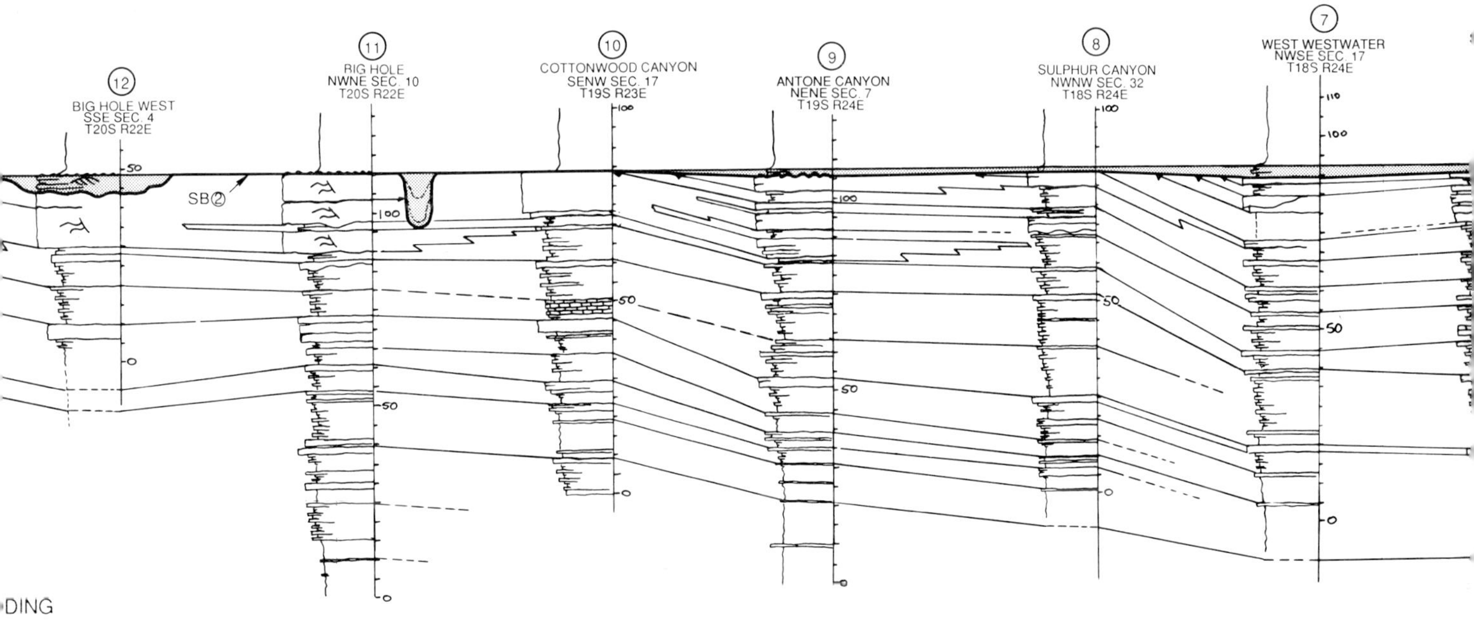

DING

S AND MINOR
VITHIN

LOWER-SHOREFACE
STAL LOWER-SHOREFACE
D MUDSTONES, AND
ES

DARIES

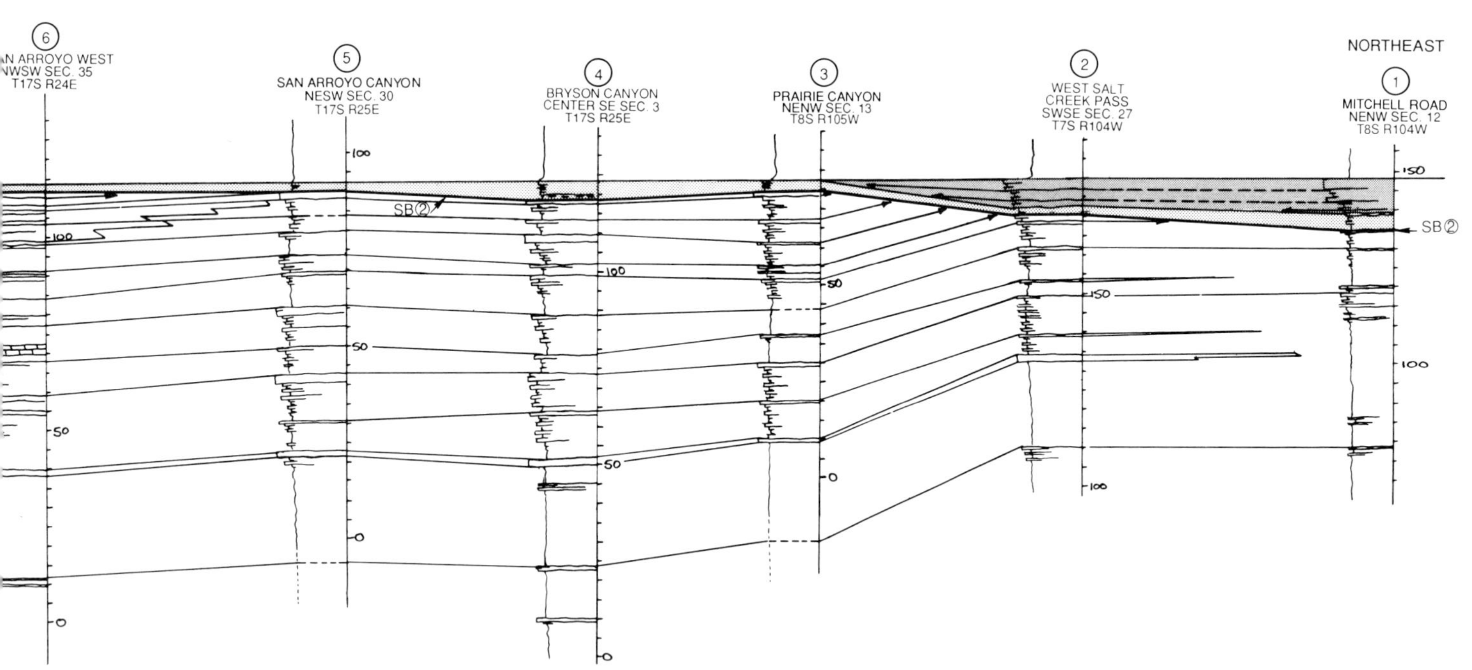

NORTHEAST
6
AN ARROYO WEST
NWSW SEC. 35
T17S R24E
5
SAN ARROYO CANYON
NESW SEC. 30
T17S R25E
4
BRYSON CANYON
CENTER SE SEC. 3
T17S R25E
3
PRAIRIE CANYON
NENW SEC. 13
T8S R105W
2
WEST SALT
CREEK PASS
SWSE SEC. 27
T7S R104W
1
MITCHELL ROAD
NENW SEC. 12
T8S R104W
SB②
SB②

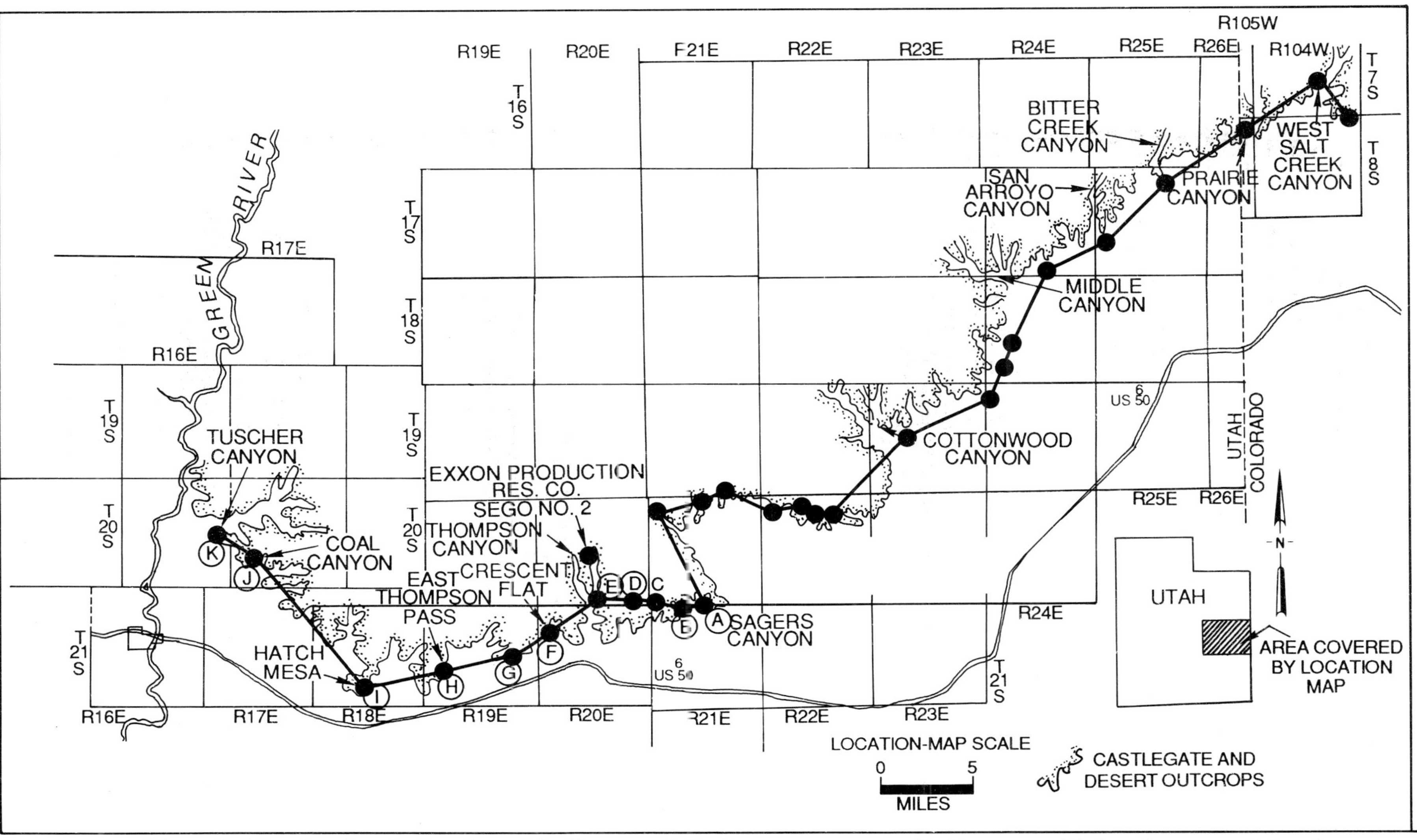

Figure 2: Map showing the location of the Desert and Castlegate outcrops studied in eastern Utah and western Colorado, and the locations of the cross sections illustrated in Figures 3 and 4. Letters A through K mark locations of sections shown on Figure 3. The cross-section line east of Sagers Canyon marks the locations of measured sections shown on Figure 4. See Figure 4 for a detailed location map.

Day Five

Grand Junction, Colorado, to
Farmington, New Mexico

ROAD LOG, DAY FIVE: NOTES ON SELECTED ASPECTS OF THE SEQUENCE STRATIGRAPHY OF THE SAN JUAN MOUNTAINS, COLORADO

by

Donald E. Owen
Lamar University, Beaumont, Texas

OBJECTIVES: The scale of sequence recognition shifts to major first-order sequences, which may be called Sloss sequences or megasequences. The Kaskaskia-Absaroka sequence boundary is seen at STOP ONE and the Zuni-Tejas sequence boundary is seen at STOP THREE. A higher order sequence boundary within the Absaroka sequence that has been enhanced by local tectonic activity producing an angular unconformity is seen at STOP TWO. An objective today is to emphasize the effects of recurrent movement of basement faults on stratigraphy and sequence boundaries. These effects may be seen on outcrop in the San Juan Mountains, but their effects continue in a more subtle manner in the subsurface of the San Juan Basin, the site of the next two days of stops.

Leaving Grand Junction, drive south to US 50 east. The road log begins at the junction of US 50 and CO 146 approximately eight miles southeast of Grand Junction.

0.0 Junction of US 50 and CO 146 near Whitewater. View of Book Cliffs to north and west, Uncompahgre uplift to west, and Grand Mesa (approximately 10,000 feet in elevation) to east, capped by Pliocene (9 Ma) lava flows. The road between here and Delta is in the lower Mancos Shale. The underlying Dakota Sandstone and Burro Canyon Formation may be seen to the west in the Gunnison River valley and close along the road in places. The rest of the Mancos Shale forms the hills to the east.

31.0 Delta. Confluence of Gunnison and Uncompahgre rivers. Continue on US 50 to Montrose. Road is on Quaternary alluvium of Uncompahgre River most of the way, but some hills of Mancos Shale rise above alluvium.

52.0 Montrose. Follow US 550 to Ridgway. Road continues on Quaternary alluvium of Uncompahgre River to vicinity of Colona. South of Colona, road enters a canyon cut into Dakota, Burro Canyon, and Brushy Basin Member of Morrison Formation strata with local K-T boundary (65 Ma) sills. See Figure 5-1 for a cross-section along the road from Montrose to Durango (Lee et al, 1976, p. 142-143).

77.0 Cross Ridgway fault. Morrison Formation in upthrown block to north faulted against Mancos Shale in downthrown block to south. This Laramide fault locally marks the southwest boundary of the Uncompahgre uplift.

78.0 Ridgway. Continue south on US 550 to Ouray. Very flat, wide valley of Uncompahgre River visible south of Ridgway is the surface of Pleistocene glacial lake sediments deposited behind a moraine. Bedrock dips north between here and Ouray, so we descend the stratigraphic column over the next few miles.

81.3 Road at Dolores-Cutler contact. Dike to east.

88.5 Ouray. Tan and gray Hermosa Group on both sides of valley. At south edge of town cross fault near Box Canyon Falls Park that bounds San Luis uplift (Sneffels block) to south. A magnificent angular unconformity between the Uncompahgre Formation (middle Proterozoic) and the upper member of the Elbert Formation (Upper Devonian) may be seen at the top of the trail in the park. Near-vertical Uncompahgre quartzites and slates with primary sedimentary structures may be seen along the road for approximately the next six miles.

94.5 Uncompahgre Formation overlain by San Juan Tuff. In San Juan volcanics for approximately next 20 miles through Silverton.

100.0 Idarado Mine.

102.5 Red Mountain Pass. Elevation 11, 018 feet.

112.8 Silverton. At south edge of Silverton caldera.

114.8 First appearance of Paleozoic (Leadville Limestone) below volcanics along road.

117.5 Turn left into Molas Lake area.

STOP ONE. MOLAS LAKE. The first two stops are in the San Juan Mountains on the San Luis uplift (see Figure 1 in the accompanying overview paper), and involve Paleozoic strata. At STOP ONE we see a paleokarst disconformity forming the Kaskaskia-Absaroka Sloss sequence boundary here where the Molas Formation (Morrowan Pennsylvanian) overlies the Leadville Limestone (Osagean Mississippian) with a lacuna of approximately 20 m.y. Locally, the Leadville has been completely removed by solution, and the Molas Formation rests directly on the Upper Devonian Ouray Limestone (a dolostone here). The Leadville Limestone is present as karst towers at this location (Figure 5-2). An active borrow pit has been dug into the west side of a paleokarst tower of Leadville Limestone surrounded by terra rossa paleosol of the Molas Formation. Maslyn (1977, p. 17) concisely described this locality:

> *"Following deposition of the Late Devonian and Early Mississippian Ouray Formation-Leadville Limestone interval, the Molas Lake area was uplifted and a karst topography developed on the emergent carbonate rocks. The Early Pennsylvanian transgression deposited the upper Molas over the area and preserved the karst. At least 3 fossil karst towers extend up to 68 ft into the overlying Molas Formation from a base on this karst surface. The towers are composed of ridge-like cores of recrystallized Leadville Limestone up to 400 ft long, oriented in a north-south direction. Mantling the sides of the towers are 3 related fragmental units: a limestone conglomerate, a Molas-cemented chert breccia, and a Molas-cemented karst breccia. Locally, cryptocrystalline silica and 1/8-in.-diameter quartz crystals both encrust and replace the carbonate rock in and adjacent to the towers. Remnants of at least 5 other karst towers are present in the area but are not so well exposed. Adjacent to the fossil karst towers are north-south trending valleys composed in part of coalescing depressions averaging about 150-200 ft in width. These small depressions probably represent paleosinkholes which must have coexisted with the karst towers. The fossil tower karst in the Molas Lake area developed as a product of strong jointing in a massive limestone in an area with adequate relief in a humid sub-tropical to tropical climate. The faulting present in the area was probably a major contributor to the degree of tower karst development."*

As may be seen at this stop, karsting develops much secondary vuggy porosity in the Leadville, and the overlying clayey Molas provides a good seal. The Molas Formation may be subdivided into three members here: (1) the lower 70 foot-thick Coalbank Hill Member, an unstratified residual terra rossa soil; (2) the middle member, an 80 foot-thick reworked poorly stratified terra rossa soil with sand and chert pebbles, and (3) the upper member, a 25 foot-thick stratified succession of siltstone, sandstone, and limestone containing marine invertebrate fossils (Spoelhof, 1976, p. 164). Some of the petroleum occurring in the Paradox basin is associated with this paleokarst unconformity. The Lisbon field in the Leadville in southeastern Utah is an example. It contains approximately 43 million recoverable barrels of oil (Clark, 1978).

Large sinkholes were also developed on this sequence boundary. Molas Lake has reoccupied one of these paleosinkholes, and a large dry one may be seen on the high plateau approximately one mile southwest of Molas Lake. Most of the karst on the Leadville is of the more normal, flatter variety where jointing is less prominent. For example, a good exposure may be seen in Rockwood Quarry 17 miles south of here. The flatter karstification could produce more potential secondary porosity than tower karst, because most of the reservoir is preserved rather than being dissolved.

Note the thin and incomplete nature of the sub-Pennsylvanian strata in this area (Figure 5-3), an excellent example of the "more gap than record" rule (Ager, 1973). The total sub-Pennsylvanian Phanerozoic section is less than 300 feet thick and represents less than 15% record and more than 85% lacuna. From Cambrian through Mississippian time this area was part of the North American craton with a passive margin to the west, so that deposition was sporadic and consisted mainly of carbonate rocks and mature quartz sandstones with some thin shales. During Pennsylvanian time the Ancestral Rocky Mountain orogeny occurred as a long-distance result of transpression from a continent-continent collision in the Ouachita mobile belt, and, consequently, the style of sedimentation changed to the thick and more complete strata of the Hermosa Group, more than 2000 feet thick in this area and recording perhaps 20 m.y. of deposition.

118.9 Molas Pass. Elevation 10,901 feet.
119.2 Engineer Mountain straight ahead. Capped by a rhyolite sill at 12,972 feet. View of Cutler (red) and Hermosa (tan and gray) below.
119.7 Contact of Tertiary intrusive with Hermosa.
121.3 Lime Creek Burn overlook on left.
122.1 Pull off on right shoulder.

STOP TWO. LIME CREEK BURN. In the roadcut at this stop a local angular unconformity at a small-scale sequence boundary in the lower Honaker Trail Formation (Hermosa Group) may be seen. The unconformity was produced by Early Pennsylvanian (Desmoinesian) upward movement of a basement feature below, the Sheep Camp horst (Figure 5-3). Structures such as these not only produce local structural highs, but may produce local erosional truncation of strata. To the north, in the Lime Creek Burn area, we can see the effect of the north-bounding fault of the horst. There, it produced a drag fold forming the south limb of a non-compressional syncline in the Hermosa that is visible (the north limb of the syncline was caused by a Tertiary intrusive, also visible). Figure 5-4 displays the geology of this area.

The area below us is capped by Upper Cambrian quartzite-boulder conglomerate on top of the basement horst. The conglomerate rests with angular unconformity on the Uncompahgre Formation, a near-vertical succession of middle Proterozoic quartzite and slate at least 8000 feet thick. The Sheep Camp horst is a small high block on the larger Grenadier fault block, which is composed of all the Uncompahgre Formation in the area (Spoelhof, 1976). The Grenadier fault block formed during Proterozoic time (as a graben!), but it has been rejuvenated as a horst many times since, producing local erosion of strata and unconformities over it during at least parts of Cambrian, Devonian, and Pennsylvanian time. The Grenadier and Sneffels (north, near Ouray) blocks and associated structures form part of the San Luis uplift, which has an en echelon orientation to the Uncompahgre uplift to the north (Figure 1 of the accompanying overview paper). The San Luis uplift extends southeastward into north-central New Mexico (Baars and Stevenson, 1984).

Examine the strata above and below the angular sequence boundary carefully, and note any change in parasequence stacking patterns. Millberry (1984) interpreted the thin shale and sandstone beds as marine shelf deposits and the thick conglomeratic sandstone as a progradational fan-delta deposit. How would you interpret her representative section (Figure 5-5) in light of current concepts of sequence stratigraphy?

You may also want to examine the effect of the structure on the detailed bed-by-bed stratigraphy here in the roadcut (Figure 5-6).

123.7 Coal Bank Pass. Elevation 10,647 feet. An outcrop of quartzite-boulder conglomerate and McCracken Sandstone Member of the Elbert Formation resting nonconformably on Precambrian gneiss may be seen in the rest area. The upper member of the Elbert is exposed in the roadcut. Approximately 2500 feet of Hermosa Group is exposed above in the steep slope toward Engineer Mountain.
126.1 Begin a series of roadcuts in various Paleozoic formations (with a Tertiary sill) for the next mile.
128.3 Cascade Creek.
129.0 For the next six miles the road is on a bench at the Kaskaskia-Absaroka sequence boundary (Molas on Leadville). Good view of Hermosa Cliffs to west. Many mixed clastic-carbonate parasequences and small-scale sequence boundaries are exposed in this cyclic succession.
141.0 Dirt road beyond gate to Rockwood Quarry to left. Excellent exposure of planar paleokarst Kaskaskia-Absaroka sequence boundary in quarry.
155.0 Good view of Mesozoic section from Dolores Formation up to Mesaverde Group at Durango. Another glacial lake floor and its terminal moraine may be seen for a few miles as we approach Durango along the Animas River.
160.0 Durango. All major towns in southwestern Colorado are built on the Mancos Shale.
164.5 Make a U-turn and park on the lateral road in front of the Texaco bulk station to view the exposure across the canyon.

STOP THREE. DURANGO SOUTH. An outcrop across the Animas River canyon displays the geometry of strata across the Zuni-Tejas Sloss sequence boundary (the K-T unconformity). Also, there is an excellent roadcut outcrop of this unconformity at Yellowjacket Pass 24 miles east of here. The K-T unconformity is difficult to define within the San Juan basin subsurface because of the similarity in well-log response of the Kirtland and Animas Formations. However, along the northern monocline of the basin near Durango, the unconformity is quite obvious on nearby seismic sections and outcrops (Figure 5-7). The present-day structure that we see here is a result of early Cenozoic deformation of the Laramide orogeny along reverse faults that may be rooted in Precambrian basement rocks.

The strata here consist of the tan-colored Kirtland Formation below, the purple-colored McDermott Member of the Animas Formation in the middle, and the tan-colored upper member of the Animas Formation above. The K-T unconformity is apparently at the McDermott-Animas upper member contact. The Animas Formation thickens basinward, and the McDermott Member is truncated just east of here by the unconformity. Note the spectacular onlap (approximately 150 feet of onlap in this outcrop) of the Animas upper member on the McDermott at this sequence boundary! Also, a thick paleosol is present in the uppermost McDermott just below the sequence boundary. The McDermott consists of approximately 125 feet of andesite-cobble breccia and tuff (Baltz et al, 1966) overlain by a 60 foot paleosol. The onlapping Animas upper member, which is more than 1000 feet thick here (Barnes et al, 1954), consists of ledge-forming andesite-cobble conglomerate and slope-forming shale. The andesite records the first (very late Cretaceous) volcanic activity in the San Juan Mountain area and corresponds to the age of the older laccoliths in the Four Corners area, such as the La Plata Mountains west of Durango. Most of the laccoliths and the San Juan Mountain volcanics are Oligocene.

165.8 Junction of US 160 and 550. Turn south on US 550 to Aztec, New Mexico. The upper member of the Animas Formation is exposed for the next several miles.
A series of Amoco coal-degasification wells in the Fruitland coal are along the road in the next two miles.
182.7 New Mexico state line.
196.7 Aztec.
208.5 Farmington.

REFERENCES

Ager, D. V., 1973, The nature of the stratigraphic record: New York, John Wiley and Sons, 114 p.

Baars, D. L., J. A. Ellingson, and R. W. Spoelhof, 1987, Grenadier fault block, Coalbank to Molas passes, southwest Colorado, *in* S. S. Bues, ed., Centennial Field Guide: Geological Society of America, Rocky Mountain Section, v. 2, p. 343-348.

Baars, D. L., and G. M. Stevenson, 1984, The San Luis uplift, Colorado and New Mexico--an enigma of the ancestral Rockies: The Mountain Geologist, v. 21, p. 57-67.

Baltz, E. H., S. R. Ash, and R. Y. Anderson, 1966, History of nomenclature and stratigraphy of rocks adjacent to the Cretaceous-Tertiary boundary, western San Juan basin, New Mexico: U.S. Geological Survey Professional Paper 524-D, p.1-23.

Barnes, H., E. H. Baltz, and P. T. Hayes, 1954, Geology and fuel resources of the Red Mesa area, La Plata and Montezuma Counties, Colorado: U.S. Geological Survey Oil and Gas Investigations Map OM-149.

Clark, C. R., 1978, Lisbon:, *in* J. E. Fassett, ed., Oil and gas fields of the Four Corners area, v. II: Four Corners Geological Society, p. 662-665.

Lee, K., R. C. Epis, D. L. Baars, D. H. Knepper, and R. M. Summer, 1976, Road log: Paleozoic tectonics and sedimentation and Tertiary volcanism of the western San Juan Mountains, Colorado, *in* R. C. Episk and R. J. Weimer, eds., Studies in Colorado field geology: Colorado School of Mines Professional Contribution 8, p. 139-158.

Maslyn, R. M., 1977, Fossil tower karst near Molas Lake, Colorado: The Mountain Geologist, v. 14, p. 17-25.

Millberry, K. W., 1984, Fan-delta and associated shelf-bars, lower member of the Honaker Trail Formation (Desmoinesian), southwestern Colorado: M.A. thesis, University of Texas, Austin, 62 p.

Spoelhof, R. W., 1976, Pennsylvanian stratigraphy and paleotectonics of the western San Juan Mountains, southwestern Colorado, *in* R. C. Episk and R. J. Weimer, eds., Studies in Colorado field geology: Colorado School of Mines Professional Contribution 8, p. 159-179.

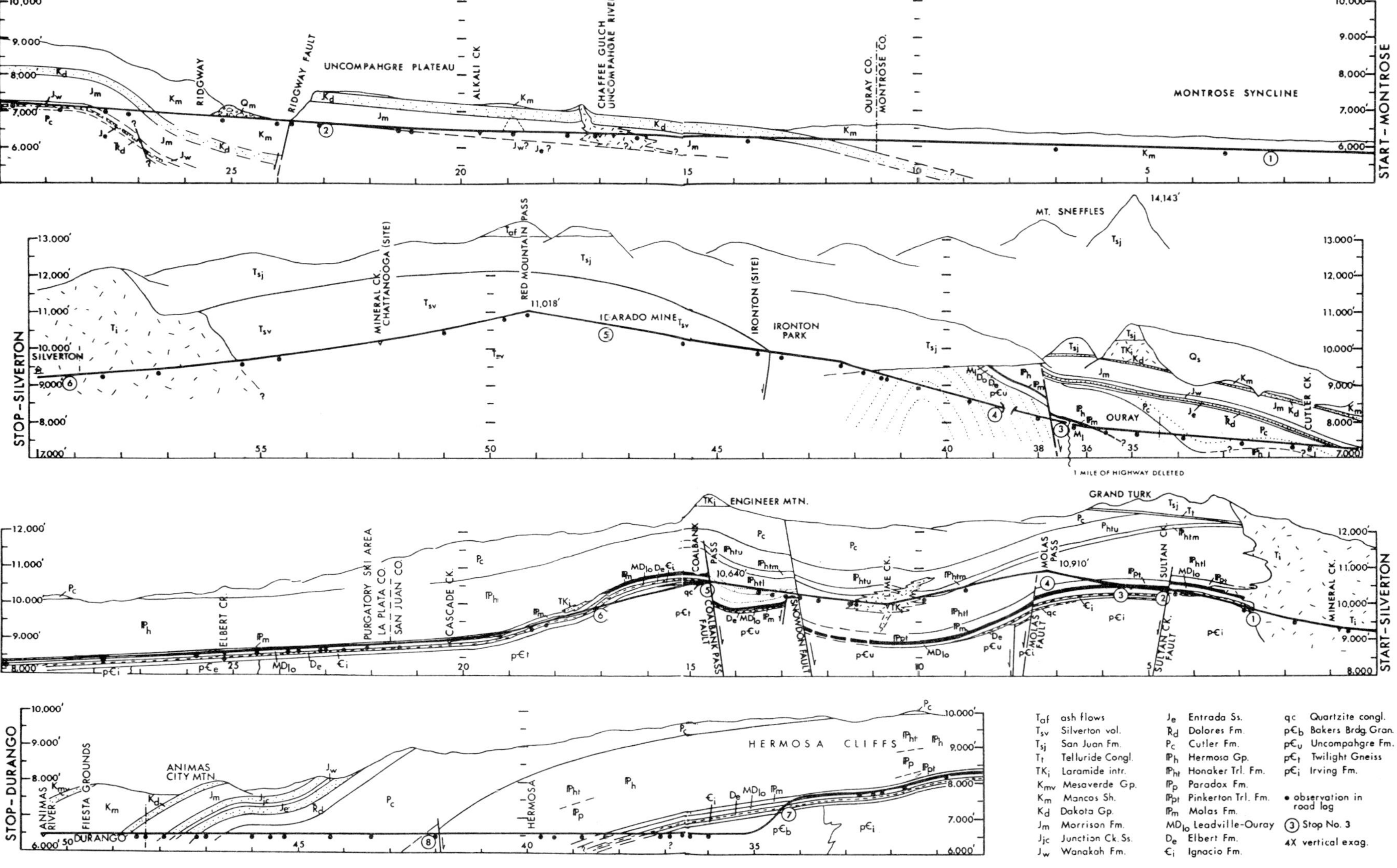

Figure 5-1. Cross-section along the west side of the road from Montrose to Durango. Not to true scale because line of section is U.S. 550. From Lee et al (1976, p. 142-143.). Junction Creek Sandstone is now member of Morrison Formation. Dakota Sandstone as shown includes Burro Canyon Formation.

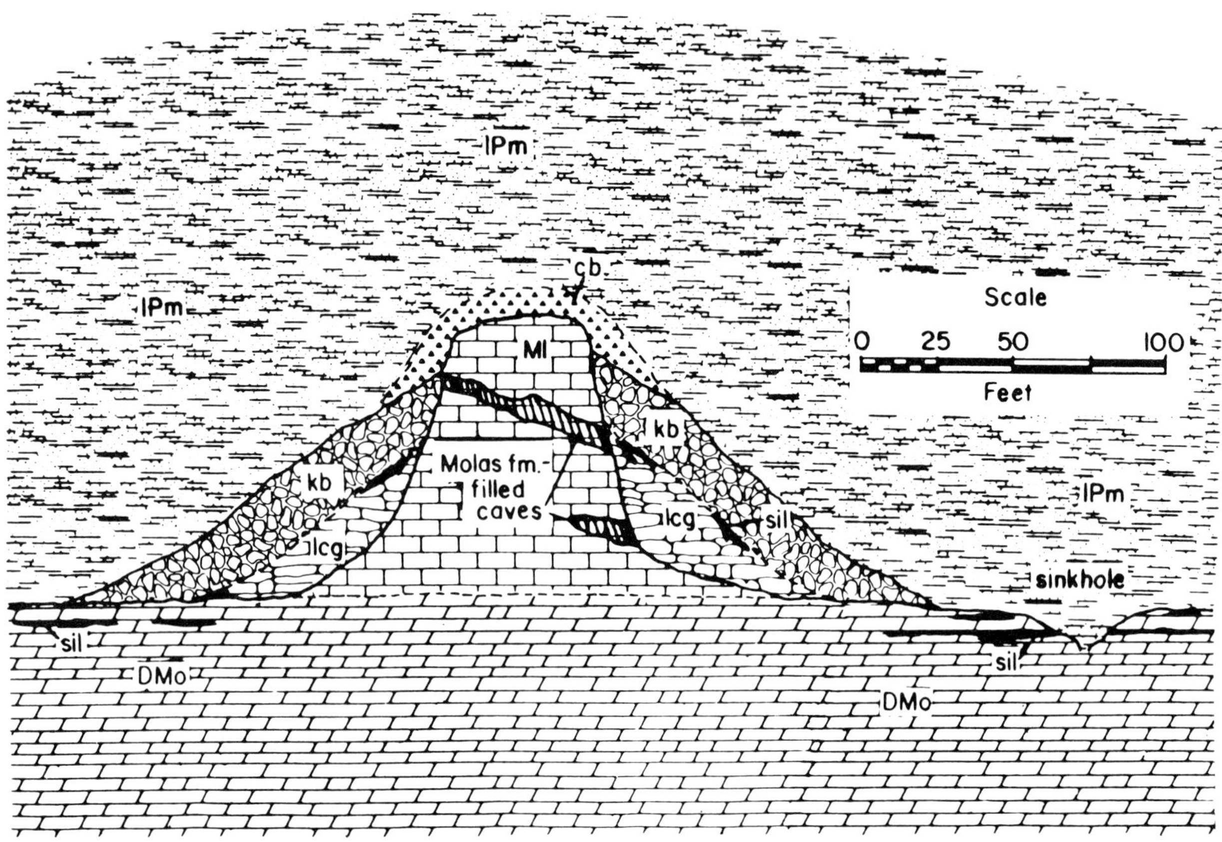

Figure 5-2. Idealized cross-section of the Molas Lake karst tower. Ml = Leadville Limestone; DMo = Ouray Limestone; IPm = Molas Formation; lcg = limestone conglomerate; kb = karst breccia; cb = chert breccia; sil = silica. From Maslyn (1977). Note that paleokarst surface between towers is on top of Ouray Limestone, a dolostone here.

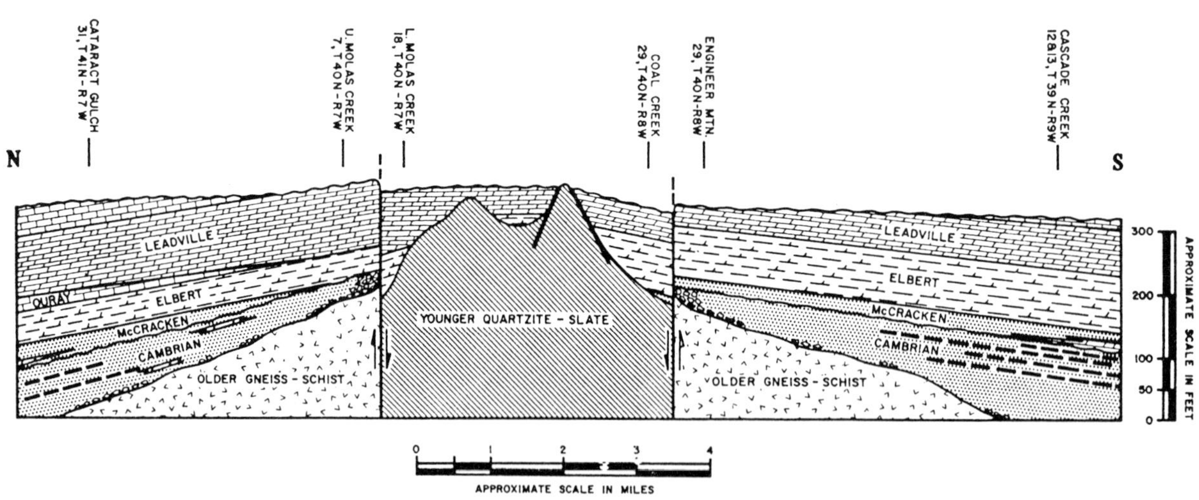

Figure 5-3. Cross-section across Grenadier block during latest Mississippian time. Small horst in center is Sheep Camp horst. From Baars et al (1987, p. 14).

AGE	FORMATION	LITHOLOGY	THICKNESS	MAP SYMBOL
TERTIARY	LIME CRK. ENGINEER MTN. INTRUSIVES	DACITE SILL	260+m	Ti
LOWER TERTIARY	SAN JUAN	ACIDIC FLOWS & TUFFS	305+	Tsj
UPPER PENNSYLVANIAN (MISSOURIAN)	CUTLER	RED BEDS	825+	Pc
MIDDLE PENNSYLVANIAN (DESMOINESIAN)	HERMOSA: UPPER HONAKER TRAIL	SANDSTONE	240-265	₱htu
	HERMOSA: MIDDLE HONAKER TRAIL	LIMESTONE	34-104	₱htm
	HERMOSA: LOWER HONAKER TRAIL	SANDSTONE	488-503	₱htl
	HERMOSA: PINKERTON TRAIL	LIMESTONE	67-113	₱pt
LOWER PENNSYLVANIAN (ATOKAN)	MOLAS	REGOLITH	0-17-46	₱m
LOWER MISSISSIPPIAN UPPER DEVONIAN	LEADVILLE-OURAY	DOLOMITE	0-46	MDlo
UPPER DEVONIAN	ELBERT-McCRACKEN	DOLOMITE SANDSTONE	0-33	De / Dm
UPPER CAMBRIAN	IGNACIO	SANDSTONE	0-46	Ꞓi
PROTEROZOIC	UNCOMPAHGRE	QUARTZITE	1525(?)	pꞒu
ARCHEOZOIC	IRVING-TWILIGHT	GRANITE GNEISS		pꞒgg

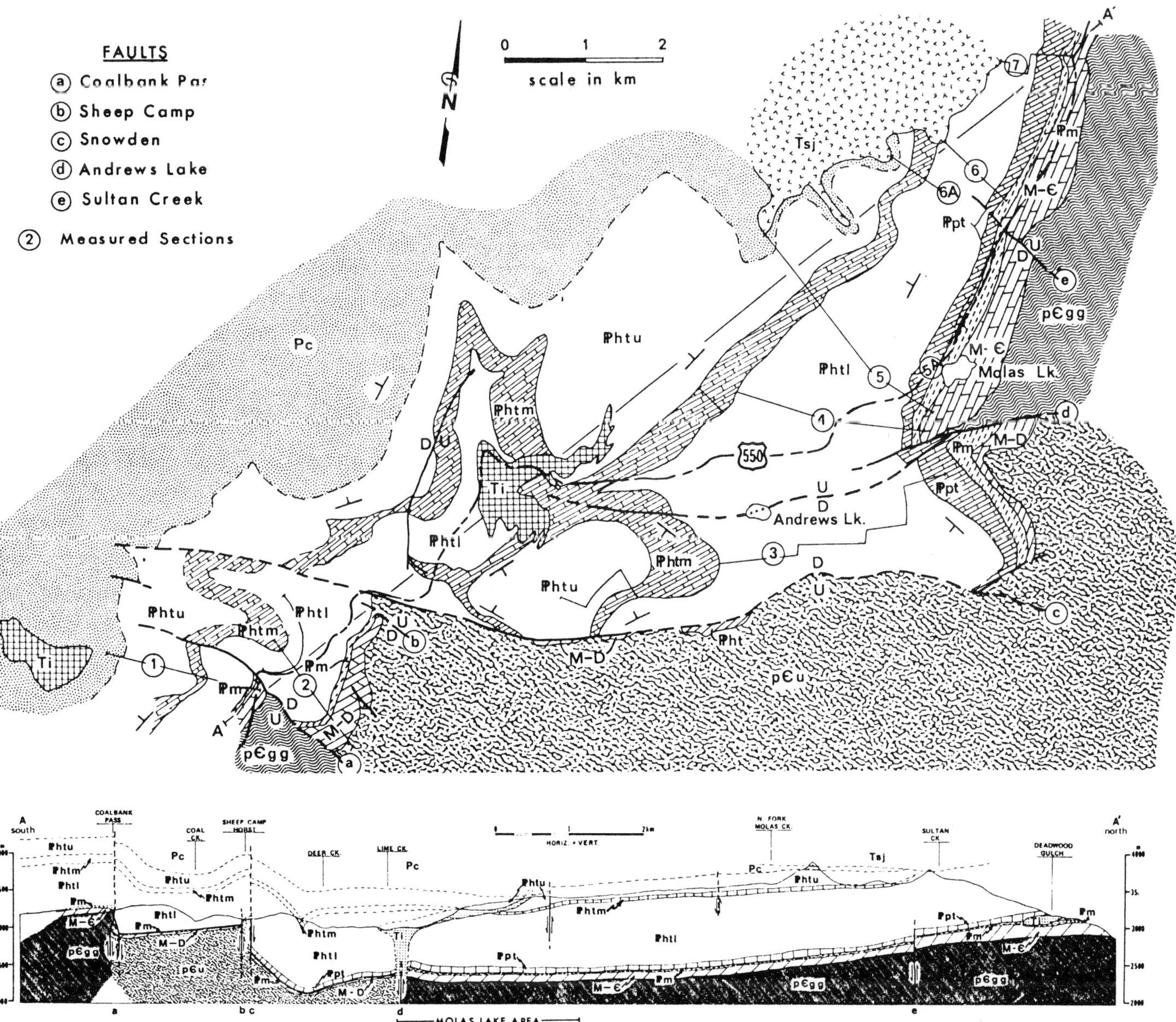

Figure 5-4. Stratigraphic column, geologic map, and cross-section of Molas Lake-Lime Creek Burn-Coal Bank Pass area. From Spoelhof (1976, p. 162-163).

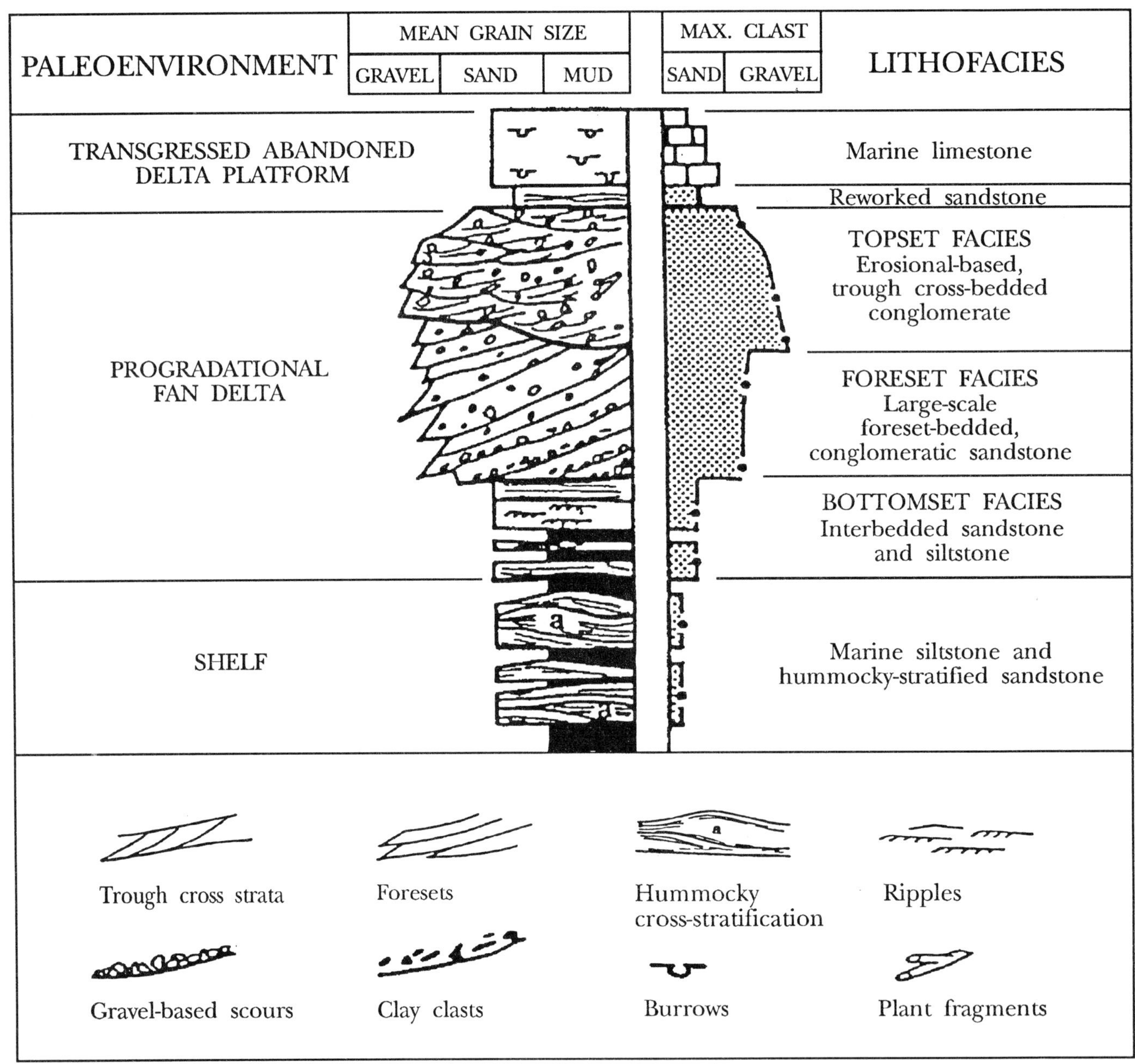

Figure 5-5. Representative section of lower member of Honaker Trail Formation (Hermosa Group) in Molas Lake-Lime Creek Burn-Coal Bank Pass area with paleoenvironmental interpretations of Millberry (1984, p. 7). Modified from Millberry (1984, p. 7).

Figure 5-6. Local intraformational angular unconformity at high-order sequence boundary in Honaker Trail Formation (Hermosa Group) at Stop 2. Pennsylvanian movement of basement Sheep Camp horst below produced angularity.

Figure 5-7. Onlap at Zuni-Tejas sequence boundary (K-T unconformity) at Stop 3. Onlapping beds and upper member of Animas Formation (tan strata to left) resting on McDermott Member of Animas Formation (Purple strata in center), which in tern rests unconformably on Kirtland Formation (tan strata on right).

LARGE-SCALE SEQUENCE STRATIGRAPHY OF THE PHANEROZOIC OF THE SAN JUAN MOUNTAIN REGION, COLORADO

by

DONALD E. OWEN
Lamar University
Beaumont, Texas

ABSTRACT

The Sauk, Kaskaskia, Absaroka, Zuni, and Tejas Sloss sequences are all represented in the San Juan Mountain region, Colorado, but the Tippecanoe sequence is absent. The tectonic setting changed from cratonic bounded by a passive margin to the west during the pre-Pennsylvanian Paleozoic to active with orogenic periods related to the Gondwana collision during the Pennsylvanian and later orogenic periods related to subduction and thrusting to the west during the Mesozoic and early Cenozoic. Sedimentation also changed from thin, carbonates and mature clastics in sub-Pennsylvanian strata to thick, more immature clastics in super-Mississippian strata. Small-scale sequences are also present, but mostly await additional investigation for recognition.

INTRODUCTION

Large-scale sequences bounded by major unconformities are well displayed in the San Juan Mountains of southwestern Colorado. Large lacunas are associated with most of the unconformities. The Sauk, Kaskaskia, Absaroka, Zuni, and Tejas Sloss sequences are all present in the area, but the entire Tippecanoe sequence is absent (Table 1).

The purposes of this paper are to review the large-scale sequence in the San Juan Mountain region and to speculate on smaller scale sequences within them in order to encourage additional sequence stratigraphic analysis.

TECTONICS

The tectonic setting of the San Juan Mountains changed during the Phanerozoic. From Cambrian through Mississippian time, while the Sauk and Kaskaskia sequences (Table 1) were deposited, the region was in a cratonic setting near the southwest extent of the Trans-Continental arch, and a passive margin adjoined to the west. Structural activity consisted mainly of vertical movement on basement faults. The stratigraphic section deposited in such a setting is thin (approximately 300 feet), consisting of carbonate rocks and mature quartz sandstones, with little shale. At least 85 percent of Cambrian through Mississippian time is represented by lacunas (Molenaar, 1989).

In contrast, tectonic activity increased during Pennsylvanian time with transpression transmitted into the area from the Gondwana collision occurring in the Ouachita-Marathon thrust belt. This resulted in the Ancestral Rocky Mountain Orogeny that uplifted the San Luis and Uncompahgre uplifts in the San Juan Mountain area (Figure 1) with related transtension causing rapid subsidence of the Paradox basin to the west. Later orogenies (Nevadan and Sevier), near the developing active margin to the west, brought increased amounts of sediments from uplifts and thrust belts into the developing Western Interior foreland basin, especially during Late Jurassic and Cretaceous time. The Laramide Orogeny of very late Cretaceous and early Cenozoic time rejuvenated many older structures in the San Juan Mountains and produced the present-day overall doming, which exposed the older basement-related structures. Rejuvenation of Precambrian basement faults has been a frequent occurrence and has had considerable effect on sedimentation and stratigraphy, both in cratonic and active tectonic settings. Super-Mississippian strata of the Absaroka, Zuni, and Tejas sequences (Table 1) total as much as 10,000 feet in thickness. Deposited in such an active tectonic setting, they consist almost entirely of clastic lithologies. The lacunas are much briefer than during the earlier passive phase, possibly as little as one-third of Pennsylvanian through Eocene time (Molenaar, 1989).

STRATIGRAPHY

Little detailed work has been done on the sequence stratigraphy of the San Juan Mountain region, although some of the stratigraphic column abounds in opportunity. Therefore, this paper emphasizes only the large-scale sequences on the order of Sloss sequences and their major subdivisions.

Sauk Sequence

The Sauk sequence contains but a single formation, the Upper Cambrian Ignacio Formation (Table 1). The Ignacio is a brachiopod-bearing, shallow-marine quartz sandstone with some shale and quartzite-boulder conglomerate, present near contemporaneous topographic highs, where the Ignacio is absent (Baars and Ellingson, 1984, p. 11). It only locally exceeds 50 feet in thickness (Spoelhof, 1976, p. 162). The Ignacio may represent a late transgressive systems tract deposit as the shoreface transgressed eastward. Sequence boundaries include a nonconformity on Precambrian igneous and metamorphic basement below and a paraconformity with an Upper Devonian quartz sandstone above.

Kaskaskia Sequence

The Kaskaskia sequence overlies the Sauk sequence in the San Juan Mountains because the entire Absaroka sequence is missing at a paraconformity, a rather curious relationship. This paraconformity represents a lacuna of approximately 120 m.y. (Kent et al, 1988)

Included in the Kaskaskia sequence in the San Juan Mountains are the Upper Devonian Elbert Formation and Ouray Limestone and the Mississippian Leadville Limestone (Table 1). Total thickness approaches 300 feet. The Elbert consists of a lower McCracken Sandstone Member and an upper member of dolostone and shale overlain by the Ouray, a dolostone in most of the region. The McCracken may be a transgressive systems tract deposit and the upper Elbert and Ouray may be high-stand deposits. There are two possible sequence boundaries within the Kaskaskia sequence. One is at a paraconformity with a lacuna of approximately 11 m.y. (Kinderhookian and most of Osagean time) recognized by Armstrong and Mamet (1976), who considered the Leadville as a transgressive limestone over the Ouray dolostone. Baars (1966), on the other hand, contended that the Ouray-Leadville contact was conformable and that an erosional disconformity separated the lower and upper members of the Leadville. This disconformity is another possible sequence boundary. Obviously, this part of the Kaskaskia needs additional work in light of modern concepts of sequence stratigraphy. The upper sequence boundary of the Kaskaskia is an extensive paleokarst surface with terra rossa soil, discussed in the next section.

Absaroka Sequence

Strata of the Absaroka sequence are thick, complex, and contain many smaller scale sequences within. The Molas Formation, Hermosa Group, Cutler Formation, and Dolores Formation make up the Absaroka (Table 1). The Hermosa and Cutler are each at least 2000 feet thick in the San Juan Mountains, and they thicken toward the Uncompahgre uplift (Figure 1). The Molas is relatively thin, reaching up to 175 feet (Spoelhof, 1966, p. 164). The Dolores is generally several hundred feet thick.

The lower sequence boundary of the Absaroka is marked by the spectacular terra rossa paleosol of the Molas Formation developed on a paleokarst surface developed in the Leadville Limestone. Although the boundary is conventionally mapped at the base of the Molas, perhaps a more accurate sequence boundary should be placed at the upper contact of the residual paleosol where it is overlain by reworked sediment within the Molas Formation.

The thick Hermosa Group is a high-frequency cyclic succession that contains many sequences. Several obvious sequence boundaries where coarse, arkosic sandstones overlie marine limestones with a thin terra rossa paleosol just below the boundary have been noted. The cyclic nature of the Hermosa, typical of Pennsylvanian strata, makes it ideal for a sequence stratigraphic approach, and some analysis is in progress (Thomas J. Feldkamp, 1991, personal communication).

A major sequence boundary is placed at the unconformity between the Hermosa and Cutler. Other sequence boundaries may be placed at the weathered tops of alluvial fan cycles within the Cutler. The Dolores Formation is bounded by sequence boundaries that are regional unconformities--the Tr-3 unconformity below and the J-2 unconformity above (Pipiringos and O'Sullivan, 1978) (Table 1).

Zuni Sequence

The Zuni sequence encompasses the entire, very thick Upper Jurassic through Upper Cretaceous succession that continues south of the San Juan Mountains into the San Juan basin. It is bounded by the J-2 unconformity below and the K-T unconformity above (Table 1). Large areas of erosional truncation and depositional onlap are associated with both of these unconformities. Several major sequences may be recognized within the Zuni.

The J-2 to J-5 interval is such a major sequence. The eolian Entrada Sandstone may represent a lowstand deposit, overlain by a lacustrine flooding surface that may be hydrologically coupled with relative sea-level rise. The Pony Express Limestone Member of the Wanakah Formation rests on the flooding surface. It is a lacustrine limestone that may represent the transgressive systems tract. The remaining upper part of the Wanakah, largely fluvial, may represent the highstand systems tract prograding into the lake.

The Morrison Formation, sandwiched between the J-5 and K-1 unconformities, forms the next major sequence. It includes fluvial, eolian, and lacustrine strata that probably represent several hard-to-recognize sequences in this non-marine setting. The fluvial Burro Canyon Formation forms the next sequence. It is bounded by the K-1 unconformity below and the K-2 above. The importance of the K-1 unconformity can be questioned as evidence accumulates that the upper Brushy Basin Member of the Morrison Formation is Lower Cretaceous (Kowallis and Heaton, 1987). The Brushy Basin was deposited in a large lake centered on Durango, Colorado (Turner-Peterson, 1987), and it would be logical to assume that sedimentation continued later in the central part of the lake near the San Juan Mountains. Perhaps the K-1 surface at the base of the Burro Canyon in this area consists of a series of channel-base diastems cut into lacustrine sediments.

There is no question about the K-2 surface being a major regional unconformity at the base of the Upper Cretaceous Dakota Sandstone. The Dakota consists of coastal plain sandstones, shales, and coals in the lower part with marine sandstones and intertongues of marine Mancos Shale in its upper part. It onlaps progressively lower strata southward across the San Juan basin (Owen and Sparks, 1989, p. 266). Above the Dakota, there is little evidence of major sequence boundaries in the rapidly subsiding foreland basin strata until the "sub-Niobrara unconformity" (Dane, 1960), associated with the Tocito Sandstone, is reached (Table 1). This relationship is the subject of the San Juan basin part of this conference (see Jennette et al and Nummedal and Riley, this volume).

The upper boundary of the Zuni sequence is at the K-T unconformity. At the south edge of the San Juan Mountains, this unconformity is well displayed where the upper member of the Animas Formation onlaps the McDermott Member of the Animas near Durango. Farther south in the San Juan basin near Farmington, New Mexico, this surface is at the disconformity where the Ojo Alamo Formation is channeled into the Kirtland Formation.

Tejas Sequence

The Tejas sequence is rather poorly represented in the San Juan Mountain region. It contains the Paleocene and Eocene nonmarine strata in the San Juan basin (Table 1) and the series of Oligocene-Miocene tuffs in the San Juan volcanics, each of which is bounded by erosion surfaces in the San Juan Mountains.

CONCLUSIONS

The Sauk, Kaskaskia, Absaroka, Zuni, and Tejas Sloss sequences are all represented in the San Juan Mountain region, but the Tippecanoe sequence is absent.

The tectonic setting in the San Juan Mountains changed from cratonic bounded by a passive margin to the west during the pre-Pennsylvanian Paleozoic to active with orogenic periods related to the Gondwana collision during the Pennsylvanian and later orogenic periods related to subduction and thrusting to the west during the Mesozoic and early Cenozoic.

Sedimentation also changed from thin, carbonates and mature clastics in sub-Pennsylvanian strata to thick, more immature clastics in super-Mississippian strata.

Small-scale sequences are also present in the San Juan Mountains, but mostly await additional investigation for recognition.

ACKNOWLEDGEMENTS

I am grateful to Don Baars for a field introduction into the stratigraphy of the San Juan Mountains. I have also learned much from his many publications on the region. I also thank Dag Nummedal for his insight into sequence stratigraphy, frequently expressed while we were in the field in this region. He has encouraged me to ask the right questions of the rocks.

I appreciate the helpful suggestions of Diane Sparks in reviewing the manuscript.

REFERENCES

Armstrong, A. K., and B. L. Mamet, 1976, Biostratigraphy and regional relations of the Mississippian Leadville Limestone in the San Juan Mountains, southwestern Colorado: U.S. Geological Survey Professional Paper 985, 25 p.

Baars, D. L., 1966, Pre-Pennsylvanian paleotectonics--key to basin evolution and petroleum occurrences in Paradox basin, Utah and Colorado: American Association of Petroleum Geologists Bulletin, v. 50, p. 2082-2111.

Baars, D. L., and J. A. Ellingson, 1984, Geology of the western San Juan Mountains, *in* D. C. Brew, ed., Field trip guidebook: Four Corners Geological Society, p. 1-45.

Baars, D. L., J. A. Ellingson, and R. W. Spoelhof, 1987, Grenadier fault block, Coalbank to Molas passes, southwest Colorado, *in* S. S. Bues, ed., Centennial Field Guide: Geological Society of America, Rocky Mountain Section, v. 2, p. 343-348.

Dane, C. H., 1960, The boundary between rocks of Carlile and Niobrara age in San Juan basin, New Mexico and Colorado: American Journal of Science, Bradley Volume, v. 258-A, p. 46-56.

Jennette, D. C., C. R. Jones, J. C. Van Wagoner, and J. E. Larsen, this volume, High-resolution sequence stratigraphy of the Upper Cretaceous Tocito Sandstone: the relationship between incised valleys and hydrocarbon accumulation, San Juan basin, New Mexico

Kent, H. C., E. L. Couch, and R. A. Knepp, 1988, Central and southern Rockies region correlation chart: American Association of Petroleum Geologists, COSUNA Chart, 1 sheet.

Kowallis, B. J., and J. S. Heaton, 1987, Fission-track dating of bentonites and bentonitic mudstones from the Morrison Formation in central Utah: Geology, v. 15, p. 1138-1142.

Molenaar, C. M., 1989, Cretaceous rocks of the San Juan basin, *in* W. I. Finch, A. C. Huffman, Jr., and J. E. Fassett, eds., Coal, uranium, and oil and gas in Mesozoic rocks of the San Juan basin--anatomy of a giant energy-rich basin: 28th International Geological Congress, Field Trip Guidebook T-120, plate 2.

Nummedal, D., and G. W. Riley, this volume, Origin of late Turonian and Coniacian unconformities in the San Juan basin

Owen, D. E., and Sparks, D. K., 1989, Surface to subsurface correlation of the intertongued Dakota Sandstone-Mancos Shale (Upper Cretaceous) in the Zuni embayment, New Mexico, *in* O. J. Anderson et al, Southeastern Colorado Plateau: New Mexico Geological Society, 40th Annual Field Conference, p. 265-267.

Pipiringos, G. N. and R. B. O'Sullivan, 1978, Principal unconformities in Triassic and Jurassic rocks, western interior United States--a preliminary survey: U. S. Geological Survey Professional Paper 1035-A, 29 p.

Spoelhof, R. W., 1976, Pennsylvanian stratigraphy and paleotectonics of the western San Juan Mountains, southwestern Colorado, *in* R. C. Episk and R. J. Weimer, eds., Studies in Colorado field geology: Colorado School of Mines Professional Contribution 8, p. 159-179.

Turner-Peterson, C. E., 1987, Sedimentology of the Westwater Canyon and Brushy Basin members, Upper Jurassic Morrison Formation, Colorado Plateau, and relationship to uranium mineralization: Ph.D. dissertation, University of Colorado, Boulder, 169 p.

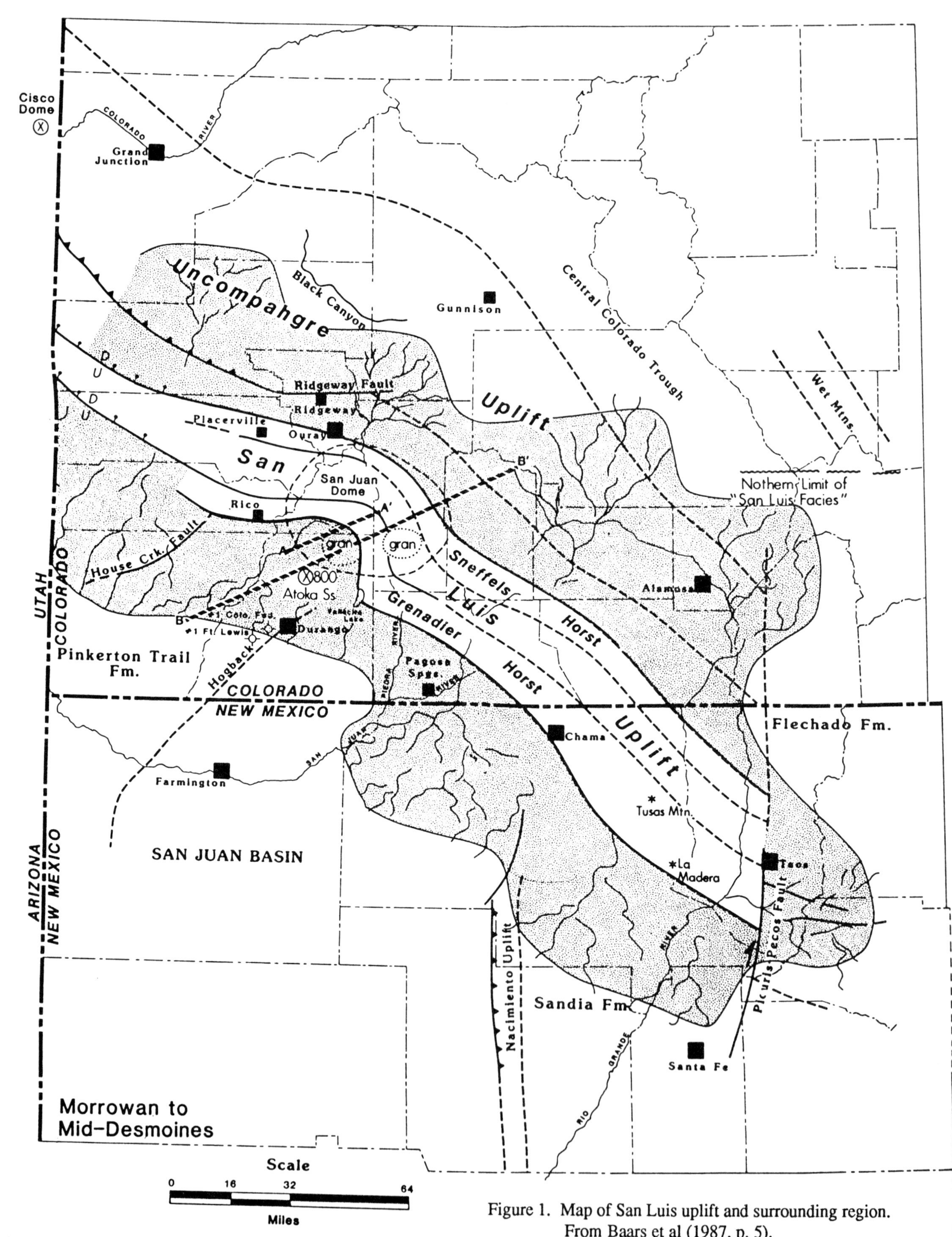

Figure 1. Map of San Luis uplift and surrounding region. From Baars et al (1987, p. 5).

Table 1: Stratigraphic column and regional unconformities of San Juan Mountains, southwestern Colorado.

CHRONOSTRATIGRAPHY		LITHOSTRATIGRAPHY	*SEQUENCE*	TYPE
QUATERNARY		alluvium & colluvium		
				DIS & ANG
EOCENE		San Jose Fm.		
				DIS
PALEOCENE		Animas Fm., upper mbr.	***TEJAS***	
				DIS & ANG
UPPER CRETACEOUS		Animas Fm., McDermott Mbr.		
				DIS
		Kirtland Fm. Fruitland Fm. (basal coal) Pictured Cliffs Fm. Lewis Sh. Mesaverde Gp. upper Mancos Sh. Tocito Ss..		
				DIS
		Gallup Ss. lower Mancos Sh. Dakota Ss.	***ZUNI***	
	K-2			DIS & ANG
LOWER CRETACEOUS		Burro Canyon Fm.		
	K-1			DIS
		Morrison Fm.		
	J-5			DIS
UPPER JURASSIC		Wanakah Fm. Entrada Ss.		
	J-2			DIS & ANG
UPPER TRIASSIC		Dolores Fm.		
	Tr-3			DIS & ANG
PERMIAN		Cutler Fm.	***ABSAROKA***	
				DIS & ANG
PENNSYLVANIAN		Hermosa Gp. Molas Fm.		
				DIS
MISSISSIPPIAN		Leadville Ls.		
				PARA
UPPER DEVONIAN		Ouray Ls. Elbert Fm.	***KASKASKIA***	
				PARA
UPPER CAMBRIAN		Ignacio Ss.	***SAUK***	
				ANG
PRECAMBRIAN		Uncompahgre Fm.		
				NON
		granitic & gneissic basement		

DIS = disconformity; ANG = angular; PARA = paraconformity; NON = nonconformity

DAY SIX

Four Corners Platform area
northwestern New Mexico

SAN JUAN BASIN SEGMENT, NEW MEXICO

by

Clive R. Jones, John C. Van Wagoner and David C. Jennette
Exxon Production Research, Houston, Texas

Dag Nummedal and Greg W. Riley
Louisiana State University, Baton Rouge, Louisiana

ROAD LOG, DAY SIX: SEQUENCE STRATIGRAPHY AND FACIES ARCHITECTURE OF THE TORRIVIO AND TOCITO SANDSTONES NEAR BEAUTIFUL MOUNTAIN, FOUR CORNERS PLATFORM, NORTHWESTERN NEW MEXICO.

OBJECTIVE: For the next two days we will examine the stratigraphy and lithofacies of Turonian- and Coniacian-age strata in northwestern New Mexico, on the lands of the Navajo Nation (see location map Figure 6-1). We will focus on the Gallup (shallow marine and coastal plain), Torrivio (predominantly braided-fluvial) and Tocito (estuarine to shallow marine) sandstone formations. An overview of the stratigraphy, lithofacies and hydrocarbon-trapping styles pertaining to these sandstones, in outcrop and from extensive subsurface well-log correlations, is presented in the accompanying paper entitled 'High- resolution sequence stratigraphy of the Upper Cretaceous Tocito Sandstone...'. A briefer overview is also presented under STOP 1 below. The stratigraphy of the San Juan Basin is summarized in Figures 6-2, 6-3 and 6-4. These are increasingly higher resolution stratigraphic columns, with Figure 6-4, being specifically designed for this field trip. This figure reflects some of the latest ideas on the distribution of significant sequence-stratigraphic surfaces in the Late Turonian and Coniacian parts of the section.

The discussion in the text at each of the field stops is designed to present two somewhat different sequence-stratigraphic interpretations of the rocks; an Exxon (Jones, Van Wagoner and Jennette) and L.S.U. (Nummedal and Riley) interpretation. The reader can draw his/her own conclusions from the observations made at outcrop and in the subsurface (see accompanying papers).

Any person wishing to conduct geological investigations on the Navajo Reservation, including visiting the stops described in this guidebook, must first obtain a permit from the Navajo Nation Minerals Department, P.O. Box 146, Window Rock, Arizona 86515.

0.0 Turn right off the parking lot of the Holiday Inn, Farmington on to Broadway (US 64 west)
0.2 Intersection of Pinon/Butler. Go straight ahead.
1.7 Intersection of Main St. and Broadway. Drive straight, heading west on US 64.
3.4 Junction with road 170, continue straight ahead.
7.5 Cliffs in the distance to the left are the Tertiary Ojo Alamo Sandstone.
13.0 Driving on uppermost Cretaceous rocks of the Kirtland Shale.
13.9 Road cuts through shoreface deposits of the Upper Cretaceous Pictured Cliffs Formation to the right (north).
20.1 The road cuts through a prominent ridge called The Hogback and passes the San Juan River to the south.We will be driving on alluvial deposits of the river or at the base of Cretaceous cliffs for the next 19 miles.

Regional outcrop geology and seismic data show the Hogback ridge to be part of an easterly dipping monocline, which forms a drape fold across an east-vergent, Laramide high-angle reverse fault. As we drive west we will cross progressively older Cretaceous rocks. The sandstone capping the Hogback comprises shoreface parasequences of the Upper Cretaceous Cliff House Sandstone. Coastal-plain deposits of the Menefee Formation outcrop at the base of the Hogback on the west side.

26.1 Parasequences within the Upper Mancos Shale Formation outcrop on the north side of the road.
29.0 We have entered the town of Shiprock, intersection of US 64 (east) and US 666 north (which carries on to Cortez, Colorado). Head south on US 64 (west) towards the San Juan River, follow signs for Gallup.

29.5 Crossing the bridge over San Juan River in Shiprock.
29.9 Intersection of US 666 with US 64 (westbound). Go straight, on US 666 out of the town of Shiprock, signposted for Gallup.
32.2 Some of the local landmarks are clearly visible. To the east is the Hogback ridge. Straight ahead is the flat-topped Table Mesa capped with sandstones of the Point Lookout Formation and Cathedral Rocks, a Tertiary volcanic neck similar to the prominent Ship Rock to the west.
32.6 The Four Corners power plant can be seen in the distance to the east.
36.2 Turn right onto road to Red Valley.
38.8 Good view of Beautiful Mountain ahead to the southwest. The elevation of the peak is 8945 feet. Beautiful Mountain is capped by Tertiary Chuska sandstone and a Tertiary sill. The Gallup and Tocito sandstones outcrop at the base of Beautiful Mountain, close to where our first field stop will be. The southerly-directed igneous dike emanating from Ship Rock is now clearly visible ahead.

43.8 PHOTO STOP. Pull off for a photo-opportunity stop. At this point the road crosses the largest radiating dike from Ship Rock. Ship Rock is a volcanic plug exhibiting three well-developed dikes and many smaller ones radiating out from the neck.

Ship Rock rises some 1600 feet from the surrounding desert floor. The top of Ship Rock has an elevation of 7178 feet above sea level. This volcanic neck was intruded into surrounding Cretaceous strata between 32 and 26 Ma. Ship Rock is a sacred place to the Navajo Nation and takes its name from the Navajo word *"tse bida' hi"* meaning "rock with wings".
47.3 To the southwest the dip slope of Rock Ridge is seen, dipping steeply to the east. Gallup shoreface parasequences comprise this prominent ridge. Steep-walled canyons dissect the ridge, creating the type of landscape for which the Colorado Plateau is famous.
49.9 The road passes through the lower-shoreface deposits of the Gallup, which form the Rock Ridge outcrops. The Gallup dips to the east.
50.8 Another volcanic plug called Mitten Rock outcrops to the north.
51.7 Upper Cretaceous Dakota Sandstone and Lower Mancos Shale outcrop to the north. They form the core of the breached anticline on which we are now driving.
52.1 Turn left off the paved Red Valley Road onto gravel road Navajo 5012, cross cattle guard, heading south. Beautiful Mountain lies directly ahead. Gallup Sandstone is the cliff former at the base of the mountain.
52.7 Reach three-way intersection. Turn left, now heading southeast. Just passed two-story house with red roof on the left.
55.2 Cliffs to the right (southwest) comprise Gallup shoreface parasequences.
55.5 Descending small hill through the Juana Lopez Member of the Lower Mancos, which occurs stratigraphically beneath the Gallup Sandstone.
56.4 Now on top of poorly exposed Juana Lopez outcrops.
57.6 The Juana Lopez forms the lowest, dark gray resistant cliffs to the right (southwest), approximately 50 feet in thickness. Lower Mancos Shale lies above and below the Juana Lopez. The major cliff former above the Juana Lopez is the Gallup Sandstone which largely comprises shoreface parasequences.

Above the marine Gallup are coastal-plain strata and very coarse-grained, fluvial sandstones of the Torrivio Sandstone Member. At the top of the succession are the tidally influenced deposits of the Tocito Sandstone, which forms brown, recessive outcrops on top of the cliffs.
61.1 Turn right onto jeep trail and drive towards "amphitheater" formed by the cliffs. Small ravine is off to the left of the road.

61.4 STOP ONE. "The Amphitheater" (SW NE Sec. 17 T26N R19W). Overview of the typical Gallup, Torrivio and Tocito succession in the Beautiful Mountain area (see stratigraphic summary and sketch section, Figures 6-5, 6-6 and 6-7A). A more comprehensive geologic overview of these deposits in outcrop and the subsurface is presented in the accompanying papers.

Juana Lopez Member of the Lower Mancos Shale

The low bench at the base of the cliffs is the top of the Juana Lopez Member of the Lower Mancos Shale. The Juana Lopez is a widely correlative unit easily followed in outcrop and in the subsurface on well logs. It comprises a series of beds composed of black shale and fine-grained calcareous sandstone. We have recovered the ammonite *Scaphites whitfieldi* in beds several feet above the top of the Juana Lopez on

Beautiful Mountain. This ammonite suggests an age of middle Late Turonian. Faunas yielding a similar age have been collected from the Juana Lopez at other localities in New Mexico (Hook and Cobban, 1980).

Gallup Sandstone

The Gallup Sandstone grades up out of the Lower Mancos Shale tongue which occurs above the Juana Lopez. The Gallup Sandstone in the Beautiful Mountain area is represented by several shoreface parasequences commonly comprising fine-grained, hummocky and wave-rippled lower shoreface deposits. They locally grade upward into upper shoreface and foreshore deposits. We will investigate the Gallup in more detail today at STOP 5 and on DAY SEVEN (STOP 2). A detailed study of the Gallup in this area by D. Valasek (Colorado School of Mines) will be completed later this year (1991).

The Gallup shoreface deposits in the Beautiful Mountain area form some of the most distal, and probably youngest Gallup parasequences in the San Juan Basin. Exxon palynological data suggests a latest Turonian age for these particular outcrops. Molenaar (1973, 1983a, 1983b) has suggested an age range of middle Late Turonian to Early Coniacian for the whole of the Gallup Sandstone Formation in the San Juan Basin based on inoceramids and ammonites.

Non-marine portion of the Gallup Sandstone

In the Beautiful Mountain area, the fine-grained, upper shoreface and foreshore deposits at the top of the marine Gallup are frequently absent through truncation by an erosional surface, which often has pronounced vertical relief (Fig. 6-8A). The erosion surface is predominantly filled with medium- to coarse-grained, tabular, trough and sigmoidally cross-bedded sandstones and current-rippled sandstones of fluvial to estuarine origin (Fig. 6-9). The dominant paleotransport direction from measured cross-bed sets is to the east and northeast. The erosive surface is correlatable throughout the Beautiful Mountain area and with the apparent downward shift in facies of coarse-grained fluvial to estuarine deposits, resting on lower-shoreface deposits of the Gallup, it is a probable candidate for a sequence boundary. It is possible that these incised valleys were feeding a co-eval shoreline system (McCubbin, 1982, 1991 pers. comm.) or lowstand delta further basinward.

Nummedal and Riley note that the base of the Gallup Sandstone at many locations near Sanostee is remarkably sharp suggesting perhaps, that there was shelf erosion associated with a downward shift. Nummedal and Riley think this implies a "forced regression" during this downward shift in the Gallup. In the accompanying paper entitled "Tectonics and eustasy in the development of the Gallup and Torrivio Sandstones", they present arguments that this sequence boundary can be correlated to the updip Gallup outcrops (near the town of Gallup) and that it may correspond to the Late Turonian sea level fall (90 Ma: UZA-2/UZA-3) on the chart by Haq et al. (1988). This is the unconformity U-1 in the accompanying paper by Nummedal and Riley.

Above the fluvial to estuarine sandstones which dissect the top of the marine Gallup section, lies an interval of fine-grained, probably brackish-water deposits, which attain a maximum thickness of about 45 feet in the Beautiful Mountain area. In the type area of the Gallup Sandstone some 100 miles to the south (Molenaar, 1973), lithofacies equivalents reach thicknesses of over 100 feet (Flores et al., 1991). These non-marine Gallup deposits show a variety of back-barrier and coastal-plain facies, including discontinuous fluvial channel sandstones, upward-coarsening and upward-thickening bay head/ lagoonal delta sandstones, crevasse-splay sandstones, carbonaceous mudstones and coals. The sandstones are predominantly fine-grained, but are locally medium- to coarse-grained. The fine-grained paludal facies are interpreted to represent marsh and coastal swamp deposits. Burrowing and rooting is common.

Four bayhead/ lagoonal delta units are recognizable in the Beautiful Mountain area. Although quite thin (about 10 feet thick), they form excellent horizons for correlation, and are laterally continuous across the Beautiful Mountain outcrops for up to 8 miles.

Torrivio Sandstone

A probable equivalent of the Torrivio Sandstone occurs above the non-marine Gallup deposits described above (Molenaar, 1973). The base of the Torrivio on Beautiful Mountain and in the area of Sanostee is erosional (Fig. 6-10G). Nummedal (1989, p.41) and Flores et al., (1991) comment that the base of the Torrivio is everywhere erosional, suggesting a widespread unconformity. The Torrivio is a coarse-grained, sheet-like, fluvial braided-channel complex. Some tidal influence may also be evident in places (Miall, 1991, pers. comm.). In the area south and west of the town of Gallup, the Torrivio reaches 100 feet in thickness (Molenaar, 1973), but thins to the north to a maximum thickness of 30 feet in the Beautiful Mountain area. The basal erosion surface is overlain by pinkish-red, feldspathic sandstones. Grains are angular to sub-angular and range mainly from medium to very coarse. Granules are common. Locally, pebbles and cobbles rest on the basal surface. The sandstones display planar-tabular, trough- and sigmoidal-cross-bedding which usually form the internal lamination and bed sets that comprise large, downstream accreting macroforms (commonly five to ten feet high and can be several hundred feet in length), bars and channels.

The high-energy facies of the Torrivio are incised mainly into fine-grained, coastal-plain deposits of the underlying Gallup (Fig. 6-8D, STOP 3), although at STOP 6 today, we will see the Torrivio resting on lower-shoreface deposits of the Gallup. This may represent a basinward dislocation of facies due to a relative fall in sea-level, and thus the surface at the base of the Torrivio is another candidate for a sequence boundary. The inferred unconformity at the base of the Torrivio is identified as U-2 in the accompanying paper by Nummedal and Riley.

An alternative hypothesis however is that the Torrivio lithofacies is diachronous, being oldest in the south and youngest in the Beautiful Mountain area. In this case, several 'Torrivio' sandstones may exist that are landward time-equivalents of the five Gallup 'tongues' identified by Molenaar (1973). The Torrivio at Beautiful Mountain could thus alternatively represent a coarse grained distributary system that has cut down into the underlying Gallup, as younger shorefaces have prograded seaward to the northeast. Unfortunately, no published biostratigraphic data is available from the Torrivio in order to constrain the synchroneity of this lithofacies. Exxon Production Research is currently attempting to establish the chronostratigraphic significance of the basal Torrivio surface and correlate this unit through the outcrop belt.

Dilco Member of the Crevasse Canyon Formation

In the Beautiful Mountain area a thin interval of coastal plain lithofacies separates the Torrivio from the overlying Tocito. The top of the Torrivio is often incised by an erosional surface with between 5-10 feet of vertical relief (Fig. 6-10G). This surface is overlain by a very coarse-grained lag (Fig. 6-8E) and occasionally, large cobbles of polycrystalline quartz (Fig. 6-8F). Chert and feldspar can also be abundant components of the lag. The coastal-plain coals, carbonaceous mudstones and fine-grained, current-rippled sandstones which infill and onlap this erosional surface, thicken markedly to the south (D. Valasek, pers. comm. 1991), because the erosion surface at the base of the Tocito rises exposing more of the underlying Dilco coastal-plain strata.

Tocito Sandstone

The Tocito forms the low, recessive cliffs above the non-marine portion of the Gallup Sandstone in the Beautiful Mountain area. The Tocito is separated from strata of the underlying Torrivio, Gallup and Juana Lopez, by a coarse-grained to pebbly basal lag, comprised of quartz clasts, phosphate and carbonate pebbles, fragments of shell, bone and sharks teeth (Fig. 6-14G). This erosive surface is overlain by tan-brown sandstones, which are medium- to very coarse-grained and frequently stacked into coarsening-upward packages (Fig. 6-14E). Trough and sigmoidal cross-bedding is common (Fig. 6-14A). The sandstones are often burrowed to bioturbated with an ichnofauna including *Skolithos*, *Paleophycus*, *Planolites*, *Thalassinoides* and occasional *Ophiomorpha*. Fragments of the robust bivalve *Inoceramus* are also present. Locally, spores, pollen and low diversity assemblages of dinoflagellates occur. Based on the palynomorphs, ichnofossils, sedimentary structures and sandbody geometry, the coarsening-upward stratal packages within the Tocito, are interpreted to represent the migration of sub-tidal bars within broad estuaries (Exxon).

The basal Tocito erosion surface in the Beautiful Mountain area based on outcrop and sub-surface work, is interpreted by Exxon to represent a sequence boundary formed during a lowstand in relative sea-level. Broad incised valley complexes were cut and subsequently filled with tidal deposits, during the late lowstand as relative sea level rose.

Nummedal and Riley alternatively interpret the basal Tocito surface to be a ravinement surface, cut by erosional shoreface retreat during transgression. The Tocito sandbodies which occur above this ravinement surface, are interpreted as transgressive tidal-sand bodies on a restricted shelf.

Based on megafossil assemblage zones, including inoceramids, ammonites and oysters, the Tocito Sandstone here at the Beautiful Mountain area is inferred to be of Middle Coniacian age. Palynomorphs suggest a Middle to Late Coniacian age.

Return to gravel road.

61.8 Turn right onto gravel road.
62.4 Rock Ridge to left (east) formed by Gallup shoreface strata.
63.7 Reach 5-way intersection. Turn left onto Navajo 5016. Pull over to side of road.

STOP TWO. "Five Points" (NW NE NW Sec. 28 T26N R19W). - Brief overview of regional structural elements of this part of San Juan Basin. Return to vehicles and drive straight ahead, now heading east. Panoramic view of Beautiful Mountain, Rock Ridge, Ship Rock, Hogback and the San Juan Mountains to the northeast in the distance.

64.4 Road bends sharply left, slow down.
64.5 Turn right on jeep trail, marked by sign 'CAUTION -WATERLINE'.
64.7 Trail bends sharply right with outcrops of Torrivio Sandstone on the right and left.

64.8 STOP THREE. "Sanostee East" (SW NW Sec. 27 T26N R19W). - Figures 6-11A (Exxon) and 6-12 (Nummedal) show measured sections at this outcrop.

Torrivio

Examine the basal contact of the Torrivio Sandstone. Note the coarse grain size, well-exposed trough, tabular and compound cross-beds (Fig. 6-11C) which occur in the Torrivio at this locality (Figs 6-10B, C, D and E). Paleotransports are to the east and northeast at this outcrop, the same as the Torrivio on Beautiful Mountain (Figure 6-11B), and in up-dip locations near the town of Gallup, 70 miles to the south. Walk up through the measured section and note the coarse, burrowed basal grit which marks the top of the Torrivio (Figure 6-10F).

Tocito

Outcrop work at Beautiful Mountain coupled with subsurface well-log correlations show this, the youngest Tocito sandbody to have variable thickness, thinning in some cases to zero on to what Exxon considers to be interfluve areas (see the accompanying paper by Exxon). The Tocito would then be deposited in broad, erosional lows on a sequence boundary which is probably represented at this locality by the contact at the base of the lowest Tocito grit bed (Fig. 6-11A).

The Tocito at this locality is comprised mainly of medium- to very coarse-grained sandstone, which is highly burrowed to churned (e.g. Skolithos). Some original sedimentary structures such as trough cross-bedding can however still be identified. The Tocito at this and other Beautiful Mountain outcrops appear to be the youngest of the four Tocito sequences identified in the subsurface study by Exxon. Palynology suggests a Middle to Late Coniacian age for these strata.

Alternatively, Nummedal and Riley interpret this basal Tocito surface as a “ravinement surface” (Swift, 1968) and the Tocito Sandstone as a transgressive shelf-sand ridge. This interpretation is based on outcrop work in this area, primarily from Bergsohn (1988). The basal Tocito contact contains chert and quartzite cobbles and displays large wave ripples where it overlies channel-fill sandstones of the Torrivio Member, otherwise it is marked by a glauconitic, coarse-grained sandstone lag. The basal surface truncates a variety of pre-Tocito facies and shows a gradual stratigraphic 'rise' to the southwest. Above this surface, the Tocito Sandstone consists of well-burrowed, glauconitic, marine sandstones and muddy sandstones and forms a northwest-southeast oriented, linear sandbody, parallel to the trend of the shoreline (Fig. 6-13). This linear, shore-parallel, marine sandbody, overlying the transgressive erosion surface is felt to represent a transgressive shelf-sand ridge, analogous to sand ridges presently forming on modern transgressive shelves (Swift, 1975, Thomas, 1990).

It is significant in this context to note that the mapped trend of the Tocito sandbody is perpendicular to that of the underlying fluvial Torrivio Sandstone. Such orthogonal orientations characterize all modern shelf-sand ridges relative to underlying, late Pleistocene, incised fluvial valley fills (Thomas, 1990, and Nummedal, unpublished data).

Return to the gravel road (Navajo 5016).

65.1 Turn left onto Navajo 5016
65.5 Pass under powerlines.

65.7 STOP FOUR. "Five Points Wash" (NE SW SE Sec. 22 T26N R19W). - Pull vehicles off to the north side of the road. Walk north about 300 yards along an old jeep trail in the direction of Ship Rock. Descend into the arroyo and examine the coastal plain deposits of the non-marine portion of the Gallup Sandstone. Marine parasequences of the Gallup lie beneath us.

In the measured section (Fig. 6-15A), four coarsening-upward units are recognized. A vertical profile through one typically begins with a coal (up to 2 feet thick), followed by light to dark gray carbonaceous mudstones which are frequently bioturbated, then very fine-grained sandstones with abundant wave ripples, burrows and thin organic-rich mudstone partings (Fig. 6-8B and C). These beds coarsen upward into fine-grained, current-rippled and burrowed sandstones containing small-scale trough cross-beds. Locally, this typical profile is cut by a coarser-grained channel. These coarsening-upward units are sharply overlain by grey organic-rich mudstones and coals. Each package is typically 10-15 feet in thickness.

These deposits are interpreted to represent small deltas, distributary channels and distributary mouth bars, which prograded into a restricted body of shallow water such as a lagoon or distributary bay. Each of the delta complexes are laterally persistent and can be walked out along the entire length of the Beautiful Mountain outcrop belt (approximately 8 miles). The geometry and orientation of the shallow water body into which these deltas were building is unknown, since these deposits always overlie the Gallup shoreface and associated coarser grained fluvial sandstones. The lack of both marine palynomorphs and significant shoreface re-working, suggests the deltas built out into a body of water which was a sheltered, low energy environment of restricted salinity. Van Heerden and Roberts (1988) have described similar deposits with comparable geometries (10-15 feet thick and 5-6 miles across), from the Holocene of the Louisiana Gulf Coast. Here small bay-head deltas (e.g. Atchafalaya) and sub-deltas (e.g. Cubits Gap) are building into bays and interdistributary bays in a back-barrier setting. By comparison, typical marine shelf-phase deltas, such as the Holocene delta lobes of the Mississippi (e.g. La Fourche), which prograded out on to an open marine shelf, are usually thicker (30-50 feet), of greater lateral extent and show evidence of significant wave re-working (Penland et al., 1988, Boyd et al., 1989).

Note that the Torrivio seen at STOPS 1 and 2 is absent at this locality. Exxon's interpretation is that in the 0.6 miles from our previous locality (STOP 2), we have crossed over the northern edge of a Torrivio Sandstone incised valley (Fig. 6-15B). We are now located on a narrow interfluve area where the Torrivio is absent through non-deposition. This 'interfluve' area can also be traced out on the southern edge of the Beautiful Mountain. The underlying coastal plain facies of the Gallup reach their maximum thickness (40 feet) at this interfluve area, where they have experienced only minor truncation by the overlying Torrivio erosive surface. On the basis of this one outcrop exposure alone, proponents of Walther's Law alternatively might suggest that we have only crossed a channel margin, if indeed the Torrivio has no sequence stratigraphic significance. In this context, it would represent a coarse-grained distributary system that fed a time-equivalent, strand-plain complex farther seaward of this locality (northeast) .

Regional truncation at the base of the Torrivio Member is also suggested by the northward convergence of the Torrivio Member and the top of the underlying Gallup Sandstone. Farther south (STOPS 7-2 and 7-3), the Torrivio is separated from the underlying Gallup Sandstone by approximately 25 meters of coastal plain facies (non-marine Gallup). North of STOP 5 and at STOP 6, the Torrivio is in erosional contact with shoreface facies of the Gallup Sandstone.

66.0 Return to 5-way intersection. Turn right onto Navajo 5012, heading north.
72.3 On top of low hill formed by the Juana Lopez.
74.7 On crest of small hill. Just passed under power line.
74.8 Take third jeep trail on the right. Keep left.
75.0 Road splits in to three. Take left-most track. Keep left. Head towards point (to the north).
75.3 Road forks again. Take right fork, trail veers left and now parallels the cliff face to the left, which comprises shoreface deposits of the Gallup Sandstone.

75.5 STOP FIVE. "Parasequence Point" (NW SE Sec. 14 T27N R20W). - LUNCH STOP and discussion on the parasequence architecture, seaward-dipping imbrication and facies changes within the Gallup shoreface. The measured sections by Exxon (Fig. 6-16) and Nummedal (Fig. 6-17) shows that two parasequences separated by marine-flooding surfaces are present in the Gallup at this locality. Sets of hummocky-bedded sandstones can be seen to facies change into wave-rippled sandstone and eventually marine mudstone, in an east-northeasterly direction within each parasequence (Figs 6-7B, C, D and 6-16). The parasequences at this locality probably comprise the 'A' or youngest tongue of the marine Gallup sandstone (terminology of Molenaar, 1973).

Nummedal and Riley propose a slightly modified parasequence architecture (Figs 6-17, 18 and 19). In this model, the sharp surface above the lower, massive sandstone is interpreted as a ravinement surface. The most recessive part of the cliff corresponds to the maximum gamma-ray count (Fig. 6-18) and a change from structureless sandstone below to weakly-bedded sandstone above. This surface is felt to represent minimal sedimentation rates and is interpreted as a "final transgressive surface", separating a thin interval of transgressive shelf sediments (lower part of unit 7, Fig. 6-17) from regressive sediments of the succeeding parasequence. In this model parasequences contain both regressive and transgressive components, separated by the initial transgressive surface. Parasequences are bounded by final transgressive surfaces. The regressive part of the lower parasequence terminates in the middle shoreface, whereas, the upper parasequence contains beach deposits at the top (Figs 6-7D and 6-17). This pattern indicates forestepping parasequences (Van Wagoner et al., 1988).

Return to main gravel road.

76.2 Turn right onto gravel road. (Navajo 5012).
77.9 'T' junction. Turn right, driving towards Mitten Rock.
78.5 Intersection of Navajo 5012 and the paved road to Red Valley. Turn right.
79.0 Good view of Mitten Rock to left (north).
80.6 Passing through the Rock Ridge. Look out for the third road on the left
82.8 Turn left at the third road on the left. Cross cattle guard that has 3 tires on either side. Road forks. Take right fork on gravel road. Heading north, Ship Rock is to the right (northeast).
84.1 Pass several hogans and a house with green roof.
91.5 Pass through small Navajo settlement.
91.9 Pull off road to right, just before small descent and past metal trash. Small outcrop to left.

STOP SIX. "Roadside Dump" (SW Sec. 5 T29N R19W). - At this locality the Torrivio Sandstone rests on middle shoreface deposits of the uppermost part of the Gallup Sandstone. The thin Tocito Sandstone interval that is present thickens rapidly to the northeast from here. Exxon interprets this locality to represent the southwestern margin of an incised valley, which has been mapped using outcrop and subsurface well log data (text-figure 15 in the accompanying paper by Exxon).

Riley and Nummedal alternatively interpret this pattern to represent the formation of a tectonic "strait" created landward (southwest) of a rising anticline (Figs 6-20 and 6-21). Figure 6-22 shows their interpretation of the Gallup through Tocito section at this stop. STOPS 6 and 7 mark the approximate seaward (northeast) margin of Gallup shoreface facies, farther seaward, only shelf facies are present.

The Torrivio Member forms a lenticular sandbody that possibly truncates the ravinement surface at the top of the Gallup (Fig. 6-22). Structures within the sandbody include low-angle trough cross-beds at the base and subhorizontal bedding above. This sandbody is interpreted as a distributary mouth bar.

In the vicinity of this stop the Gallup contains a thin (1 m thick) bed of hummocky-stratified shoreface deposits overlying a sharp base (Fig. 6-23A). The sharp base of this bed and its juxtaposition with underlying inner-shelf facies contrasts with the more typical, gradual coarsening-up shoreface profile. These sharp-based shorefaces have been interpreted as reflecting a rapid seaward shift of the shoreline in response to relative sea-level fall (Plint, 1988, Posamentier and Vail, 1988). This process has been referred to as a forced regression (Posamentier et al., 1990). This basal surface is referred to as a submarine, regressive surface of erosion (H.E. Clifton, pers. comm.). The upper surface of this hummocky bed is heavily burrowed and locally contains scattered chert pebbles (Fig. 6-23B). This upper surface is interpreted as a sequence boundary and flooding surface. A gamma-ray profile and measured section (courtesy of R. Cole) near this location have been used to tie these outcrops to the subsurface (Fig. 6-24).

Overlying the Torrivio Member is the Tocito Sandstone, which is here subdivided into two informal members in outcrops in these areas; the lower sandstone member and the upper sandstone member. The lower sandstone member consists of a coarsening-upward section with bioturbated, carbonaceous, glauconitic sandstones at the base and cross-bedded, glauconitic sandstones near the top (Fig. 6-21). At the top of the lower sandstone is a continuous bed of bioturbated, muddy sandstone containing dispersed phosphate nodules. The coarsening-upward trend, upward increase in scale of physical structures and abundance of terrestrial organic matter is interpreted as indicating a regressive (allochthonous) setting for the lower sandstone. The distribution of the lower sandstone member (Fig. 6-21) suggests that it formed behind a rising northwest-southeast oriented structural high (Fig. 6-20). The regionally continuous bed of bioturbated, phosphatic, mudstone and muddy sandstone, at the top of the lower sandstone member, marks the end of progradation. The base of this unit is interpreted as an "initial transgressive surface" and the phosphatic muddy sandstone bed is interpreted as forming in an embayment during transgression.

Above the bioturbated muddy sandstone bed is the upper sandstone member. The base of the upper sandstone often contains a lag of intrabasinal clasts. Intraclasts are primarily reworked phosphate nodules and cemented "burrow clasts" (Singh, 1985), derived from erosion of the lower sandstone member. The upper sandstone primarily consists of tabular-tangential cross-bedded sandstones and varies significantly in thickness, ranging from the thickness of an individual cross-bed set to 15 meters thick (Fig. 6-26). The upper sandstone forms northwest-southeast oriented isopach thicks, with paleocurrents primarily to the southeast. This member is interpreted as transgressive shelf sand ridges and sand waves, overlying a transgressive erosion surface. The characteristics of the upper sandstone member will be discussed more fully at STOP 1, DAY SEVEN .

Surface to subsurface correlations (Fig. 6-24) are used to tie the outcrop sections into regional subsurface cross sections (Figs 6-25 and 6-26; foldouts). The location of these sections are shown on Figure 6-20. The M-2 bentonite (terminology of McCubbin, 1969), above the Tocito, is used as the horizontal datum in these sections. The anticline is well defined by the underlying Juana Lopez marker. A distinct unconformity is present at the base of the Tocito Sandstone on the crest of the anticline and on the gentle northeastern flank, with most of the Gallup/Mancos interval truncated in these areas. Landward of this structure, no significant truncation is evident. Most of the uplift occurred during the Early Coniacian as indicated by the presence of Early Coniacian fauna in the Gallup Sandstone (pre-uplift) and Middle to Late Coniacian fauna in the post-uplift, upper Tocito Sandstone. The progradational lower sandstone member is correlated to interval 2 and possibly interval 1 in the subsurface, and the transgressive upper sandstone member correlates to interval 3 and part of interval 4 (see Figures 6-25 and 6-26).

Subsequent downward movement along the anticline is indicated by expansion of interval 4 along the crest (primarily on section A-A', Fig. 6-25). Correlations suggest that this downward movement (block faulting) occurred during Middle to Late Coniacian. Similar reversals in fault motion have also been described from this area by Huffman and Taylor (1991).

Turn around, head back the way we came along the dirt road about 0.1 miles.

92.0 Turn left off main 'gravel' road onto dirt track.
92.3 Fork right. Two small old corrals are to the right.
93.9 Track crosses an earthen dam (Ship Rock Wash).
94.1 Track splits into four just across the earthen dam. Take leftmost fork.
94.2 Track forks. Bear left. Ship Rock is to the southeast.
95.2 'T' junction. Turn left. Old wrecked white car is to the left. Now travelling northeast up a low rising hill. Small dwellings visible ahead at top of the hill. Road separates into two going up the hill, select the left one.
96.3 Narrow gauge water pipe crosses road. Metal trash to left.
96.5 Crossing fence line with a cattleguard. Small settlement now to the left in the distance. Road here is partly composed of Tocito Sandstone outcrop.
97.0 Turn left at four-way intersection, with pink house to the left.
97.1 Passing water tower to left.
97.6 Turn left across cattleguard with white painted fencing and posts. Now driving due west between widely spaced houses of the small "Rattlesnake" settlement.
97.7 Reach a four-way intersection, go straight.
97.8 Reach another four-way intersection, go straight.
98.0 Road forks, keep left.
98.1 Cross the fence line, pass through the gate. Ship Rock is now straight ahead.
98.3 Follow main dirt track which bears right. Now heading northeast.
98.5 Descend the hill and cross over one of the youngest parasequences in the Tocito (metal trash to right).
98.7 Road forks. Take the dirt track to the right. Abandoned wells of the Rattlesnake oil field are all around. The Rattlesnake oil field produces form the Dakota sandstone.
98.9 Pass through 'oasis' where flowing water has filled a pond on the left. Keep right of the watering troughs.
99.1 Ascending a hill. Reach a 'Y'-intersection, take the right fork. A deep gully exposing the Tocito sandstone is on the right.
99.3 Turn left (fence line is straight ahead).
99.5 Reach another 'Y'-intersection, go right. Old corral is to the right.

99.8 Road forks. Take right fork.
100.4 Keep right.

100.6 STOP SEVEN. "Rattlesnake" (SW SE Sec. 34 T30N R19W). - Walk down into Ship Rock Wash to observe the Tocito resting directly on top of distal Gallup shoreface (see measured section and photograph, Figure 6-27). The base of the Gallup is quite sharp at this locality. Hummocky cross-stratification and 'ball and pillow' structures are abundant. The loading at the base of the Gallup, coupled with frequent water escape structures and convolute lamination, suggests that the basal hummocky beds were emplaced rapidly, presumably as storm deposits. Exxon interpret these deposits to represent deposition during latest highstand (available accommodation space is decreasing), when the shoreface system would be prograding rapidly basinward, probably over relatively unlithified shelfal mudstones. This may also account for some of the soft sediment deformation observed at the base of the Gallup here.

Between this locality and the stops we visited this morning (STOPS 1 to 4) , the basal erosive surface of the Tocito truncates out the Torrivio and nonmarine portion of the Gallup beneath an angular unconformity (Figure 6-28). As the Tocito outcrops are traced to the north and into Colorado, the base of the Tocito continues to erode into older strata (Fig. 6-14F), ultimately cutting down to the top of the Juana Lopez (Figs 6-14G and 6-29).

Drive to the paved road US 64, by retracing the route in from the white painted posts and cattle guard near the "Rattlesnake" settlement at roadlog marker 97.6.

103.5 Cross the white painted cattleguard and turn left on to the main dirt road which goes north to US 64.
105.3 Reach US 64, turn left and drive to the west. Ship Rock is visible to the south, ahead are the Chuska Mountains, the high peaks to the north are the Ute Mountains.
108.5 Crossing Ship Rock Wash. Rocks outcropping in the Wash are Tocito.
110.1 Crossing a small bridge across a tributary to Ship Rock Wash, more Tocito outcrops in the wash.
112.1 Turn right through the fence line onto a dirt track heading north. Stay left.
112.2 The road forks, take the right fork to the northeast. Stay on this main dirt road.
113.9 Intersection of two dirt roads, one veers right; take the road that goes straight ahead. Small cream colored house to left.
114.0 Pass a white house with a green roof on the right. Track splits into three, go straight ahead on the middle track.
114.1 Road forks. Keep right on the small dirt track.
114.2 Deserted house with red roof to left. Carry straight on. A corral is on the right.
114.3 Pass boarded up hogan on the right, under power line. Road forks, continue ahead then turn sharp right, 50 feet past the hogan. Drive out to the edge of the Tocito outcrop and park.

114.4 STOP EIGHT. "Lichii Wash" (NW NE Sec. 8 T30N R19W). - This stop provides an opportunity to examine the tidal bedding of the Tocito Sandstone (see measured section, Figure 6-30). Excellent sigmoidal cross-beds and tidal bundles are developed. A tidal-bar complex is well exposed in cross section and plan view (Figs 6-14A and B. Also see the accompanying paper by Exxon for a discussion on tidal influence and bedforms in the Tocito). A paleosol (?) is developed at the base of the uppermost tidal bar at this locality, suggesting either periodic emergent conditions during deposition of the Tocito tidal bar complex here, or an exposure surface that may have more regional distribution (sequence boundary? See Figure 6-14C). Carbon/oxygen isotopic analysis of the carbonate nodules indicates the calcite was precipitated in the meteoric zone (see the accompanying Exxon paper for a fuller discussion). At the end of STOP 8 return to US 64 by retracing the route in.

116.7 Reach US 64, turn left. Return to Farmington.

REFERENCES

Bergsohn, I., 1988, Lithofacies architecture of the Tocito sandstone, northwest New Mexico: M.S. Thesis, Louisiana State University, Baton Rouge, 170 pp., 1988.

Boyd, R., Suter , J.R., and Penland, S., 1989, Relation of sequence stratigraphy to modern sedimentary environments, Geology, V. 17, p. 926-929.

Haq, B. U., Hardenbol, J., and Vail, P. R., 1987, Chronology of fluctuating sea levels since the Triassic: Science, V. 235, p. 1156-1167.

Flores, R.M., Hohman J.C., and Etheridge, F.G., 1991, Heterogeneity of Upper Cretaceous Gallup sandstone regressive facies, Gallup Sag, New Mexico, *in* Nations, J.D., and Eaton, J.G., eds., Stratigraphy, depositional environments and sedimentary tectonics of the western margin, Cretaceous Western Interior Seaway: Geological Society of America Special Paper 260, 209 pp.

Hook, S.C., and Cobban, W.A., 1980, Some guide fossils in Upper Cretaceous Juana Lopez Member of Mancos and Carlile shales, New Mexico: New Mexico Bureau Mines & Mineral Resources Annual Report. 1978-79, p.38-49.

Huffman, A.C., Jr., and Taylor, D., J., 1991, Basement fault control on the occurrence and development of Juan Basin energy resources, Abstracts with programs, Rocky Mountain Section and South Central Section Geological Society of America meeting, Albuquerque, New Mexico, April 22-24, 1991.

McCubbin, D. G., 1969, Cretaceous strike valley sandstone reservoirs, northwestern New Mexico: American Association of Petroleum Geologists, V. 53, p. 2114-2140.

McCubbin, D.G., 1982, Barrier Island and Strand Plain Facies, *in* Scholle, P. A. and Spearing, D., eds., Sandstone Depositional Environments: American Association of Petroleum Geologists, Memoir 31, 410 pp.

Molenaar, C.M., 1973, Sedimentary facies and correlation of the Gallup Sandstone and associated formations, northwestern New Mexico, *in* Fassett, J.E., ed., Cretaceous and Tertiary rocks of the southern Colorado Plateau: Four Corners Geological Society Memoir, p. 85-110.

Molenaar, C. M., 1983a, Major depositional cycles and regional correlations of Upper Cretaceous rocks, southern Colorado Plateau, *in* Reynolds, N.W., and Dolly, E.D., eds., Mesozoic Paleogeography of West Central United States: Rocky Mountain Section of Society of Economic Paleontologists and Mineralogists Symposium, No.2, p. 201-224.

Molenaar, C.M., 1983b, Principle reference section and correlation of the Gallup Sandstone, northwestern New Mexico, Contributions to mid-Cretaceous paleontology and stratigraphy of New Mexico, Part II: New Mexico Bureau of Mines and Mineral Resources Circular, V. 185, p.29-40.

Molenaar, C. M., 1988, Cretaceous and Tertiary rocks of the San Juan Basin, *in* Sloss, L.L., ed., Sedimentary Cover - the North American Craton, Column US, The Geology of North America, DNAG, V. D-2, p. 129-134.

Molenaar, C.M, 1989, San Juan Basin stratigraphic correlation charts, *in* Finch, W.I., Huffman, A.C., and Fassett, J.E., eds, Fieldtrip Guidebook T120; Coal, uranium, oil and gas in Mesozoic rocks of the San Juan Basin: anatomy of a giant energy-rich basin, 28th International Geological Congress, p. xi.

Nummedal, D., Wright, R., Swift, J.P., Tillman, R.W., and Wolter, N.R., 1989, Depositional systems architecture of shallow marine sequences, *in* Nummedal, D., and Remy, R.R., eds., Cretaceous shelf sandstones and shelf depositional sequences, Western Interior Basin, Utah, Colorado, and New Mexico: International Geological Congress Field Trip 119, American Geophysical Union, p. 35-66.

Penland, S., Boyd, R., and Suter, J. R., 1988, Transgressive depositional systems of the Mississippi delta plain: a model for barrier shoreline and shelf sand development, Journal of Sedimentary Petrology, V. 58, No. 6, p. 932-949.

Plint, A.G., 1988, Sharp-based shoreface sequences and "offshore bars" in the Cardium Formation of Alberta; their relationship to relative changes in sea level, *in* Wilgus, C.K., Hastings, B.S., Kendall, C.G.St.C., Posamentier, H.W., Ross, C.A. and Van Wagoner, J.C.,eds., Sea-Level Changes - An Integrated Approach; Society of Economic Paleontologists and Mineralogists Special Publication, No. 42, p. 357-370.

Posamentier, H.W. and Vail, P.R., 1988, Eustatic controls on clastic deposition II - Sequence and systems tract models, *in* Wilgus, C.K., Hastings, B.S., Kendall, C.G.St.C., Posamentier, H.W., Ross, C.A. and Van Wagoner, J.C., eds., Sea-Level Changes - An Integrated Approach; Society of Economic Paleontologists and Mineralogists Special Publication, No. 42, p. 125-154.

Posamentier, H.W., James, D.P., and Allen, G.P., 1990, Aspects of sequence stratigraphy: Recent and ancient examples of forced regressions: American Association of Petroleum Geologists Annual Meeting, San Francisco, California, American Association of Petroleum Geologists Bulletin, V. 74, No. 5, p. 742.

Singh, I. B., 1985, Burrow clasts - a source of intrabasinal coarse clastic grains: Abstracts, SEPM mid-year meeting, Golden, Colorado, V. 2, p. 83.

Swift, D.J.P., 1968, Coastal erosion and transgressive stratigraphy, Journal of Geology, V. 76, p. 444 - 456.

Swift, D.J.P., 1975, Tidal sand ridges and shoal retreat massifs: Marine Geology, V. 18, p. 105-134.

Thomas, M.A., 1990, The impact of long-term and short-term sea level changes on the evolution of the Wisconsin-Holocene Trinity/Sabine incised valley systems. Ph.D. diss. Rice University, Houston, Texas, 248 pp.

van Heerden, I.L., and Roberts, H.H., 1988, Facies development of Atchafalaya Delta, Louisiana: A modern bayhead delta. American Association of Petroleum Geologists Bulletin,V.72, No.4 (April 1988), p.439-453.

Van Wagoner, J.C., Posamentier, H.W., Mitchum, R.M. Jr., Vail, P.R., Sarg, J.F., Loutit, T.S., Hardenbol, J., 1988. An overview of the fundamentals of sequence stratigraphy and key definitions, *in* Wilgus, C.K., Hastings, B.S., Kendall, C.G.St.C., Posamentier, H.W., Ross, C.A., and Van Wagoner, J.C., eds., Sea-Level Changes - An Integrated Approach; Society of Economic Paleontologists and Mineralogists Special Publication, No. 42, p.39-45.

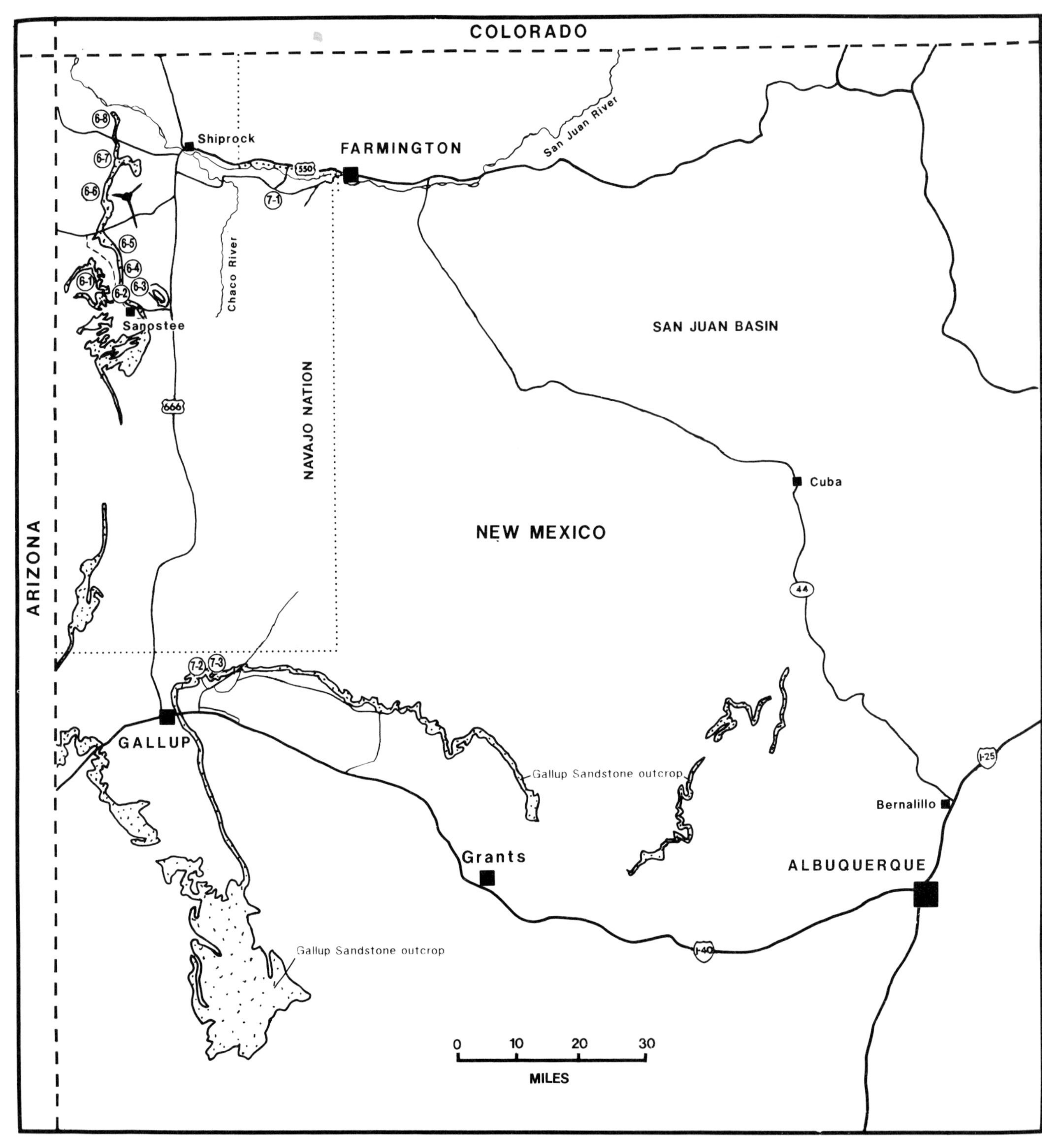

Figure 6-1. Map of northwestern New Mexico showing geological field stops on DAYS SIX and SEVEN.

SAN JUAN BASIN STRATIGRAPHIC CORRELATION CHART

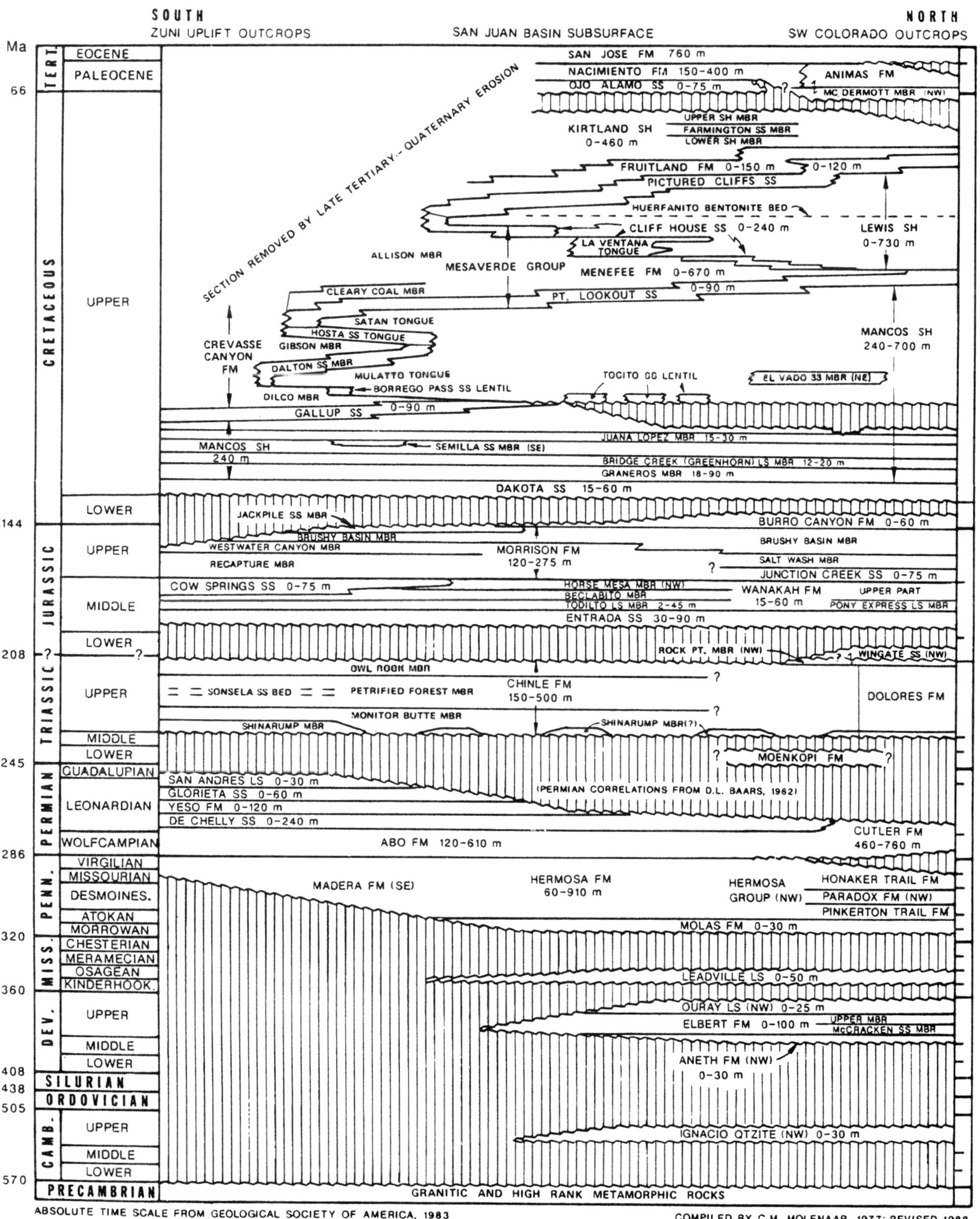

Figure 6-2. Chronostratigraphic chart of Phanerozoic section in the San Juan Basin. Lacuna of major unconformities is represented by vertical lines. From Molenaar (1988).

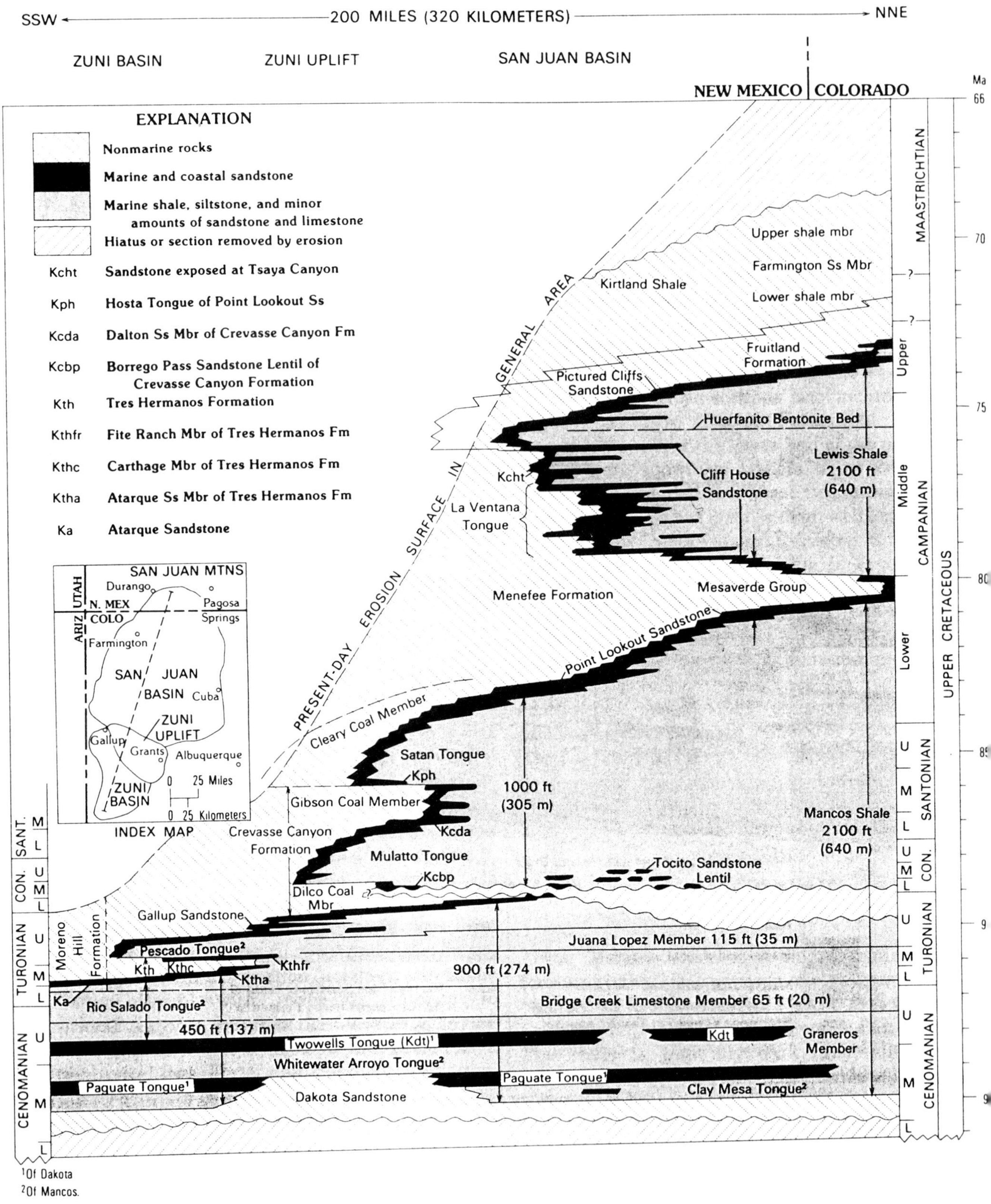

SSW ← 200 MILES (320 KILOMETERS) → NNE

ZUNI BASIN | ZUNI UPLIFT | SAN JUAN BASIN | NEW MEXICO | COLORADO

EXPLANATION

- Nonmarine rocks
- Marine and coastal sandstone
- Marine shale, siltstone, and minor amounts of sandstone and limestone
- Hiatus or section removed by erosion

Kcht	Sandstone exposed at Tsaya Canyon
Kph	Hosta Tongue of Point Lookout Ss
Kcda	Dalton Ss Mbr of Crevasse Canyon Fm
Kcbp	Borrego Pass Sandstone Lentil of Crevasse Canyon Formation
Kth	Tres Hermanos Formation
Kthfr	Fite Ranch Mbr of Tres Hermanos Fm
Kthc	Carthage Mbr of Tres Hermanos Fm
Ktha	Atarque Ss Mbr of Tres Hermanos Fm
Ka	Atarque Sandstone

INDEX MAP: SAN JUAN MTNS, UTAH, COLO, ARIZ, N. MEX, Durango, Pagosa Springs, Farmington, SAN JUAN BASIN, Cuba, ZUNI UPLIFT, Gallup, Grants, Albuquerque, ZUNI BASIN, 0 25 Miles, 0 25 Kilometers

Upper shale mbr; Farmington Ss Mbr; Lower shale mbr; Kirtland Shale; Fruitland Formation; Pictured Cliffs Sandstone; Huerfanito Bentonite Bed; Lewis Shale 2100 ft (640 m); Cliff House Sandstone; Kcht; La Ventana Tongue; Menefee Formation; Mesaverde Group; Point Lookout Sandstone; Cleary Coal Member; Satan Tongue; Kph; Gibson Coal Member; 1000 ft (305 m); Kcda; Crevasse Canyon Formation; Mulatto Tongue; Mancos Shale 2100 ft (640 m); Kcbp; Tocito Sandstone Lentil; Dilco Coal Mbr; Gallup Sandstone; Moreno Hill Formation; Pescado Tongue[2]; Juana Lopez Member 115 ft (35 m); Kth; Kthc; Kthfr; Ktha; 900 ft (274 m); Ka; Rio Salado Tongue[2]; Bridge Creek Limestone Member 65 ft (20 m); 450 ft (137 m); Twowells Tongue (Kdt)[1]; Kdt; Graneros Member; Whitewater Arroyo Tongue[2]; Paguate Tongue[1]; Dakota Sandstone; Clay Mesa Tongue[2]

PRESENT-DAY EROSION SURFACE IN GENERAL AREA

UPPER CRETACEOUS: MAASTRICHTIAN; CAMPANIAN (Lower, Middle, Upper); SANTONIAN; CON.; TURONIAN; CENOMANIAN (L, M, U)

Ma: 66, 70, 75, 80, 85, 9, 9

[1] Of Dakota
[2] Of Mancos.

Figure 6-3. Southwest-northeast chronostratigraphic cross section of Upper Cretaceous strata in the San Juan Basin. Location of section shown on inset map. Control for section is based on measured outcrop sections and interpretations of subsurface well log data. From Molenaar and Baird (1991).

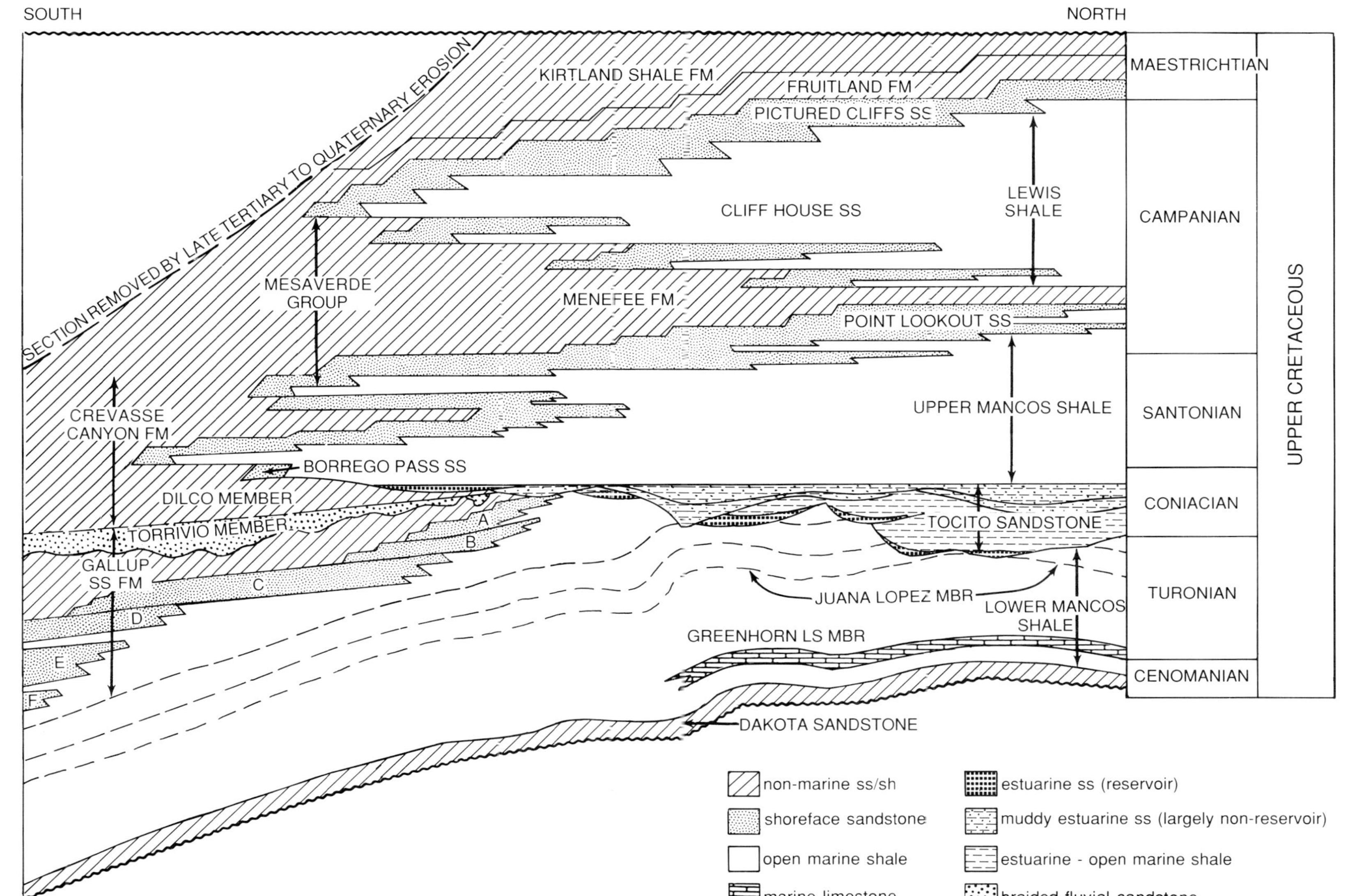

Figure 6-4. Schematic south to north cross-section of the Cretaceous stratigraphy in northwestern New Mexico (partly from Fassett 1977), with emphasis on the Late Turonian - Coniacian interval.

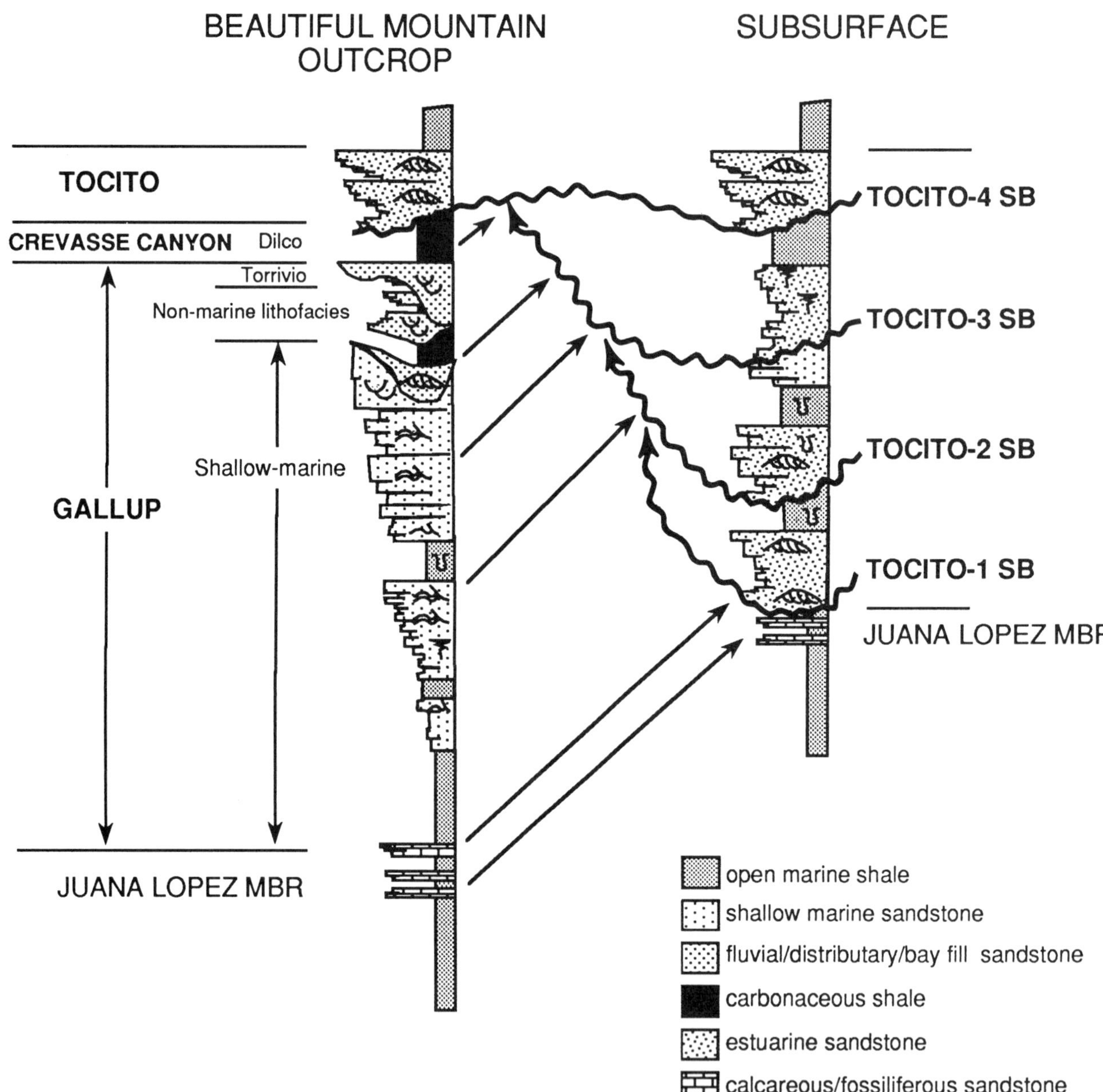

Figure 6-5. Schematic stratigraphic summary of the Gallup, Torrivio and Tocito sandstones, from outcrop at Beautiful Mountain into the subsurface in the San Juan Basin. Along Beautiful Mountain a relatively complete Gallup section from the Juana Lopez to the Torrivio Sandstone occurs beneath the Tocito Sandstone. To the east and northeast in the subsurface, four sequences comprise the Tocito interval with the lowermost sequence boundary erosionally resting on Juana Lopez strata. The total missing section is probably up to several hundred feet at this point. The four Tocito sequence boundaries identified by Exxon merge toward the outcrop and form a composite surface which everywhere separates Tocito strata from the underlying Gallup.

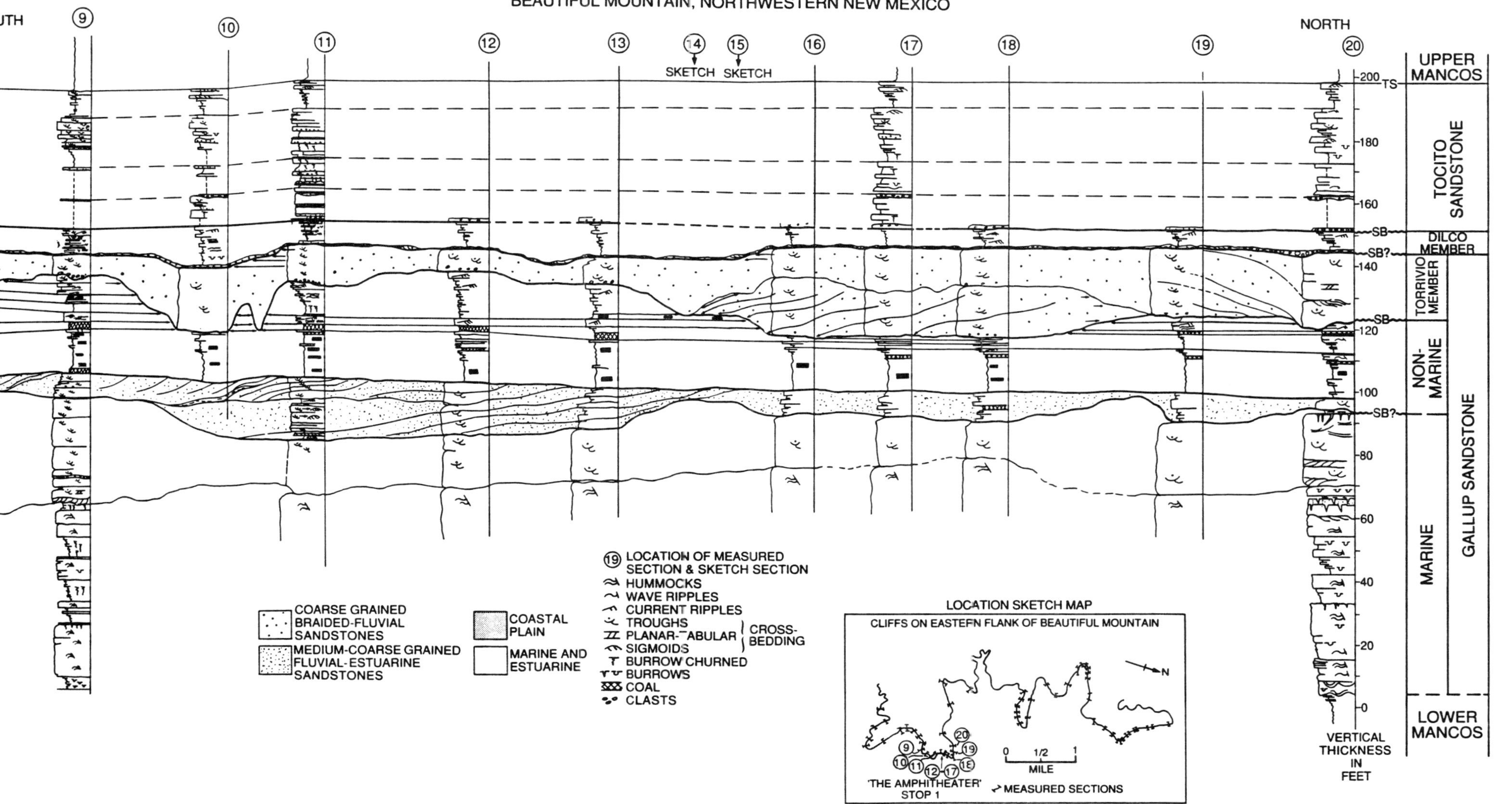

Figure 6-6. Sketch section of the cliffs at the "Amphitheater" (Day 6, Stop 1), showing the typical Turonian - Coniacian stratigraphic succession found in the Beautiful Mountain area, near the town of Sanostee, northwestern New Mexico. Fine grained shoreface deposits of the Gallup Sandstone are often incised by an erosional surface (possible sequence boundary), above which medium to coarse grained, fluvial and tidal sandstones occur (interpreted as lowstand). These deposits are then overlain by fine grained coastal plain sediments which were incised by a widespread erosion surface (sequence boundary). Coarse to very coarse grained, braided-fluvial sandstones comprising the Torrivio Member of the Gallup were deposited on this surface (lowstand deposits). The top of Torrivio is usually overlain by a very coarse-grained lag deposit. A thin interval of coastal plain, comprising fine-grained sandstones, coals and shales occur above this lag, over much of Beautiful Mountain. This is the Dilco Member of the Crevasse Canyon Formation, which thickens markedly to the south. The coarse, polymict, basal lag of the Tocito Sandstone rests unconformably on top of the coastal plain shales of the Dilco. This unconformity is thought to represent a composite sequence boundary comprised of several erosional surfaces, based on subsurface studies by Exxon, (see accompanying paper). The Tocito sandstones which overlie this erosion surface are generally coarse grained, cross bedded, burrowed and were deposited in a tidally influenced environment (late lowstand/early transgressive). The Tocito is overlain by transgressive marine mudstones of the Upper Mancos.

Figure 6-7: **A**: Photograph of the cliffs at the "Amphitheater" (Day 6, Stop 1). Labels 1 = Tocito, 2 = Torrivio, 3 = Gallup coastal plain facies, 4 = Gallup fluvial facies, 5 = Gallup shoreface. The shear Gallup cliff is about 90 feet high (27 meters).

Photographs of the Gallup shoreface at "Parasequence Point" (Day 6, Stop 5).

B: Photo mosaic showing the basinward facies change of two Gallup shoreface parasequences. Black vertical bars represent location of measured sections presented in Figure 6-16, cliff face is about 70 feet high (21m). **C**: Top of first parasequence (PS 1). Arrow indicates flooding surface.

D: Trough crossbedding in the upper shoreface which occurs on top of, and behind the main outcrop face. Paleotransport direction is to the east-southeast. Staff is 5 feet in height (1.5m).

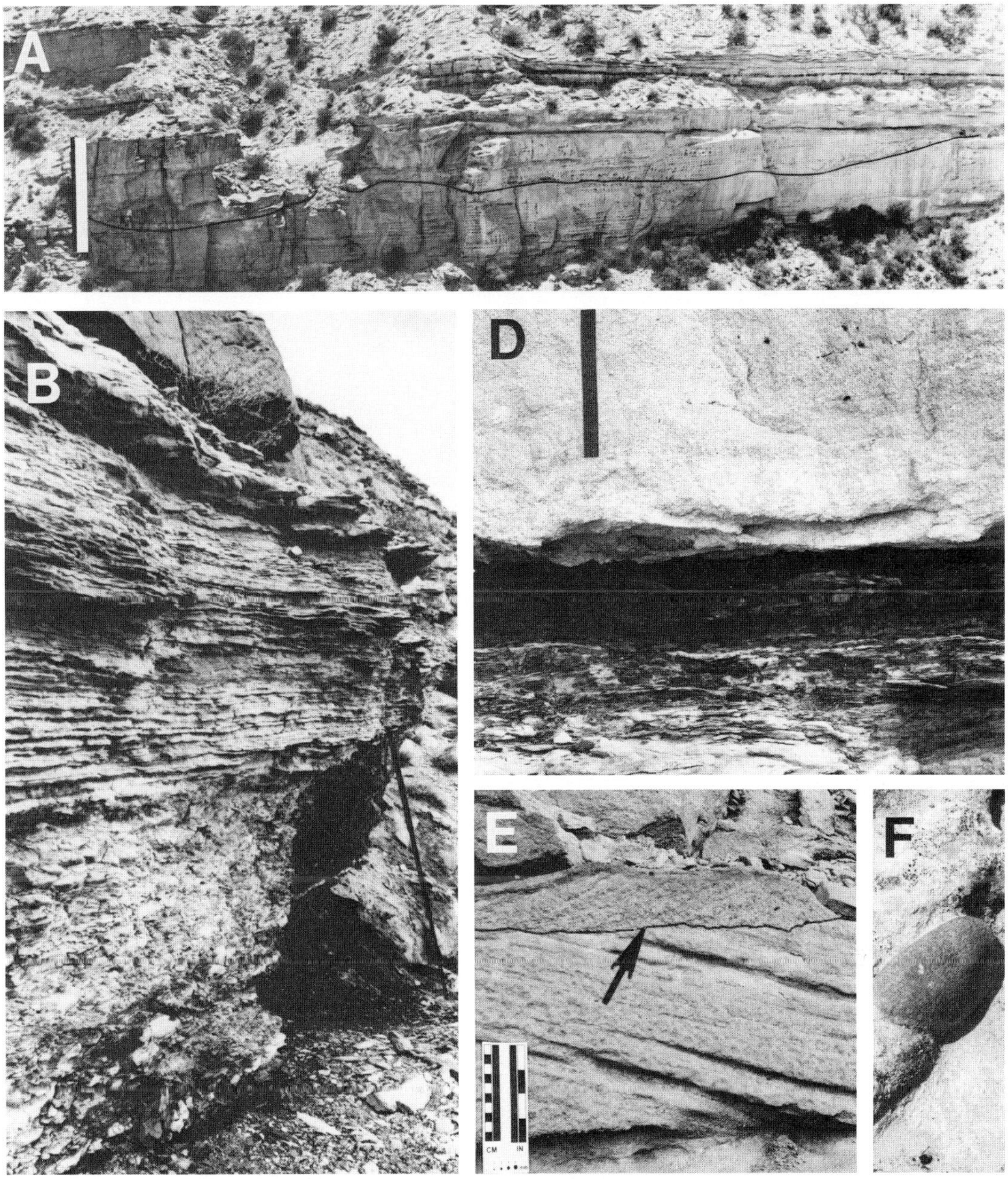

Figure 6-8. **A**: Photograph of an incised valley or large distributary channel filled with trough and sigmoidally bedded, medium to coarse sandstone (fluvial/ estuarine), which has eroded deeply (~65 feet) into the underlying Gallup shoreface deposits (NW NE Sec.12 T26N R20W). The erosion surface at the base of the channel can be traced across much of the Beautiful Mountain area .Vertical scale bar is about 80 feet (24m), it is the location of the measured section in Figure 6-9. Photographs of the Gallup Sandstone coastal plain lithofacies and Torrivio Sandstone in the area of Beautiful Mountain and Rock Ridge (T26N R20W). **B**: Typical coarsening upward profile through one of the four small delta/crevasse systems that are correlatable throughout the Beautiful Mountain area. Note the coal outcropping at the base of this photograph. See text for description of these facies. Staff is 5 feet long (1.5m). **C**: Combined flow ripples within the fine grained rippled and wavy bedded sandstone facies are typically present in the four coarsening upward packages that comprise the Gallup coastal plain facies.
D: Coarse grained Torrivio, overlying fine grained coastal-plain facies of the Gallup. Scale bar is 6 inches (7.5cm). **E**: Erosive surface at the top of the Torrivio. The erosive surface is frequently overlain by a coarse grained lag (up to cobble size, see Figure F) and sometimes mega-ripples, as in this photograph. **F**: Cobble in the coarse grained lag which overlies the Torrivio.

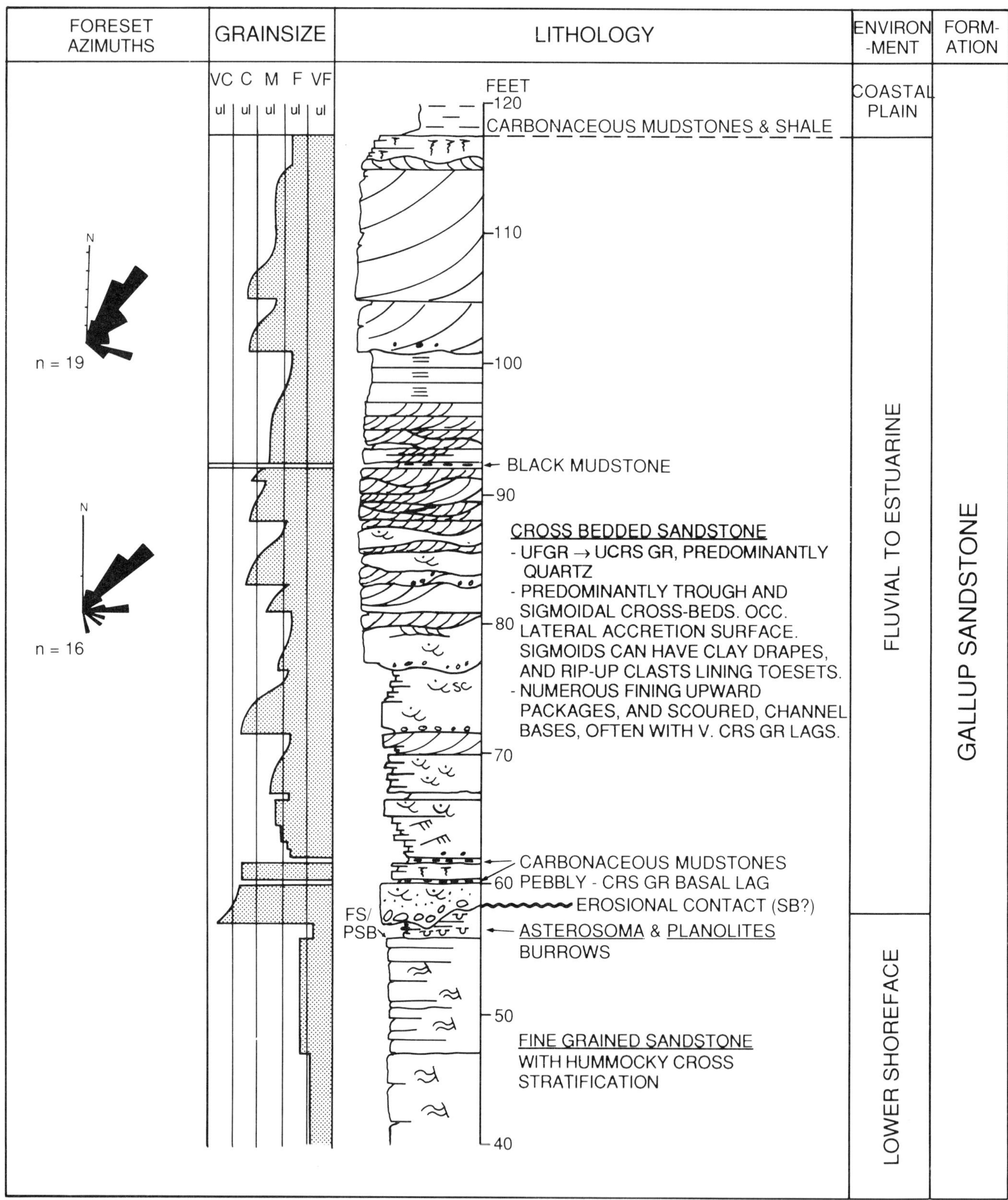

Figure 6-9. Measured section is through the predominantly fluvial fill of the deeply eroded 'valley' in the photograph, Fig. 6-8A (NW NE Sec. 12 T26N R20W). Note the medium-coarse grain size, and trough and sigmoidal cross bed sets which have foreset azimuths oriented to the northeast. The underlying Gallup shoreface deposits are finer grained and when troughs are developed where the upper shoreface is preserved, they are oriented to the southeast.

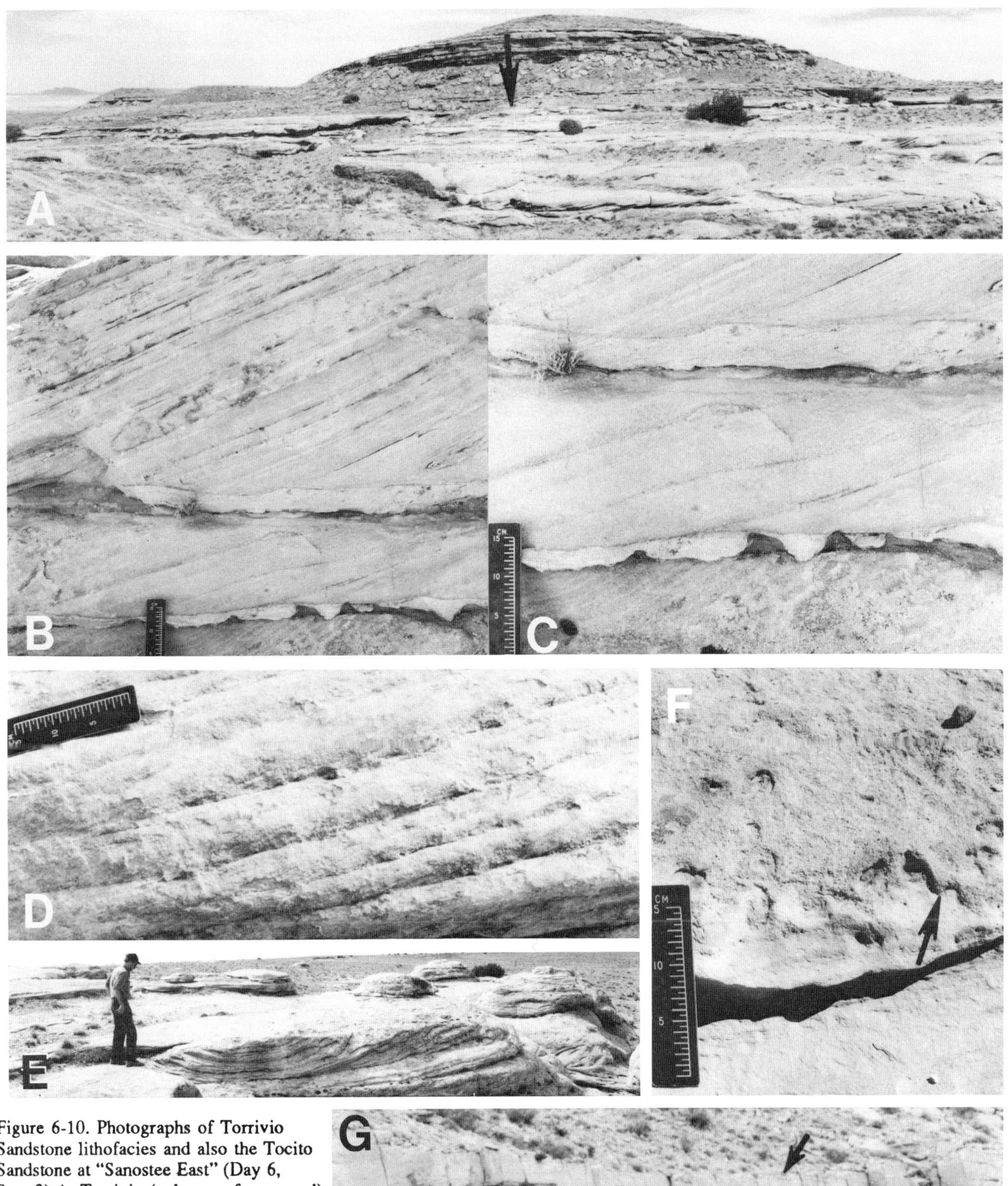

Figure 6-10. Photographs of Torrivio Sandstone lithofacies and also the Tocito Sandstone at "Sanostee East" (Day 6, Stop 3) A: Torrivio (pale gray foreground) and Tocito (dark gray strata forming the hill top) sandstones outcrop at Sanostee East. Tocito cliff is 40 feet high (14m).

B: Upper cross bed set is planar-tabular with each lamina fining upward. Lower crossbed set is a complex compound bedform which in places has a sigmoidal geometry and is composed of numerous northeasterly dipping laminar sets which fine upward.

C: Close up of the compound bedform seen in photograph B.

D: Close up of stacked lamina sets in the compound bedform sketched in Figure 6-11C. They are frequently overlain and underlain by the more typical tabular, and planar and trough crossbed sets seen in Figure 6-10B above.

E: Festoon of a northeasterly directed trough crossbed set.

F: Coarse, burrowed grit on top of the Torrivio Sandstone at Sanostee East (see arrow on Figure A above).

G: Photograph showing the erosional base of the Torrivio Sandstone, Beautiful Mountain (T 26N R20W). Note that at the top of the Torrivio there is an erosional surface. A very coarse grained lag overlies this surface. Sandstone ledge at its thickest point in this photograph is 10 feet (3m).

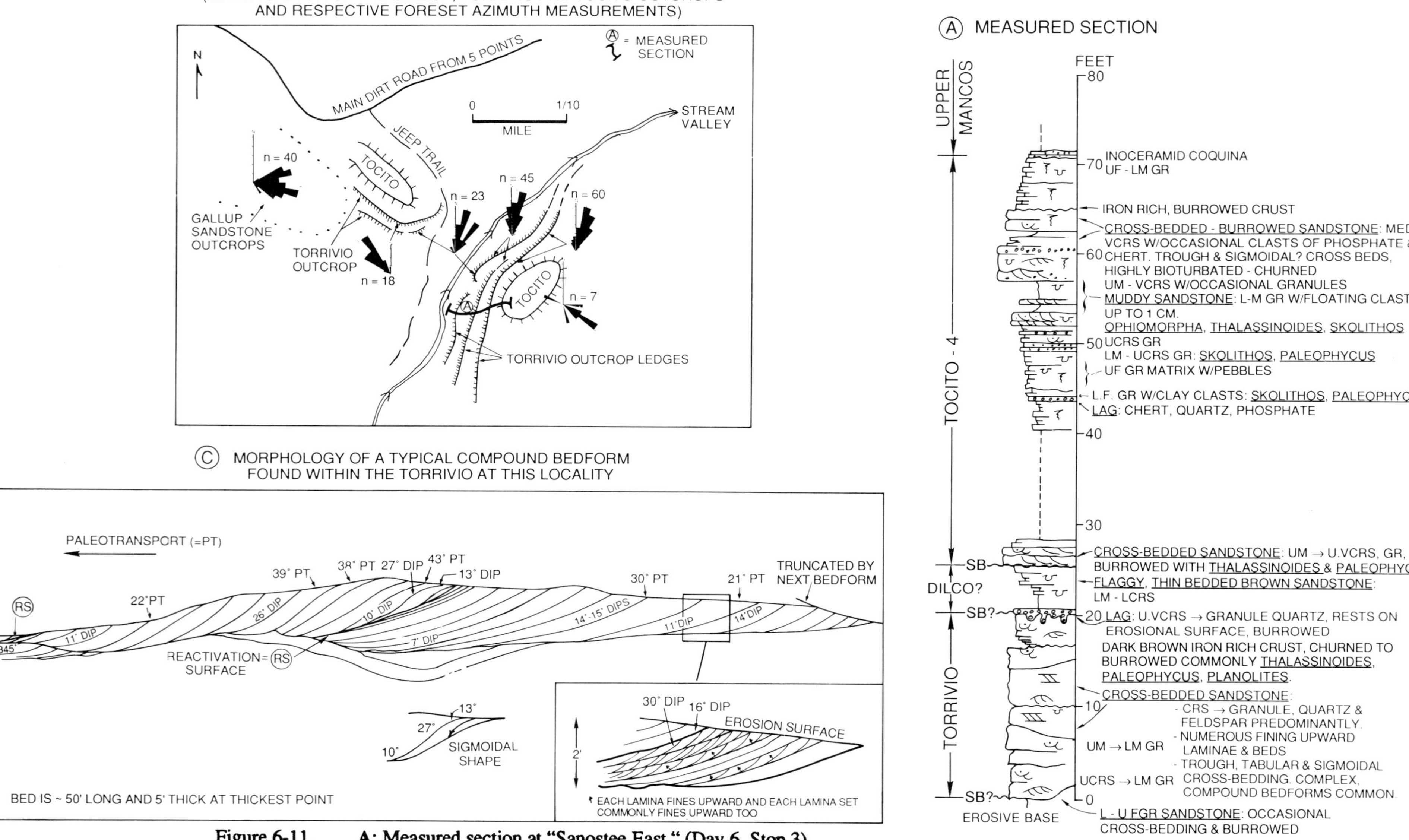

Figure 6-11. **A: Measured section at "Sanostee East" (Day 6, Stop 3)**
B: Paleotransport directions for the Tocito (southeast), Torrivio (north-northeast) and underlying Gallup (east-southeast) are plotted on the location sketch map.
C: An example of one of the complex, compound bedforms found in the Torrivio at this stop is sketched above. Opinion varies as to whether this type of lamination may indicate tidal influence.

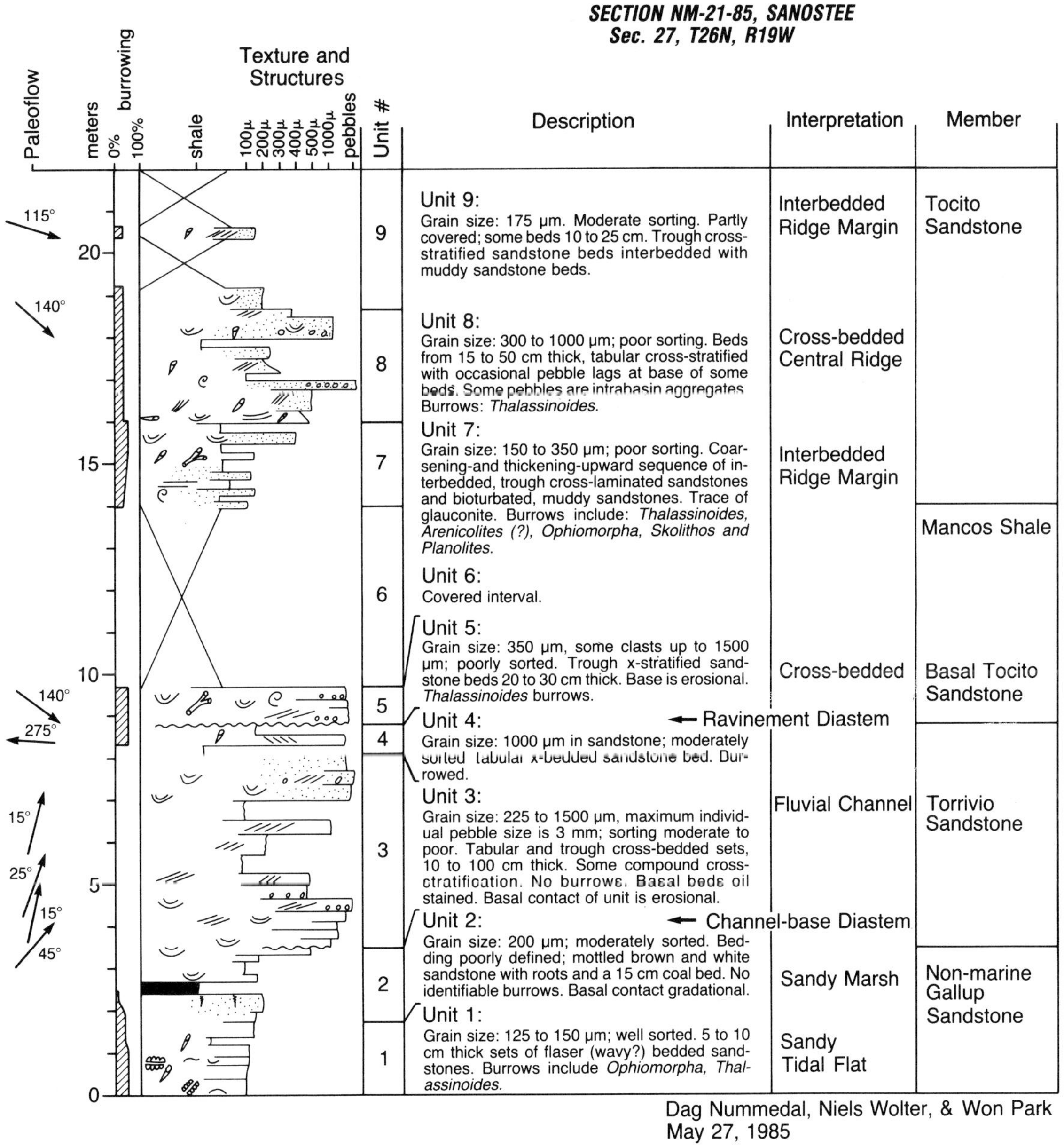

SECTION NM-21-85, SANOSTEE
Sec. 27, T26N, R19W

Paleoflow | meters | 0% burrowing 100% | Texture and Structures: shale, 100μ, 200μ, 300μ, 400μ, 500μ, 1000μ, pebbles

Unit #	Description	Interpretation	Member
9	Unit 9: Grain size: 175 μm. Moderate sorting. Partly covered; some beds 10 to 25 cm. Trough cross-stratified sandstone beds interbedded with muddy sandstone beds.	Interbedded Ridge Margin	Tocito Sandstone
8	Unit 8: Grain size: 300 to 1000 μm; poor sorting. Beds from 15 to 50 cm thick, tabular cross-stratified with occasional pebble lags at base of some beds. Some pebbles are intrabasin aggregates. Burrows: *Thalassinoides*.	Cross-bedded Central Ridge	
7	Unit 7: Grain size: 150 to 350 μm; poor sorting. Coarsening-and thickening-upward sequence of interbedded, trough cross-laminated sandstones and bioturbated, muddy sandstones. Trace of glauconite. Burrows include: *Thalassinoides, Arenicolites (?), Ophiomorpha, Skolithos and Planolites.*	Interbedded Ridge Margin	
6	Unit 6: Covered interval.		Mancos Shale
5	Unit 5: Grain size: 350 μm, some clasts up to 1500 μm; poorly sorted. Trough x-stratified sandstone beds 20 to 30 cm thick. Base is erosional. *Thalassinoides* burrows.	Cross-bedded	Basal Tocito Sandstone
	← Ravinement Diastem		
4	Unit 4: Grain size: 1000 μm in sandstone; moderately sorted tabular x-bedded sandstone bed. Burrowed.	Fluvial Channel	Torrivio Sandstone
3	Unit 3: Grain size: 225 to 1500 μm, maximum individual pebble size is 3 mm; sorting moderate to poor. Tabular and trough cross-bedded sets, 10 to 100 cm thick. Some compound cross-stratification. No burrows. Basal beds oil stained. Basal contact of unit is erosional.		
	← Channel-base Diastem		
2	Unit 2: Grain size: 200 μm; moderately sorted. Bedding poorly defined; mottled brown and white sandstone with roots and a 15 cm coal bed. No identifiable burrows. Basal contact gradational.	Sandy Marsh	Non-marine Gallup Sandstone
1	Unit 1: Grain size: 125 to 150 μm; well sorted. 5 to 10 cm thick sets of flaser (wavy?) bedded sandstones. Burrows include *Ophiomorpha, Thalassinoides*.	Sandy Tidal Flat	

Paleoflow: 115°, 140°, 140°, 275°, 15°, 25°, 15°, 45°

Meters: 0, 5, 10, 15, 20

Dag Nummedal, Niels Wolter, & Won Park
May 27, 1985

Figure 6-12 . Measured section at Sanostee East.

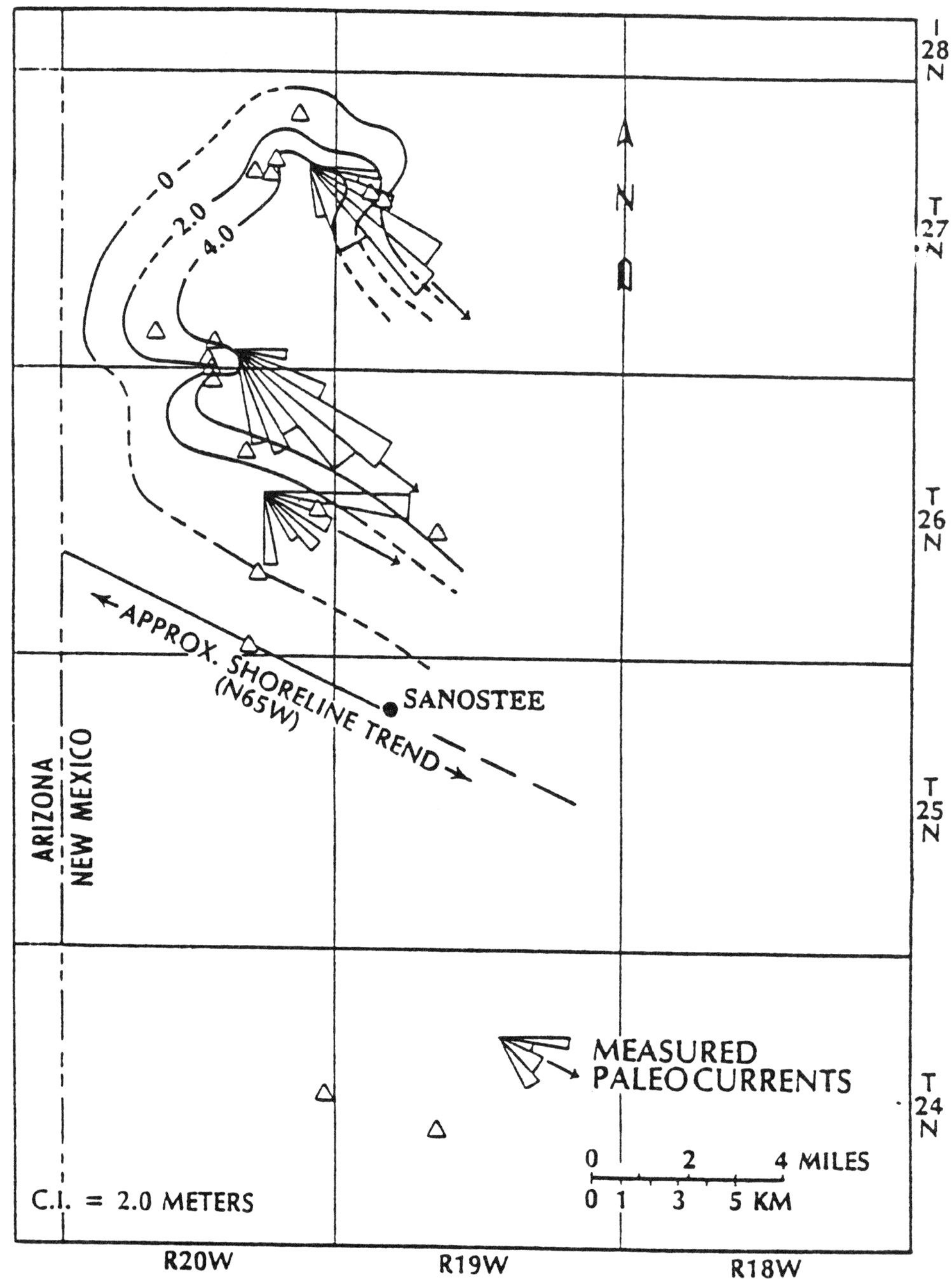

Figure 6-13. Isopach map of Tocito Sandstone, interbedded and crossbedded facies. Also shown are paleocurrents from tabular-tangential crossbed foresets. Paleocurrents show oblique, across-ridge flow on landward flank of the sandbody. Orientation of shoreline from trend of large wave-ripple crests. From Bergsohn (1988).

All scale bars are in centimeter units

Figure 6-14. Photographs of the Tocito sandstone at "Lichii Wash" (Day 6, Stop 8: these sandstones are part of Tocito sequence 4 , according to Exxon in the accompanying paper). **A**: Cross section through a subtidal bar. Excellent sigmoidal cross bedding, occasional clay drapes and tidal bundles are developed in the Tocito. **B**: Crests of foreset laminae are straight to sinuous, and can be traced out for many yards into the desert. **C**: Paleosol ? Nodular carbonate layers developed at the base of the subtidal bar shown in Figure A above. **D**: The Tocito is often very coarse grained.
E. Typical upward coarsening/shoaling profile of the Tocito Sandstone in the Beautiful Mountain area. See geologist for scale to bottom left of photograph. **F**: Erosive surface at base of Tocito north of Lichii Wash (Sec.6 T31N R19W). This surface continues to cut down through the underlying Mancos, eventually reaching the top of the Juana Lopez farther to the north (Figure G) and in Colorado.**G**: Basal Tocito lag resting on top of the Juana Lopez member in a plungepool to the north of Lichii Wash (SE SW Sec.5 T32N R19W).

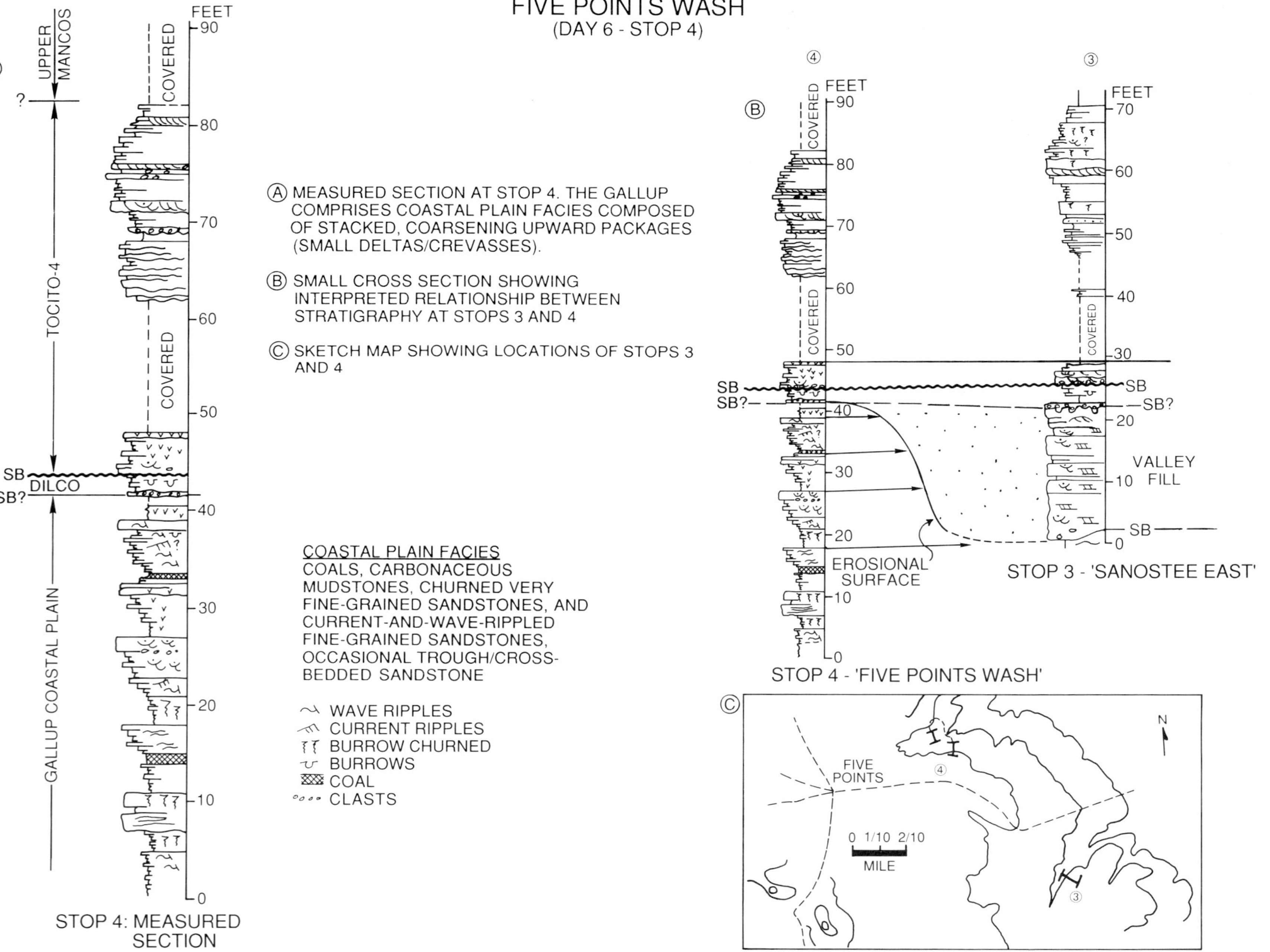

Figure 6-15. **A**: Measured section at "Five Points Wash" (Day 6, Stop 4). The Torrivio Sandstone seen at the previous stop is absent here. **B**: This sketch shows our interpretations that the Torrivio has thinned to virtually zero at Five Points Wash. Regional outcrop observations suggest we are located on an interfluve area for the Torrivio braided-stream system, that Exxon interpret to represent a lowstand incised valley system. **C**: Location sketch.

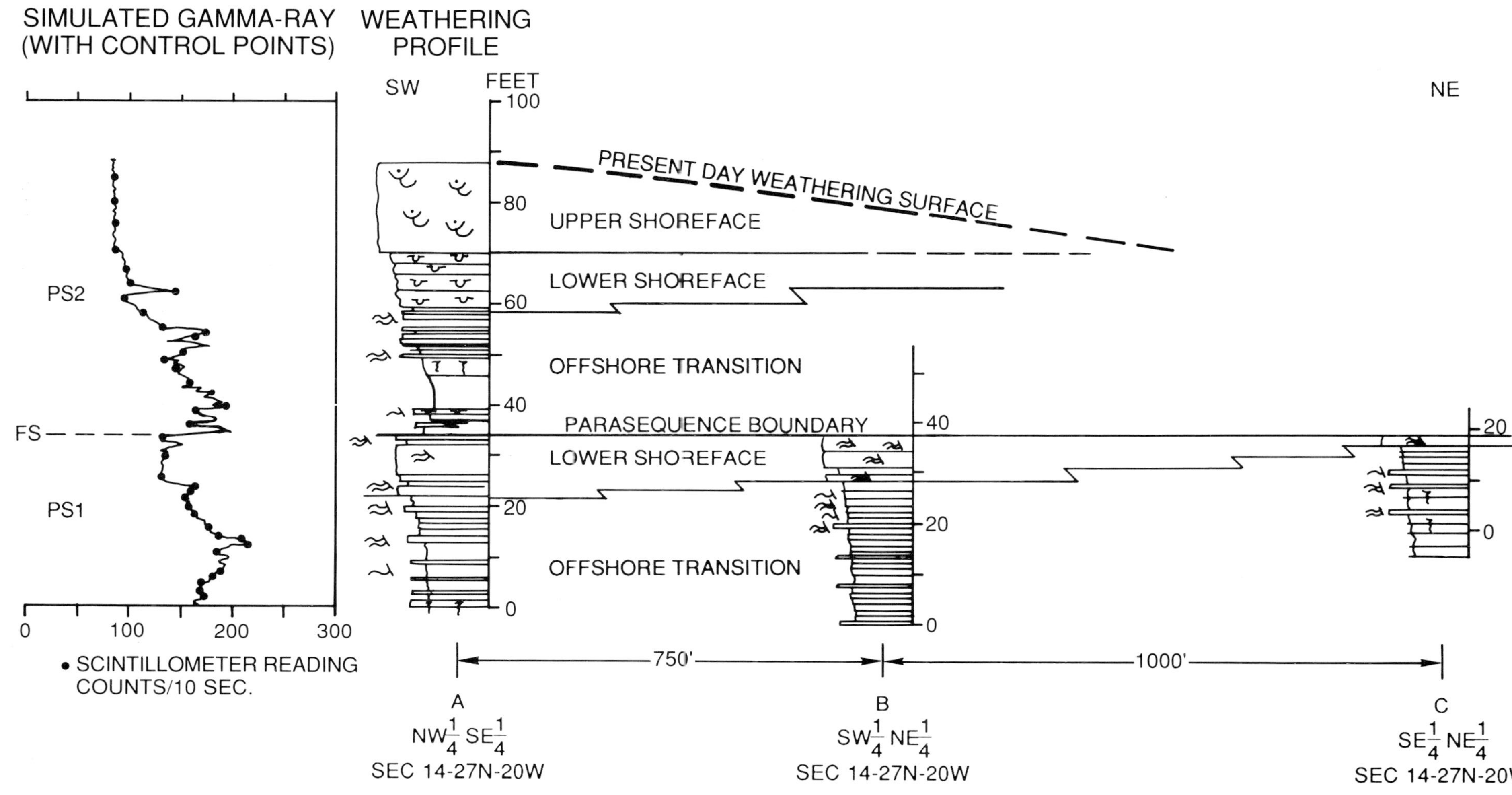

Figure 6-16. Measured sections at "Parasequence Point" (Day 6, Stop 5), with hand held gamma ray profile. Two parasequences are present within the Gallup at this locality, each facies changing in a seaward, and basinward direction to the northeast. These parasequences form part of the Gallup 'A' tongue as designated by Molenaar (1973), they are comprised of some of the youngest shoreface deposits identified in the Gallup Sandstone.

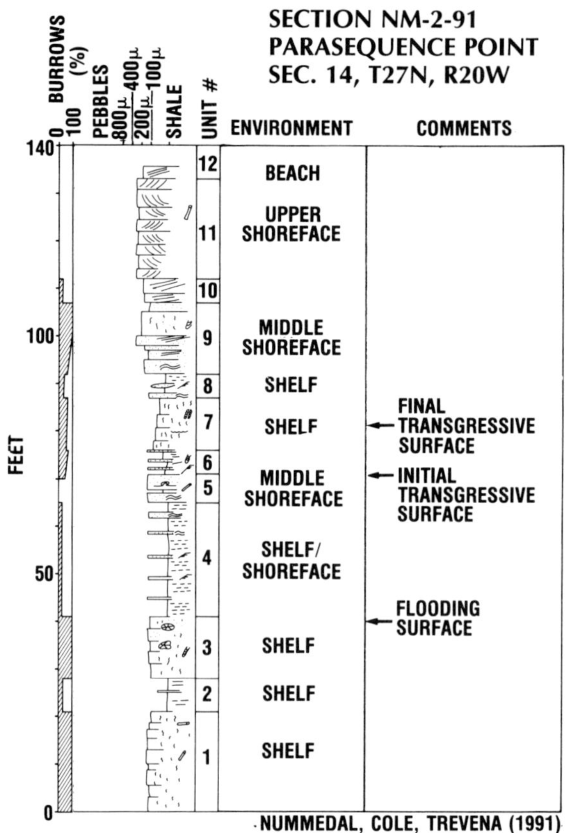

Figure 6-17. Measured section at Parasequence Point. Section shows parts of three parasequences in Gallup Sandstone. Lower parasequence extends from base of section (or below) to FS at the top of unit 3. The second parasequence extends from the base of unit 4 to a FTS in the middle of unit 7. The third parasequence begins at the FTS in unit 7 and continues through the top of the section. FTS - final transgressive surface, ITS - initial transgressive surface, FS - flooding surface.

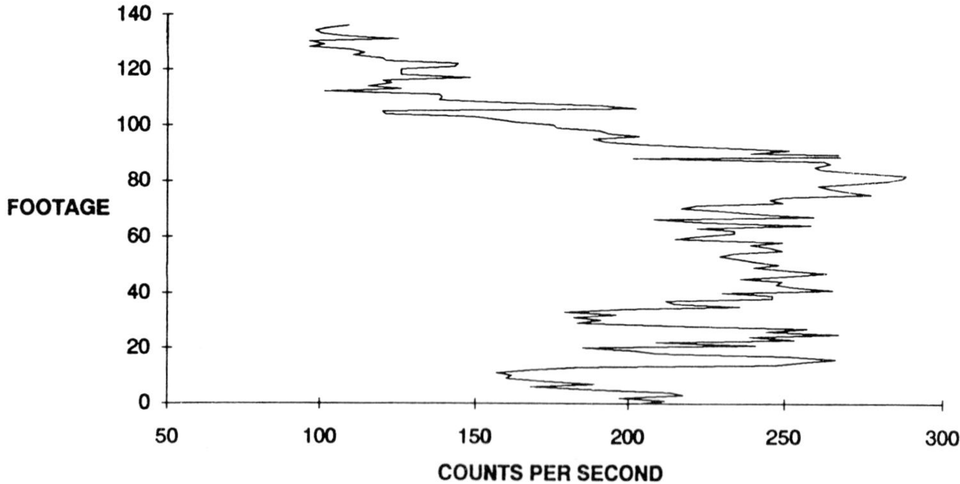

Figure 6-18. Outcrop gamma-ray log of Parasequence Point. Maximum gamma-ray count is interpreted as indicating the FTS. Marked increase in gamma-ray count is also noted at the ITS. FTS - final transgressive surface, ITS - initial transgressive surface, FS - flooding surface.

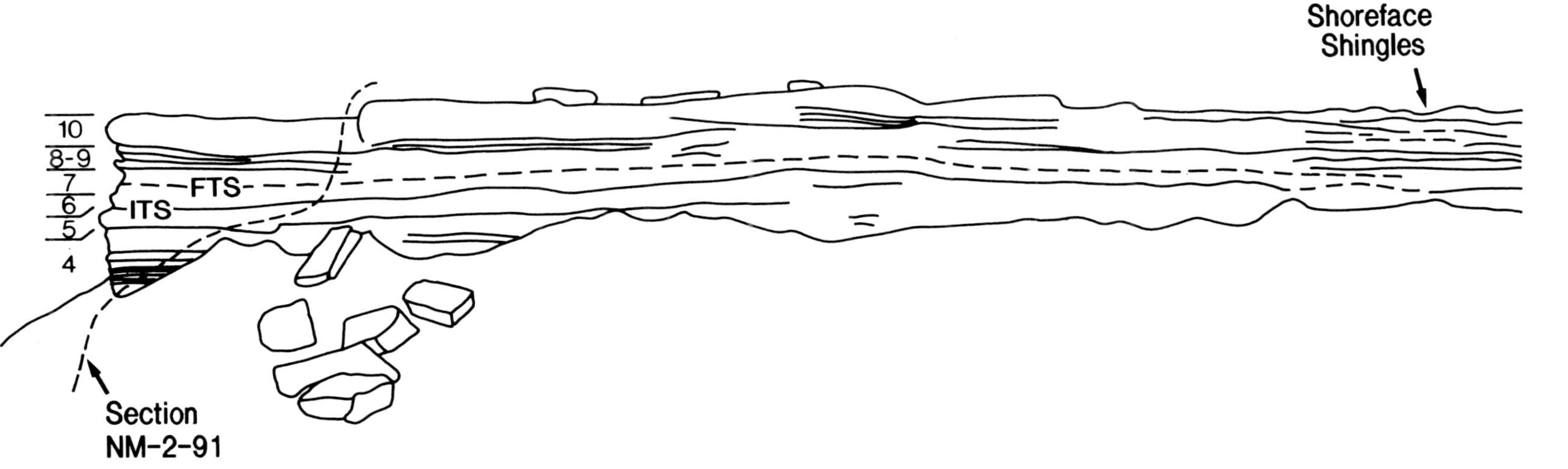

Figure 6-19. Photo and interpreted sketch of outcrop at Parasequence Point. Transgressive component of parasequence (unit 6 and lower part of unit 7) consists of bioturbated, structureless muddy sandstones. Overlying regressive facies are burrowed (upper part of 7) but display distinct bedding. The steep depositional dip of this shoreface is demonstrated by the dipping "shingles" in upper right of photo.

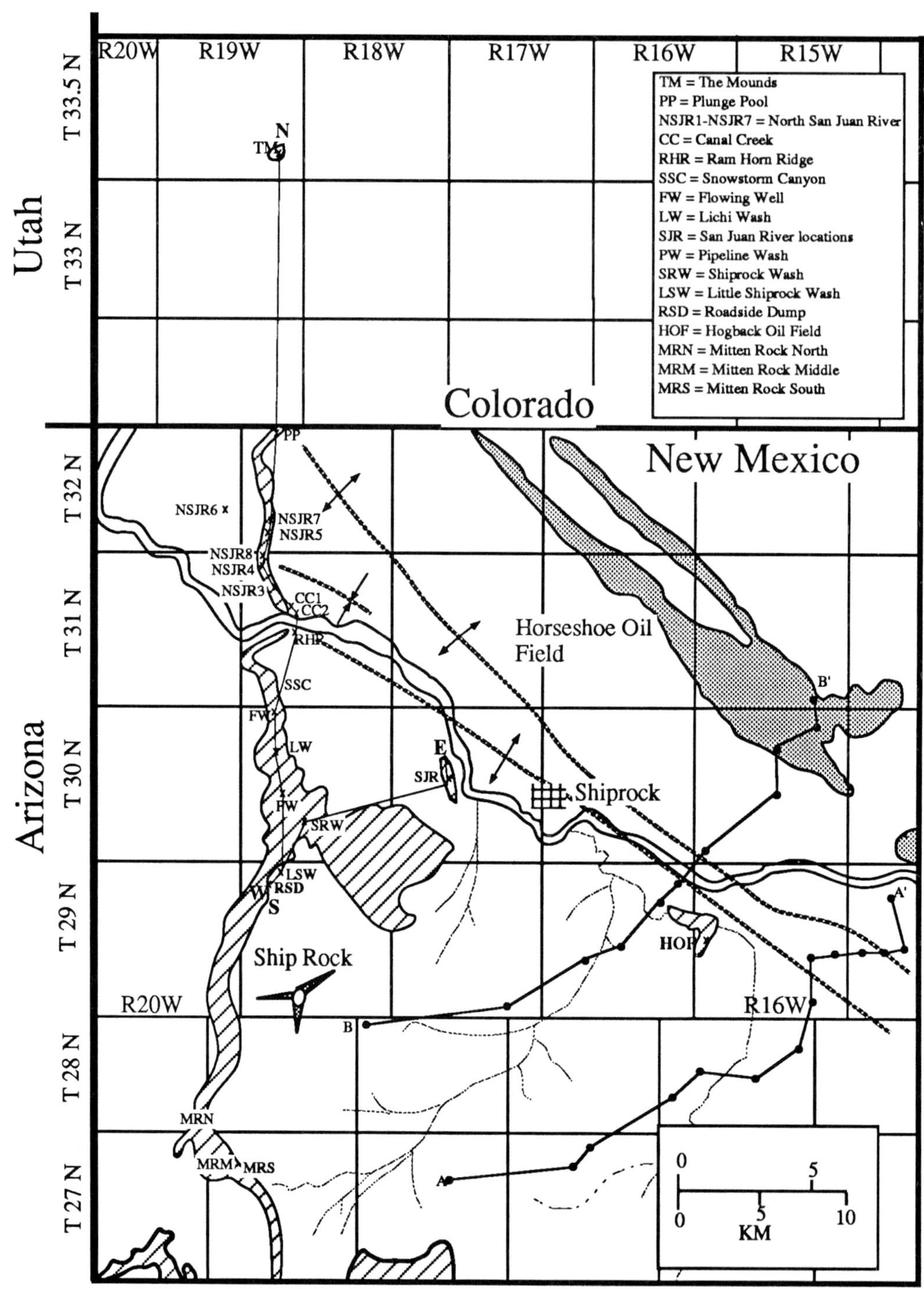

Figure 6-20. Detailed location map of Four Corners Platform. Sections S-N and W-E are surface cross sections. Sections A-A' and B-B' are subsurface cross sections. Position of northwest-southeast trending anticlines are taken from McCubbin (1969).

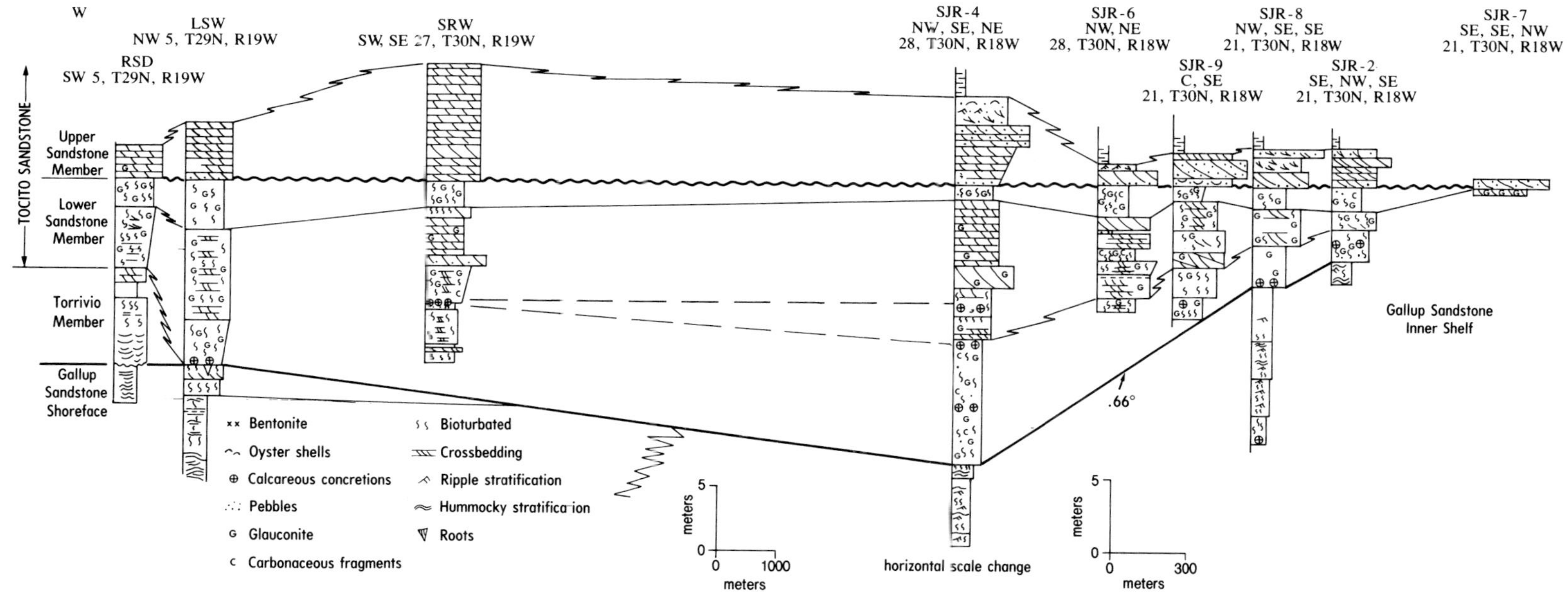

Figure 6-21. Cross section across tectonic "strait". Seaward (northeastern) thinning is interpreted as a result of thinning onto a rising anticline. A possible facies equivalence is indicated between the Torrivio Member and the Tocito Sandstone.

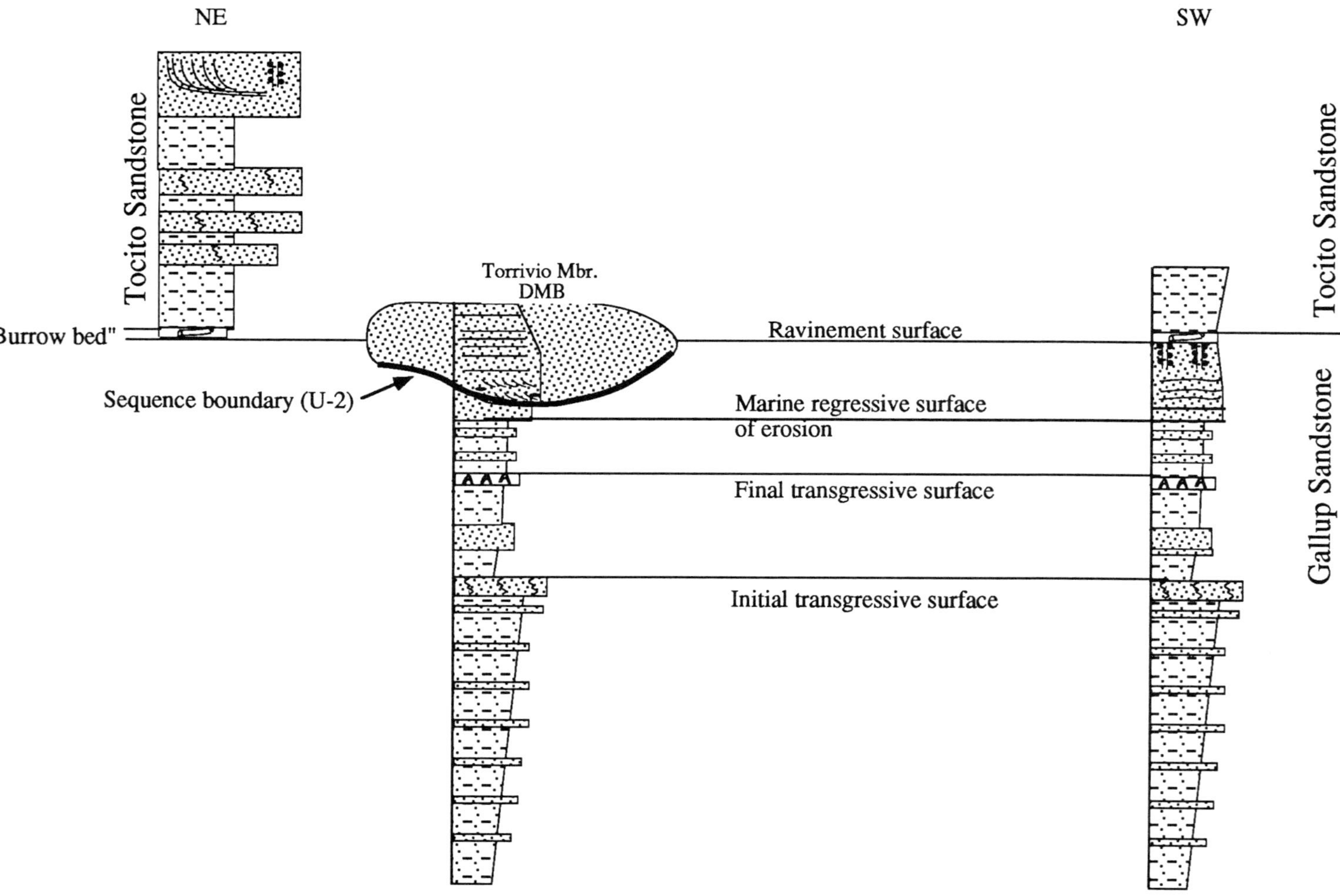

Figure 6-22. Local cross-section at Roadside Dump. Datum is final transgressive surface in Gallup Sandstone. Base of Torrivio truncates ravinement surface near top of Gallup. Lenticular Torrivio sandbody is interpreted as a distributary mouth bar of a river that probably fed part of Tocito Sandstone.

Figure 6-23A. "Sharp-based" shoreface of the upper Gallup Sandstone. Photo shows truncation at base of hummocky-stratified sandstone. Photo from near LSW location. RSE: regressive surface of erosion, Ra: ravinement surface.

Figure 6-23B. Heavily burrowed sequence boundary and ravinement surface (FSSB) at top of sharp-based shoreface. Surface is strewn with chert and quartzite cobbles.

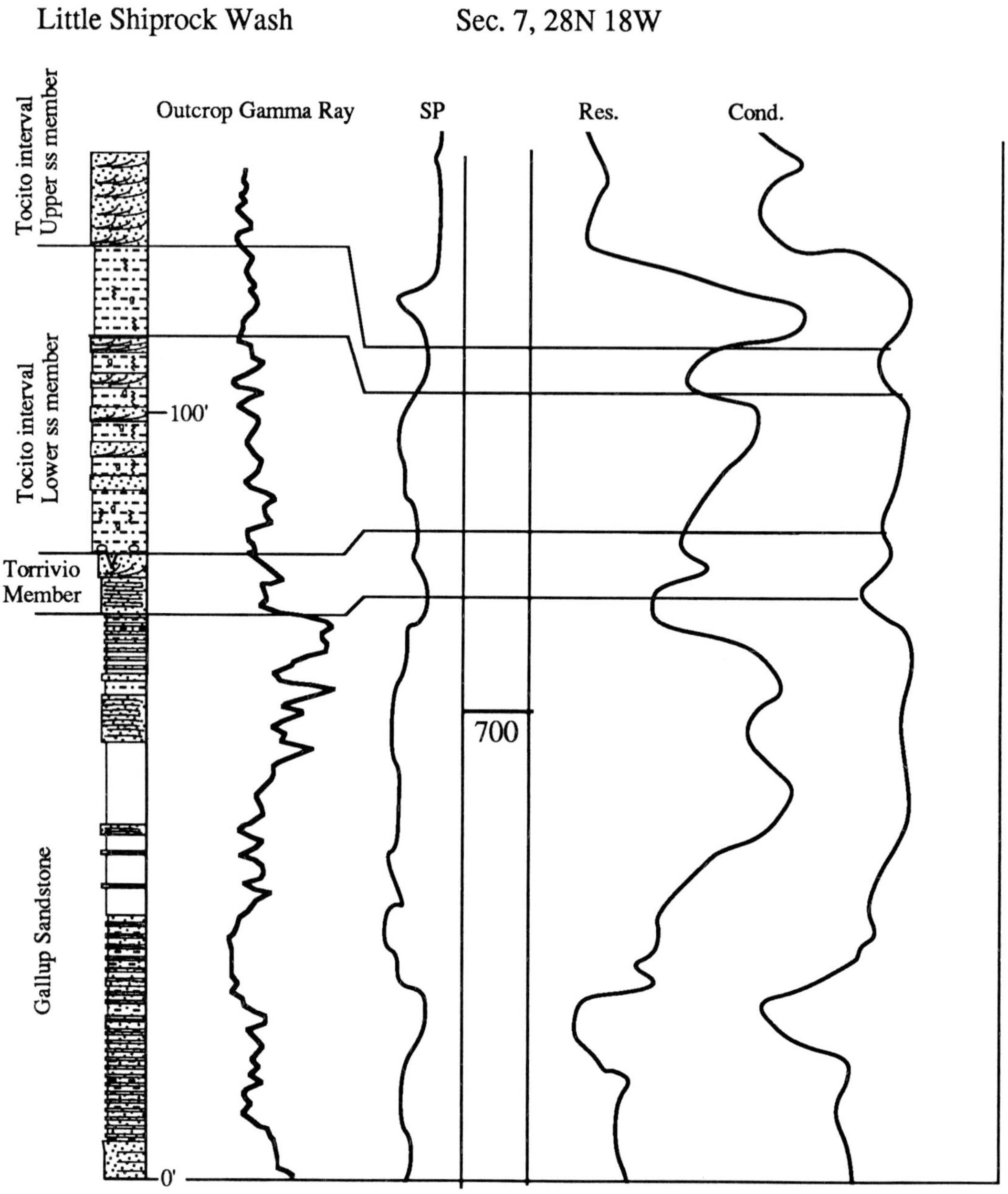

Figure 6-24. Correlation of outcrop section and outcrop gamma-ray at Little Shiprock Wash Location to nearby well. Well is most landward (southwest) well on section B-B'.

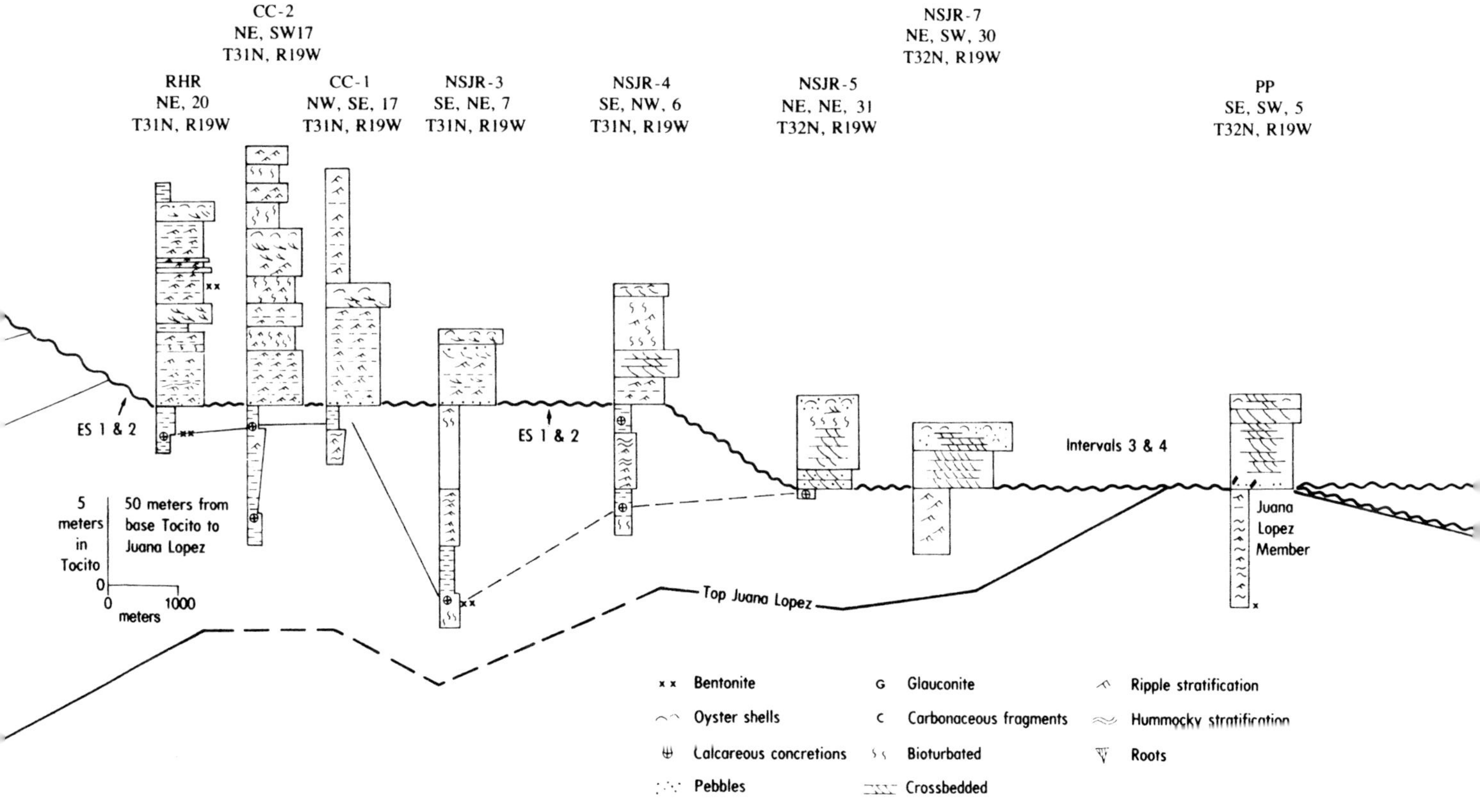

CC-2
NE, SW17
T31N, R19W

RHR
NE, 20
T31N, R19W

CC-1
NW, SE, 17
T31N, R19W

NSJR-3
SE, NE, 7
T31N, R19W

NSJR-4
SE, NW, 6
T31N, R19W

NSJR-5
NE, NE, 31
T32N, R19W

NSJR-7
NE, SW, 30
T32N, R19W

PP
SE, SW, 5
T32N, R19W

ES 1 & 2

ES 1 & 2

Intervals 3 & 4

Juana Lopez Member

Top Juana Lopez

5 meters in Tocito
0
50 meters from base Tocito to Juana Lopez
0 1000
meters

Bentonite
Oyster shells
Calcareous concretions
Pebbles
Glauconite
Carbonaceous fragments
Bioturbated
Crossbedded
Ripple stratification
Hummocky stratification
Roots

is
'e

N

TM
NW, NW, 25
T33½N, R19W

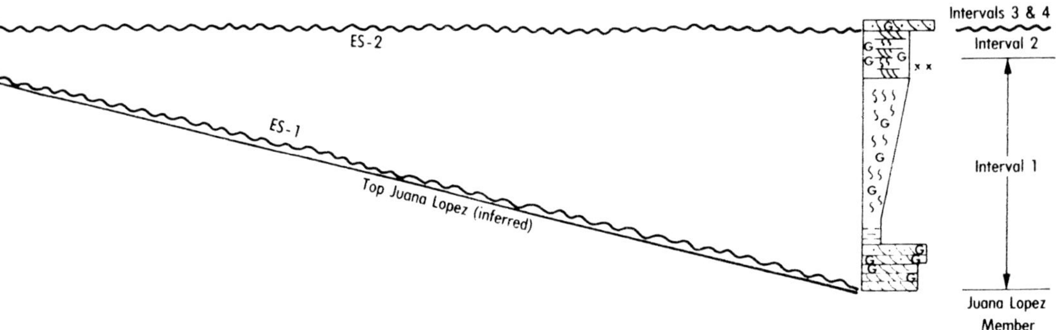

RATTLESNAKE
(DAY 6 - STOP 7)

At Rattlesnake, the Tocito rests directly on the lower shoreface of the marine Gallup. All of the Torrivio and nonmarine Gallup is missing at this location. Based primarily on subsurface correlation, the base of the Tocito is actually several intercutting sequence boundaries. This composite surface truncates strata below forming a regional angular unconformity. Strata below the unconformity have been tilted by earliest Coniacian deformation.

Figure 6-27. Measured section and photograph of the Tocito resting unconformably on lower shoreface deposits of the Gallup. This locality represents one of the most northerly outcrops of the Gallup Sandstone. Further north, these shoreface deposits continue to change facies into shelfal mudstones and muddy sandstones or are progressively truncated by the erosive surface at the base of the Tocito, as seen at this outcrop (Tocito sequence boundary 4 in the accompanying paper by Exxon).

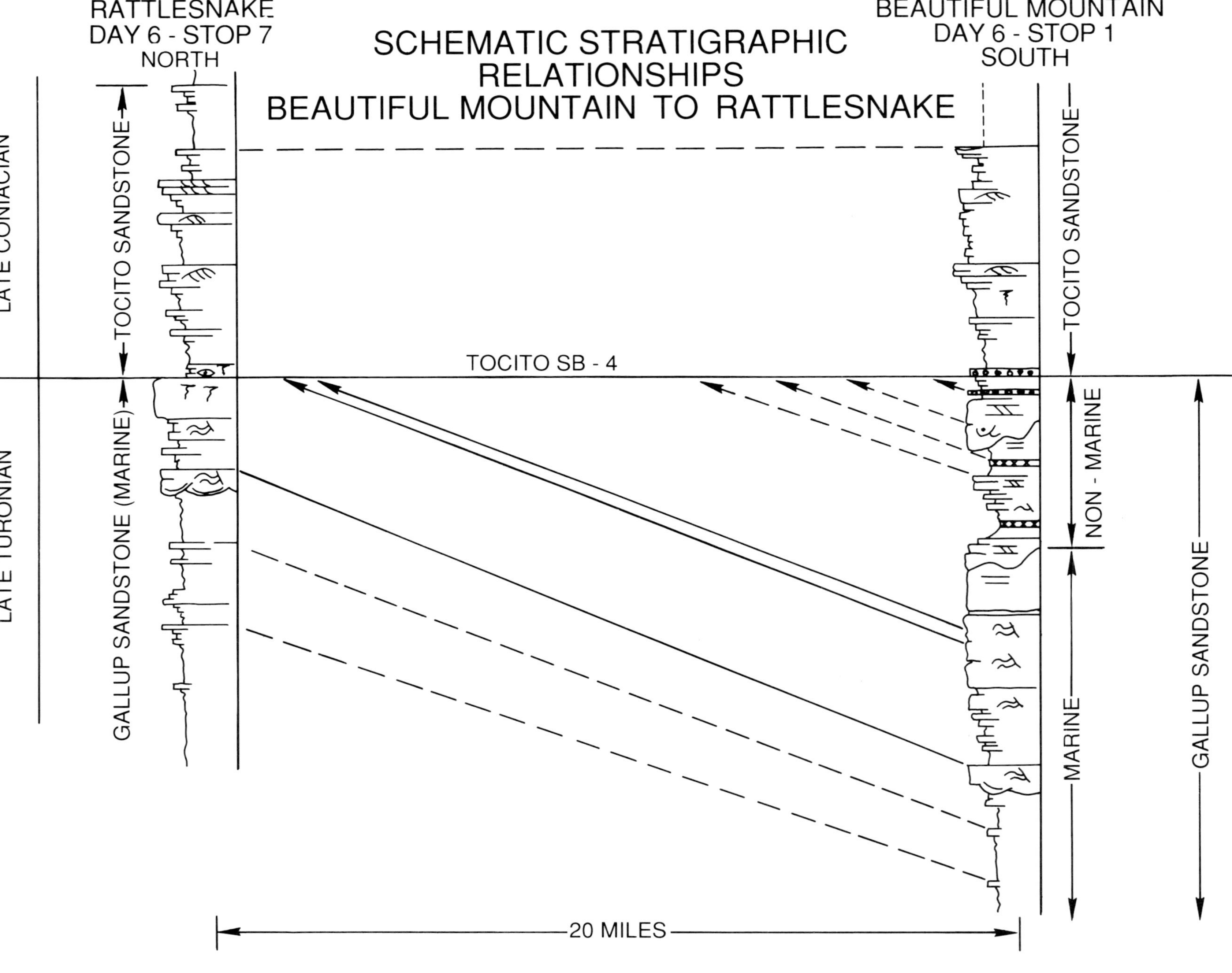

Figure 6-28. Schematic diagram showing the progressive truncation that occurs beneath the base of the Tocito sandstone at outcrop. The Torrivio sandstone and Gallup coastal plain, fluvial, and shoreface deposits, which were present at Beautiful Mountain have been removed

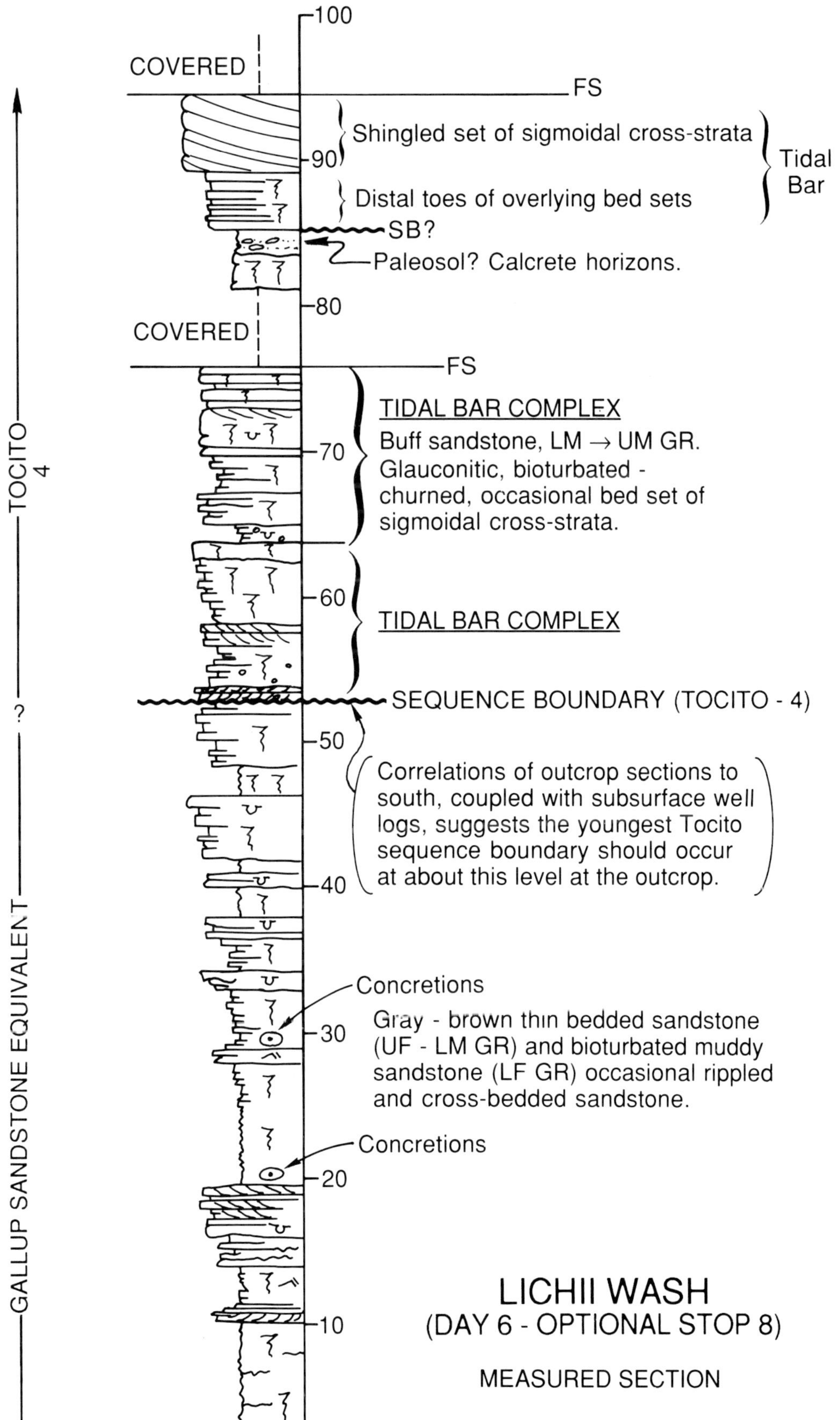

GALLUP SANDSTONE EQUIVALENT ——— ? ——— TOCITO 4

100
COVERED
FS
Shingled set of sigmoidal cross-strata
90
Distal toes of overlying bed sets
Tidal Bar
SB?
Paleosol? Calcrete horizons.
80
COVERED
FS
TIDAL BAR COMPLEX
Buff sandstone, LM → UM GR. Glauconitic, bioturbated - churned, occasional bed set of sigmoidal cross-strata.
70
60
TIDAL BAR COMPLEX
SEQUENCE BOUNDARY (TOCITO - 4)
50
Correlations of outcrop sections to south, coupled with subsurface well logs, suggests the youngest Tocito sequence boundary should occur at about this level at the outcrop.
40
Concretions
Gray - brown thin bedded sandstone (UF - LM GR) and bioturbated muddy sandstone (LF GR) occasional rippled and cross-bedded sandstone.
30
Concretions
20
10
0

LICHII WASH
(DAY 6 - OPTIONAL STOP 8)

MEASURED SECTION

Figure 6-30. Measured section at "Lichii Wash" (Day 6, Stop 8 - Optional Stop). Excellent tidal bedding in the Tocito is exposed at this outcrop in the low cliffs along the wash (see photographs, Figures 6-14A, B and C, taken at this stop). Exxon interpret these deposits to represent sub-tidal bars, deposited in a broad estuary created by the incision of the basal Tocito sequence boundary seen at DAY 6, STOP 7 (Tocito sequence boundary 4 in the accompanying paper by Exxon).

DAY SEVEN

Farmington to Albuquerque via Gallup,
New Mexico

ROAD LOG, DAY SEVEN:

Morning; SEQUENCE STRATIGRAPHY AND FACIES ARCHITECTURE OF AN EXHUMED TOCITO SANDBODY.

Afternoon; PARASEQUENCE ARCHITECTURE OF THE GALLUP SANDSTONE IN THE GALLUP AREA.

0.0 Leave Holiday Inn parking lot, turn right on Broadway.
0.2 Intersection with Butler/Pinon, continue straight on Broadway.
1.7 Intersection of Main street and Broadway. Drive straight heading west on US 64.
3.4 Junction with road 170. Straight on.
11.8 Turn left on San Juan County Road 6675 to Fruitland and Four Corners Power Plant.
12.7 Intersection with San Juan County Road 6677. Go straight on 6675.
13.1 Cross the San Juan River.
15.7 Turn right on a wide dirt road which shortly becomes paved and is signposted for Da Na Has Ta (Full Gospel Church). Now heading west.
16.1 The road becomes paved.
16.8 Pass a sign reading Da Na Has Ta.
17.4 Pass water tower on left. Four Corners Power Plant clearly visible to the left (south). The Hogback is straight ahead.
21.5 The road crosses through the Cliff House Sandstone cropping out along the Hogback. Ship Rock can be seen in the distance to the left.
23.2 Pass a dirt road to the right. There are pump jacks on either side of the road. These pump jacks are part of the Hogback oil field producing from the Dakota Sandstone at relatively shallow depth and the much deeper Pennsylvanian Hermosa Group.
24.2 Crossing the bridge over Chaco River.
24.6 Turn left across cattleguard on to dirt road. The road forks, take the left fork.
25.1 Road splits. Continue straight ahead on main dirt road. Ahead is a good view of the Hogback.
25.2 Reach a four-way intersection. Proceed straight. A farmhouse with out buildings is to the right.
25.3 Reach another four-way intersection. Continue straight.
25.7 Road veers left. Descend the hill and drop onto the valley floor.
26.0 You are now in the center of the Hogback oil field in the middle of the San Juan Basin. Rocks outcropping in the cliffs to your left are Tocito, deposited predominantly as sub-tidal bars. This is the most easterly exposed Tocito that has been found in this area.

STOP ONE. "Hogback Oil Field" (Center Sec. 19 T29N R16W). - Examine the facies and sequence stratigraphy of the Tocito around the Hogback oil field. Tidal-bar parasequences and tidal bedding are well exposed here. Both the lower and upper Tocito sandstone bodies at this locality show evidence of channelization. The tidal bedforms were probably deposited in a body of water with restricted lateral extent, which could have focused tidal energy (eg. an estuary).

This is an excellent area to make an outcrop to subsurface tie because this is the most easterly outcrop of the Tocito in this area. The Tocito was brought to the surface along the Hogback uplift and is now in the middle of the subsurface control used to make the Tocito sandstone isochore maps that are presented in the accompanying paper by Exxon. Examine the facies and sequence stratigraphy of the Tocito in the Hogback oil field area, using the measured section and sketches provided (Figs 7-1 and 7-2).

Based on outcrop and subsurface well-log correlation, Exxon interprets the Tocito outcrop at this locality to represent two stacked lowstand deposits (incised-valley fills). The lower, blocky Tocito sandstone (Fig. 7-1) probably represents a feeder channel to the main valley complex (second Tocito sequence or Tocito 2 as described in the accompanying paper by Exxon), which lies to the north of the Hogback oil field in the subsurface. The upper, flaggy Tocito sandstone which overlies the blocky basal sand, represents the youngest Tocito sequence (Fig. 7-1, Tocito 4 sequence), which correlates with the outcrops in the Beautiful Mountain area we visited yesterday (DAY SIX, STOPS 1 - 8). This upper sandstone filled broad erosional lows. The basal sequence boundary, associated with the youngest Tocito, truncates out most of the older, underlying Tocito reservoir sandstones before they reach the outcrop belt to the west. At the Hogback, this surface can be seen to gradually truncate the underlying Tocito strata towards the north (Fig. 7-2; sections 1 through 4).

Riley and Nummedal interpret the two sandstones as representing a lower regressive unit and an upper transgressive one (Figs 7-3 and 7-4A and B). The channel-fill sandstone and associated Tocito facies represent the regressive lower sandstone member. The transgressive upper sandstone member is represented by the cross-bedded sandstones above a sharp erosional surface. A basal intraclast lag is associated with this surface. This transgressive erosion surface is unconformity U-3 in the accompanying paper by Nummedal and Riley. Correlation of closely spaced measured sections in this area reveal an angular relationship between the base of the upper sandstone member and the lower sandstone member (Figs 7-5 by LSU, and 7-2 by Exxon). This location is very near the landward flank of the Hogback anticline described by McCubbin (1969), see Figure 7-6. The angular unconformity between the lower and upper sandstones (Figs 7-5, and 7-2) is inferred to result from continued uplift of this anticline.

Well logs from shallow core holes at the Hogback Oil Field have been correlated to the regional cross sections (Fig. 7-7). The lower sandstone is correlated to the coarsening-upward progradational Tocito interval (interval 2), and the upper sandstone is correlated to intervals 3 and 4. The base of the channel-fill sandstone is interpreted as merging with the pre-Tocito unconformity along the crest of the anticline (Fig. 7-7). The base Tocito unconformity is interpreted to be at the change in texture and color observed near the base of this outcrop. At the end of STOP 1 retrace your steps, drive up the hill, out of the valley, and back to the paved road.

27.4 Reach the paved road, turn left.
34.9 Reach the intersection with US 666, turn left and head towards Gallup for the final stops of the fieldtrip.
38.0 Road to Red Valley on right.
43.5 Passing the flat-topped Table Mesa, comprised of shoreface parasequences of the Point Lookout Formation and the Cathedral Cliff, an Oligocene volcanic plug.
44.7 Passing Barber Peak on the left, another volcanic plug.
51.2 Tocito Dome to the right. Gallup is exposed here. The pump jacks are producing from the Dakota and Pennsylvanian Hermosa Group.
56.0 Passing road to Sanostee on right.
59.3 Now driving between two volcanic plugs; the one on the right is Bennett Peak (elevation 6652 feet), the one on the left is Ford Butte. The road is built on the Upper Mancos Shale Formation.
62.0 Driving on coastal plain deposits of the Cretaceous Menefee Formation.
65.0 The Chuska Mountains form the prominent ridge to the right (west) and is comprised mainly of Tertiary andesitic volcanics.
65.8 Passing the town of Newcomb.
75.8 Although the road is on the Menefee, the shaley slopes to the east are comprised of the Mancos mudstones.
86.2 Crossing the McKinley County line.
93.7 The El Paso Natural Gas, Gallup pumping station is to the left (east).
99.0 Passing through the town of Tohatchi.
113.5 Now leaving the lands of the Navajo Nation. Good Menefee outcrops just right (west) of the road.
115.9 Junction with the road to Window Rock. Road forks. Go straight ahead towards Gallup.
119.6 Road cuts through trough cross-bedded, fluvial channel sandstone outcrops, probably within the Menefee.
121.8 To the east are more outcrops of the Menefee Formation, composed of a thick series of fluvial sandstones and interbedded mudstones.
122.6 Now entering Gallup city limit.
123.3 Bridge over interstate I-40, get into left lane.
123.5 Turn left onto the on-ramp of I-40 EAST, heading towards Grants and Albuquerque.
127.4 Now passing through a steep, west-dipping hogback of Gallup and Torrivio sandstones. In this area the Gallup Sandstone is composed of marine parasequences stacked in a progradational parasequence set. The top of the Gallup contains coal beds and coastal plain sandstones and mudstones. These coal deposits provided fuel for the early Santa Fe railroad and were one of the reasons why this railroad originally passed through Gallup.
128.6 Passing Exit 26. To the north, the Gallup Sandstone is outcropping as a buff-colored cliff.
130.3 To the left a small refinery
133.5 Take exit 31 to Red Rock state park and old Fort Wingate.

133.7 "T' junction, turn left to Red Rock state park.
133.8 Pass under interstate and turn left on frontage road towards Red Rock.
135.1 Turn right at junction with road 566 to Red Rock state park.
135.3 Bridge over railway tracks.
135.8 Red cliffs of Jurassic Entrada Sandstone (aeolian dunes) outcrop to left and right of road.
136.8 Now driving through Wanakah Fm. (Jurassic aeolian and lacustrine deposits)
137.5 Large hoodoos of the Recapture Member of the Morrison Formation to the right
138.3 Turn left onto gravel road McKinley County 43: Superman Canyon Rd. White Rock Mesa to the north consists of the Westwater Canyon Member of the Morrison capped by the Dakota Sandstone.
138.5 Large cross-bedded sets in the Recapture Member of the Morrison Member outcrop on the right.
138.7 Keep right and continue ahead on main dirt road.
139.1 To right good view of basal Dakota unconformity (K_2) cutting into the underlying Westwater Canyon Member of the Morrison.
139.2 Road forks, go left.
139.3 Cross one lane bridge
140.0 Cross larger steel bridge
140.5 To left good outcrops of the Westwater Canyon Member (white fluvial sandstones).
141.1 Road forks. Go right, on Hard Ground Canyon road.
141.6 Low outcrops to the left are the Two Wells Tongue of the Dakota Sandstone.
142.4 Pull onto the shoulder to the right of the road. Walk up to the outcrops on the point.

STOP TWO. "Nose Rock Point" (SW SW SW sec. 14 T16N R17 W, see location map Fig. 7-10).

The objectives of this stop are to: (a) examine the internal architecture and stacking patterns of parasequences in the Gallup Sandstone, (b) evaluate evidence for a sequence boundary at the contact between Gallup tongues D and C, and (c) discuss facies characteristics of a wave-dominated delta system. This excellent exposure owes its existence to a landslide which occurred here probably a century or more ago. The many boulders on the side of the road here where we have parked are Gallup Sandstone blocks which came down in that slide.

Individual Gallup tongues were labeled A (the youngest) to F by Molenaar (1983). Tongues F through C are exposed in this outcrop (Figs. 7-8, 7-9, 7-10). The C tongue forms the pink, vertical cliff at the top of Nose Rock Point (Fig. 7-10A). Individual parasequences in tongues F through D form coarsening- and thickening-upward successions with sharp upper boundaries (flooding surfaces). Individual sandbodies are tabular and laterally continuous across the extent of the outcrop. Typically, the parasequences are about 10 to 15 m thick. Hummocky cross stratification and marine burrows, including *Ophiomorpha*, *Zoophycos* and *Chondrites* demonstrate that most of this sandstone was deposited in wave-influenced shoreface settings. Mapping of the regional extent of the shoreface sandstone, and the fairly straight NW-SE shoreline trend (McCubbin, 1982; Molenaar, 1983), further demonstrate that the depositional environment responsible for the Gallup Sandstone was a wave-dominated delta prograding from the SW to the NE.

The section here at Nose Rock Point shows a distinct set of forestepping parasequences (Figs. 7-8, 7-9). The forestepping geometry implies either a lowstand or a highstand systems tract. It can be demonstrated in the subsurface that Gallup markers downlap onto the Juana Lopez. Moreover, the top of the Juana Lopez is a fairly coarse shell "hash", locally with abundant ammonites, oysters and inoceramid shells. We interpret this to be a maximum flooding surface acting as the downlap surface during Gallup progradation. Consequently, the forestepping set of Gallup parasequences in this outcrop (tongues F through D) are interpreted as a highstand systems tract.

Gallup Sandstone Tongue C is interpreted as estuarine because of abundant mud-drapes, an ichnofauna including *Ophiomorpha, Thalassinoides, Rosselia* and *Teredolites* (occasionally very large), total lack of hummocks and other wave structures, and lenticular, often channel-form sandbody geometries. This estuarine facies is incised into the underlying Gallup D-tongue shoreface (at 108 m in Fig. 7-9). The local relief on this surface is many meters across this outcrop. We interpret the C-D contact as a sequence boundary and, for reasons discussed in more detail in the accompanying paper by Nummedal and Riley, we suggest that this is the same unconformity observed at the base of estuarine/fluvial Gallup channels in the Beautiful Mountain area (U-1).

For those who do walk through this whole section we would like to point out the deformed bed in the lower part of the F tongue (Fig. 7-8; 31 m), the slab of exceptional trace fossils at the SE corner of the ledge atop the E-tongue (Fig. 7-8; 72 m), and the very large Teredolites "plugs" right at the sequence boundary at the base of Gallup C (Fig. 7-9, 108 m; Fig. 7-10D).

Turn around and go back (south) on Hard Ground Canyon Rd.

142.5 Gallup boulder field
142.9 Outcrop of Twowells Tongue of Dakota SS.
143.3 Turn sharp right on Superman Canyon Road
143.7 Cross bridge, turn right
144.2 Turn right onto Rock Flat Road
146.0 Cross fence line
146.8 Pull off on shoulder and stop

STOP THREE. "White Rock Canyon" (Center, sec. 16, T16N, R17W, see location map in Fig. 7-11). The measured section is located on the north side of the canyon, across a fairly deep arroyo in modern alluvium. Most of the features described, however, can also be observed along the road on the south side.

The purposes of the stop are to (a) examine in greater detail the inferred Late Turonian sequence boundary (U-1) at the contact between Gallup Tongues C and D, (b) discuss the gamma-ray expression of this boundary, and (c) describe the petrographic differences between rock types on opposite sides of this contact.

The inferred sequence boundary is located at the base of unit 2a in Fig. 7-12. It is a sharp, erosional contact traceable across the whole outcrop area and back to the 108-m-level (Fig. 7-9) at stop 2. Lithostratigraphically, this is the contact between Gallup D (below) and Gallup C. Locally, over a distance of a few hundred meters within White Rock Canyon, the unconformity has a relief of more than 5 meters. The facies contrast across the boundary is the same here as at Nose Rock Point: hummocky cross stratified shoreface deposits below and mud-draped, trough cross stratified estuarine deposits above. Large *Teredo* borings line the boundary, and some are found also within the overlying estuarine fill. Paleocurrent directions within the estuarine fill are bimodal with directions to the north and the southeast. A series of different subenvironments (estuarine bars, abandoned channel fill, active channel fill, etc.) can be identified within the Gallup C tongue in this area.

We infer that this juxtaposition of estuarine fill above lower shoreface across a wide area implies a downward shift in facies associated with a relative sea level fall.
Because of biostratigraphic constraints discussed in the accompanying paper by Nummedal and Riley, we infer that this boundary is the same as the Upper Zuni A2/A3 "supersequence" boundary of Haq et al., (1988).

The following text on the gamma-ray profiles and petrography is prepared by Rex Cole and Art Trevena, Unocal Science and Technology, Brea, CA.

To simulate subsurface well-log data we took gamma-ray readings at three different sections across the inferred sequence boundary here at White Rock Canyon. The middle profile in Fig. 7-13 (NM 4-91 B) corresponds to the measured section in Fig. 7-12, with the exception that the zero point is transferred to the sequence boundary. Measurements were taken using a gamma spectrometer every 1 ft up the outcrop. Count times ranged from 10 seconds for the routine (total counts) traverses to 30 seconds for spectral readings. All readings were reduced to counts per second. Spectral data were reduced to equivalent uranium (from 214 Bi), thorium (from 208 Th), and potassium (from 40 K) values using equations provided by the instrument's manufacturer (Scintrex).

The gamma-ray values demonstrate considerable variation upwards through all sections. Various facies show different, but overlapping gamma-ray ranges. Readings from the lower parts of the shoreface (Fig. 7-13, below - 10 ft) range from 193 to 271 cps whereas those in the estuarine complex range from 138 to 349 cps. A major radioactivity anomaly is associated with the sequence boundary which separates the lower shoreface interval and the estuarine complex. This anomaly was noted in each of the three sections measured at White Rock Canyon and also in the gamma-ray profile measured at Nose Rock Point (Fig. 7-9, 105 m). In section 4-91 A (Fig. 7-13) the anomalous zone is about 2.2 m thick and the highest total count was 750 cps. In section 4-91 B a maximum count of 805 cps was reached in the upper part of the anomaly, less than 1 ft below the sequence boundary.

In section NM 4-91 B, spectral gamma data were collected across the anomalous zone. Spectral data were reduced to equivalent parts-per-million U and Th, and equivalent percent K. No systematic trends were noted in the potassium values, however, trends in U and Th were very distinct. A plot of the U/Th together with a total gamma count profile (Fig. 7-14) demonstrates an increase in the U-Th ratio from about 0.4 in the lower shoreface to 1.68 in the most radioactive part of the weathered (?) interval. The ratio then declines to about 0.5 in the lower part of the estuarine complex. This analysis indicates that the anomalous radioactivity signal is due to uranium enrichment.

Petrographic modal analyses of three sandstone samples from section NM 4-91 B were carried out at Unocal. An unsuccessful attempt was made to document the distribution of uranium within these samples using electron microprobe analysis. The three samples are from the unaltered (low radioactivity zone) lower shoreface section of the Gallup D tongue (sample 1, Fig. 7-14), the weathered (?), radioactive part of the shoreface (Sample 2) and the lower part of the estuarine fill above the sequence boundary (Gallup tongue C, sample 3).

The shoreface samples are very fine-grained lithic arkoses (Table 1; Fig. 7-15). The upper of the two samples (in weathered ? zone) has poorer sorting, lower porosity and greater clay content than the lower one. Based on optical characteristics most of the clay appears to be kaolinite.

Sample 3 (estuary fill) is a fine-grained, moderately well-sorted arkose to quartz arenite that contains substantially more quartz and fewer lithic fragments than the lower shoreface samples below the sequence boundary. This sandstone sample contains abundant large pores and only a trace of clay (Fig. 7-16). Also, the estuarine sandstone has a higher ratio of K-feldspar to plagioclase than the shoreface sand (Table 1).

We hypothesize that this petrographic difference between the shoreface and overlying estuarine sandstones reflects the combined physical and chemical destruction of less durable components within the tidal-influenced estuary. Specifically, the presence of skeletal plagioclase grains and common oversized pores in the estuarine sandstone may indicate that dissolution of large amounts of plagioclase has occurred. It is evident in Fig. 7-16 that the estuarine fill sandstone is the best reservoir rock in this section.

Table 1. Petrographic summary of thin sections from measured section NM 4-91 B

Sample	Median grain size	Sorting	Q/F/L	K-spar/ Plag	Clay %	Porosity %
3	200 um	m-well	73/26/1	3.4	tr.	15
2	110 um	m-poor	51/36/13	0.9	19	6
1	110 um	medium	55/33/12	1.3	10	12

Turn around and backtrack to highway NM 566 and I-40.

149.4 Turn left onto Superman Canyon Road and follow this to intersection with paved road (NM 566)
150.3 Merge with Hard Ground Flat road from the left
153.1 T-junction with NM 566, go right (south)
156.3 T-junction with frontage road, turn left for I-40 East, follow signs to Albuquerque.

A recent guidebook by Finch et al. (1989), covers outcrops along the interstate between Gallup and Albuquerque. We recommend that you consult this for an in-depth discussion of the geology visible from the interstate. The following summary road log covers some of the highlights.
157.8 Pass signpost and junction for old Fort Wingate. Continue on frontage road to EXIT 33.
160.2 Turn right and cross bridge over I-40
160.4 Turn left at the entrance for I-40 EAST, head onto the interstate.
166.5 A small oil refinery can be seen to the north. Now passing Exit 39. This long stretch of the road running between Grants and Gallup is built on relatively soft Triassic Chinle Formation.
170.8 Spectacular outcrops of Jurassic and Cretaceous strata occur to the north of the interstate. The Entrada is the lowest massive red cliff. It is overlain by the recessive, greenish and red Jurassic. The cliff is capped with brown Dakota sandstone.
174.6 Passing Exit 47.

174.9 We have just crossed the continental divide at 7275 feet.
179.8 Passing Exit 53 for Thoreau. The elevation is 7200 feet.
190.0 Passing Exit 63 for Prewitt.
191.2 Descending small hill. To the north massive red Entrada cliff ends abruptly because of a fault. Entrada appears again approximately 1/2 mile to the east. The fault zone is a small graben with Morrison and Dakota strata on the hanging wall juxtaposed against the Entrada.
199.0 Passing Exit 72 to Blue Water Village. To the north is a mesa of Entrada and Morrison strata with a capping of Dakota beds.
203.4 The road descends through a road cut in Permian strata.
204.7 Approximately 10 miles north is the western edge of the Holocene lava field lapping out against the Jurassic and Cretaceous strata.
209.0 To the left is a mesa capped with thick, columnar-jointed basalt. The white water tank labelled "Grants" is built in the middle of a slope composed of lava and Cretaceous rocks.
216.3 Passing Quemado Exit 89. Lava is everywhere. These flows form part of the Malpais lava field through which we are driving. Some of the youngest flows are about 1000 years old.
225.2 Flat-topped mesas of Cretaceous strata capped with columnar-jointed basalt flows are seen north of the highway. This area is a major center of uranium exploration, mostly in the Morrison Formation. This is the heart of the Grants Uranium District. Uranium generally accumulates in association with organic matter which is deposited within fluvial channel sandstones in the Morrison. Many of the uranium mines in this area were worked in the 1950s through the 1970s, but are mostly closed now.
229.0 Passing Exit 102 to Acoma Pueblo on the Acoma Indian reservation.
230.5 Driving down a hill through the Dakota Sandstone. A rest area is 1/2 mile ahead.
234.8 Passing Exit 108. The Dakota and Morrison formations are exposed to the north of the highway along the continuous cliffs. The Morrison forms the recessive red and green unit.
239.8 Mt. Taylor is to the north. The mountain is built of many successive lava and ash flows, the oldest of basalt and the youngest of dacite and andesite. Volcanic activity began 4 million years ago. Eruptions continued repeatedly for nearly 2 million years.
241.1 Laguna Pueblo rests on a small hill to the north.
241.3 Just passed Exit 114 for Laguna. Cretaceous strata in the distance to the north form the "reference area" for the Dakota.
242.6 Driving through the Wanakah Formation.
243.3 A collapse feature in the Horse Mesa Member of the Wanakah Formation can be seen to the south.
244.6 Passing Exit 117 to Mesita.
253.7 Pass under overpass at Exit 126. To the north is a mesa capped with Dakota sandstone. The recessive Morrison Formation rests beneath the Dakota and on top of a brick red cliff formed by the Wanakah Formation, which is approximately 100 feet thick. Triassic fluvial deposits are nearest to the road. Excellent exposures of fluvial channels can be seen in these Triassic rocks to the north for the next mile.
254.7 Low-sinuosity fluvial channels can be seen in the Triassic Chinle Formation north of the highway.
255.9 Red sandstones and mudstones of the Chinle Formation can be seen 200 yards north of the highway.
258.1 Pass Exit 131 for Canoncito. Mesas to the north are Dakota rocks underlain by the Morrison. Resistive red and tan rocks at the base of the mesa are the Wanakah Formation.
259.1 Crossing Bernalillo County line.
265.7 At the top of the gentle hill is an excellent outcrop of the Santa Fe Group on the north side of the highway. These strata were deposited in the Late Tertiary by the Rio Grande River.
267.2 Passing exit I-40 for Rio Puerco. Lava flows are quarried along the river.
267.6 Cross the Rio Puerco river. This marks the approximate western edge of the Rio Grande Rift.
274.6 Ascending a gently sloping terrace of the Sante Fe Group. Ahead is a striking view of the Sandia Mountains on the eastern edge of the Rio Grande rift.
276.3 Continue east on I-40 (signposted for Santa Rosa).
279.0 In the distance to the left (north) are three of Albuquerque's five volcanoes. The volcanoes are aligned north to south along a fault or fissure. The oldest lava flows associated with these volcanoes have been dated at 190,000 years. You are driving on a broad, flat terrace that was a small sand dune field during the Pleistocene. Surrounding the volcanoes and across the dune field is a broad volcanic flow.
282.9 Cross the Rio Grande River. The west bank of the river is a terrace built of alluvium. Albuquerque is located in the north-south oriented Rio Grande rift. The rift is 28 miles wide at Albuquerque

and establishes the drainage for the Rio Grande river to the Gulf of Mexico. Parallel fault zones bounding the rift began forming 30 million years ago. In places, rifting has down dropped the central graben as much as 30,000 feet. At the same time, explosive volcanic eruptions spread ash over New Mexico. Today, great craters occur where the volcanoes once stood. Smaller volcanic eruptions within the last 200,000 years have distributed cinder cones and basaltic flows across the surface of the rift and rift margins.

284.0 Take Exit 157A for Rio Grande Boulevard (and Old Town). Turn right (south) on Rio Grande. The Sheraton Old Town hotel is visible ahead to the left.

284.4 Turn left on Bellamah.

284.5 Turn right on to the parking lot of the Sheraton.

This is the end of the fieldtrip, we hope you enjoyed the the geology.

REFERENCES

Finch, W. I., Huffman, A. C., and Fassett, J. E., 1989, Coal, Uranium, and Oil and Gas in Mesozoic Rocks of the San Juan Basin: Anatomy of a Giant Energy-rich Basin; 28th International Geological Congress, Field Trip Guidebook T120, 99 p.

Haq, B. U., Hardenbol, J., and Vail, P. R., 1987, Chronology of fluctuating sea levels since the Triassic: Science, V. 235, p. 1156-1167.

McCubbin, D. G., 1969, Cretaceous strike valley sandstone reservoirs, northwestern New Mexico: American Association of Petroleum Geologists, V. 53, p. 2114-2140.

McCubbin, D. G., 1982, Barrier-island and strandplain facies; in, Sandstone Depositional Environments; P. A. Scholle and D. Spearing (eds.), AAPG Mem. 31, p. 247-279.

Molenaar, C.M., 1973, Sedimentary facies and correlation of the Gallup Sandstone and associated formations, northwestern New Mexico, *in* Fassett, J.E., ed., Cretaceous and Tertiary rocks of the southern Colorado Plateau: Four Corners Geological Society Memoir, p. 85-110.

Molenaar, C. M., 1983, Principal reference section and correlation of Gallup Sandstone, northwestern New Mexico; in, Contributions to mid-Cretaceous paleontology and stratigraphy of New Mexico - part II; S. C. Hook (compiler); New Mexico Bureau of Mines and Mineral Resources Circ., 185, p. 29-40.

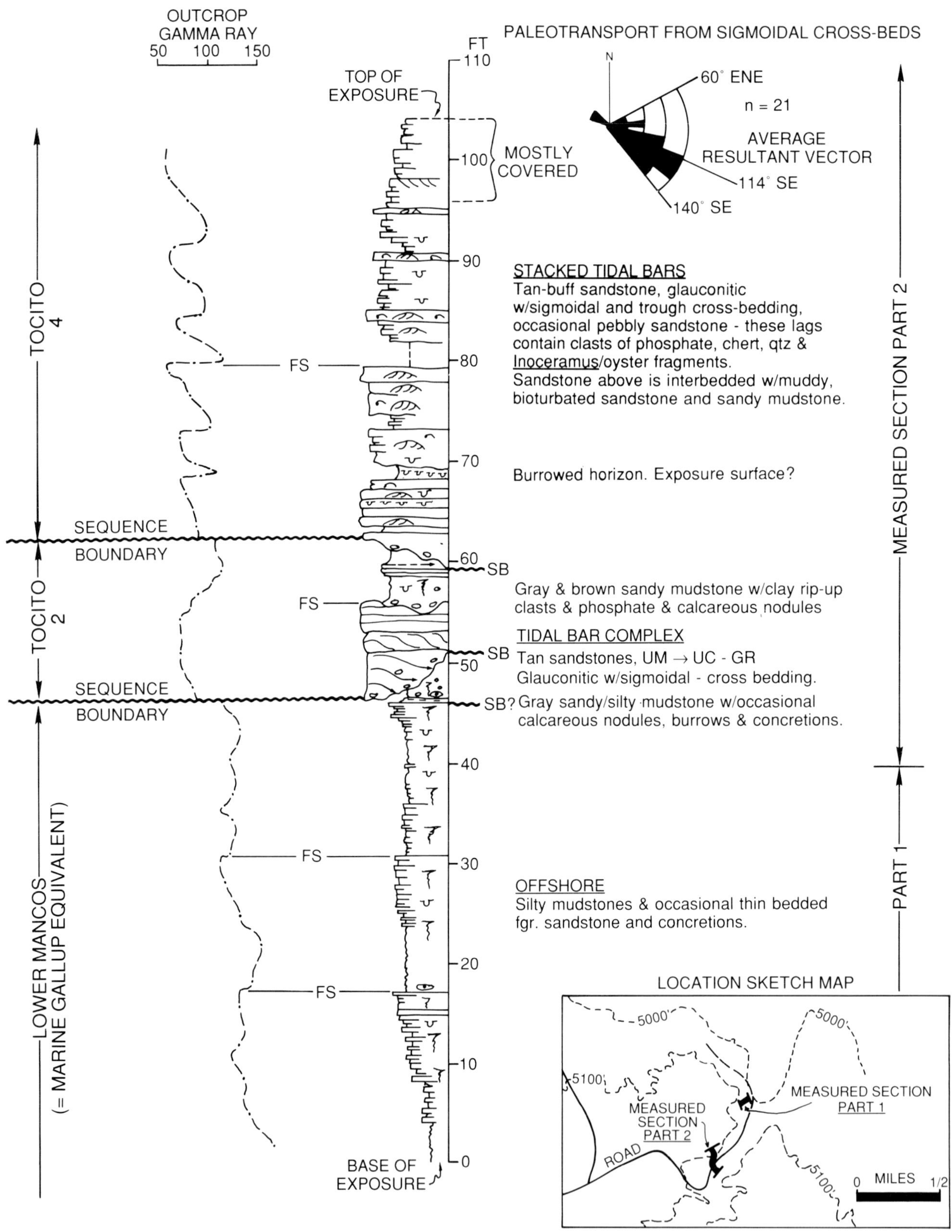

Figure 7-1. Measured section through the Tocito at Hogback Oilfield, Chaco River, near Fruitland (center Sec. 19 T29N R16W).

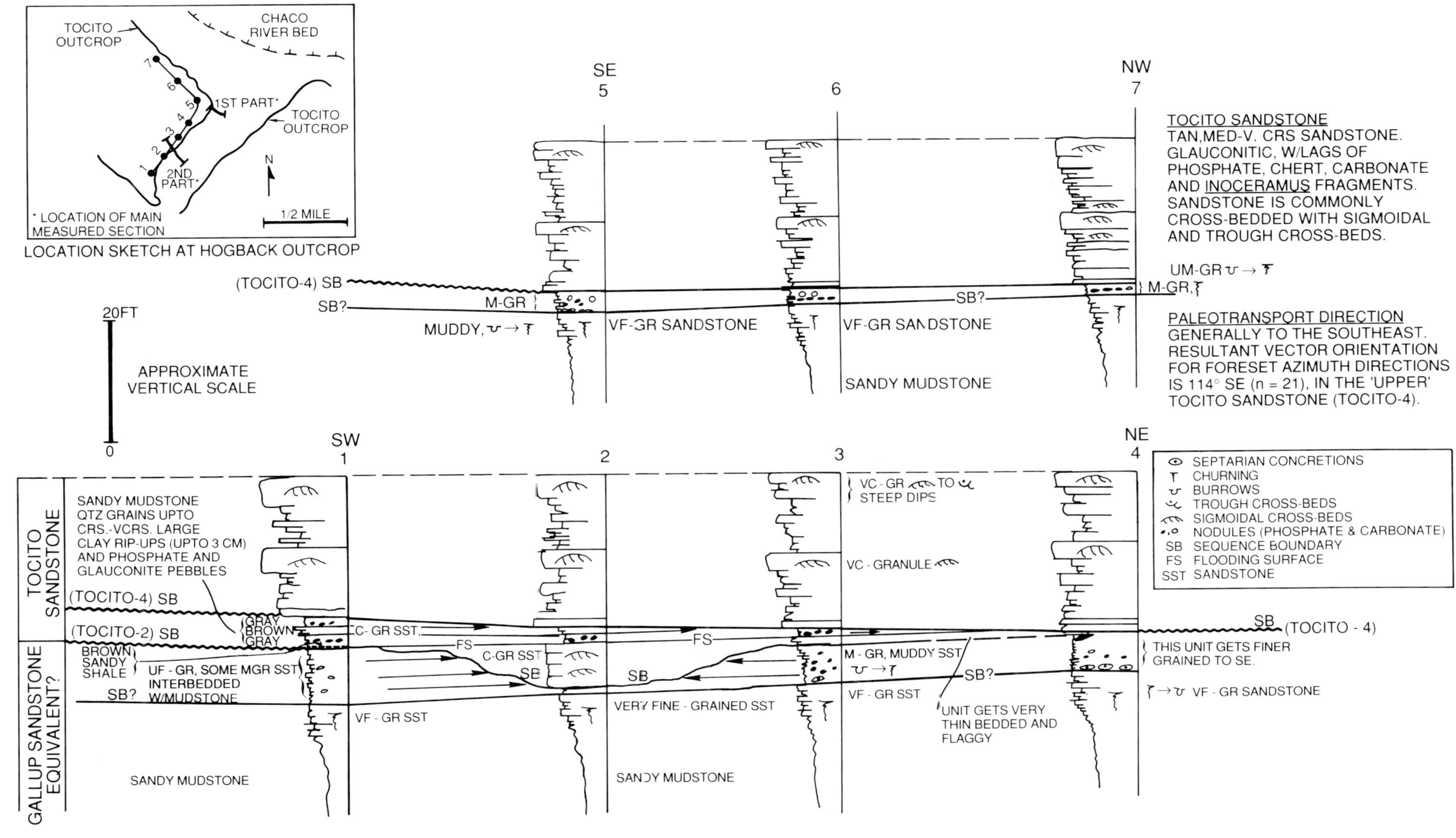

Figure 7-2. Schematic cross sections through the Tocito at the Hogback Oilfield (Day 7, Stop 1), showing the erosive based, blocky sandstone (Tocito sequence 2) and the overlying flaggy sandstone (Tocito sequence 4), the base of which gradually truncates underlying strata (see accompanying paper by Exxon for a description of Tocito sequences 1 through 4).

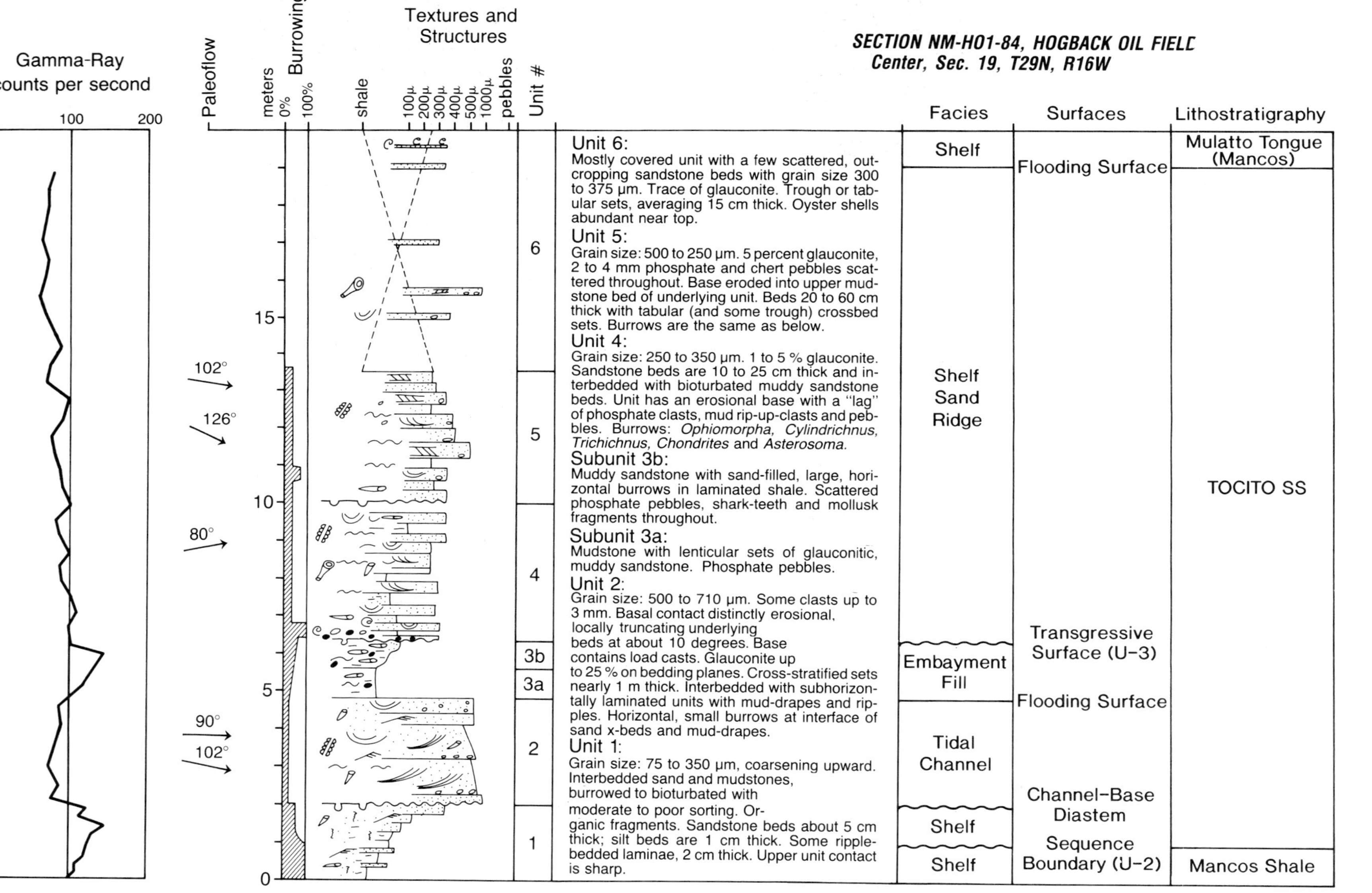

Figure 7-3. Measured section at Hogback Oil Field. Lower sandstone member is interpreted as regressive shelf facies. Channel sandstone is interpreted as a tidal channel that was a feeder to lower sandstone equivalents in subsurface. Phosphatic muddy sandstone overlies initial transgressive surface and formed in transgressive embayment. Upper sandstone overlies tectonically enhanced transgressive erosion surface and is interpreted as a transgressive shelf sand ridge.

Figure 7-4A. Photo of measured section HO1. SB is the sequence boundary at the base of the Tocito Sandstone (SB). Note sharp initial flooding surface (ITS) at top of lower sandstone member. Base of upper sandstone member is erosional and contains lag of intrabasinal clasts.

Figure 7-4B. Pseudoperna congesta layer at top of Tocito Sandstone. This bed marks the top of the upper sandstone throughout the Four Corners region. This layer is interpreted as a final transgressive surface.

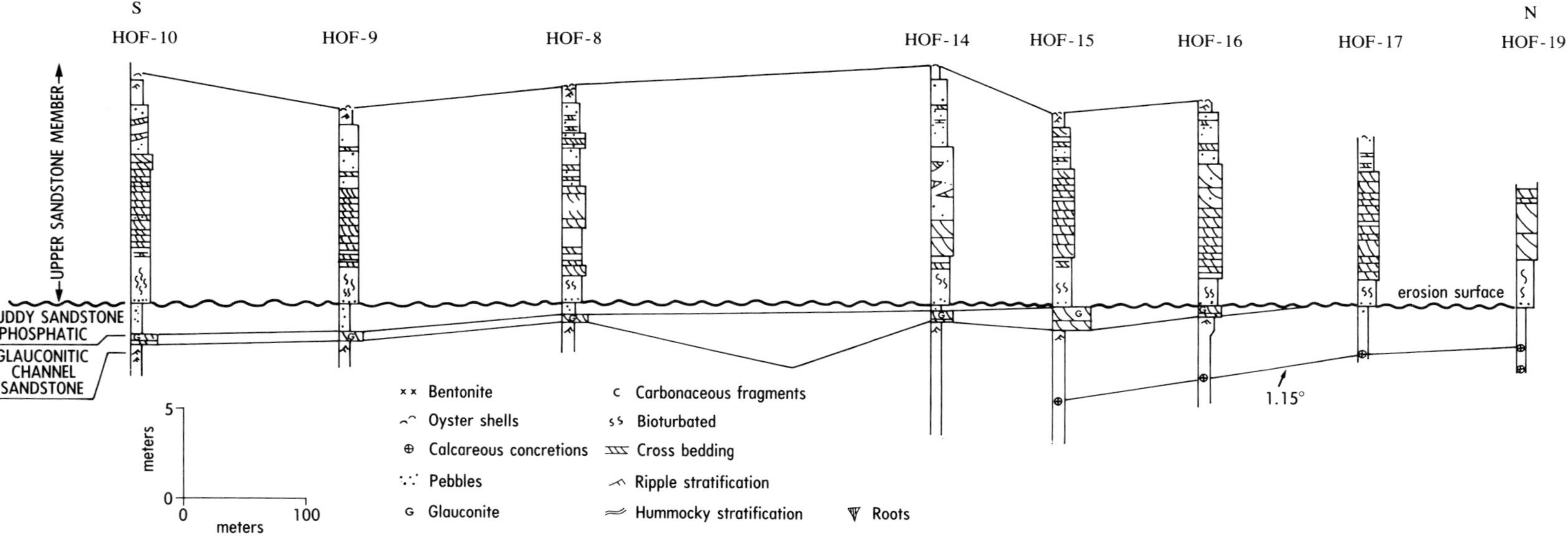

Figure 7-5. N-S cross section at Hogback Oil Field. Section shows angular truncation at base of upper sandstone member, with lower sandstone member dipping to the southwest.

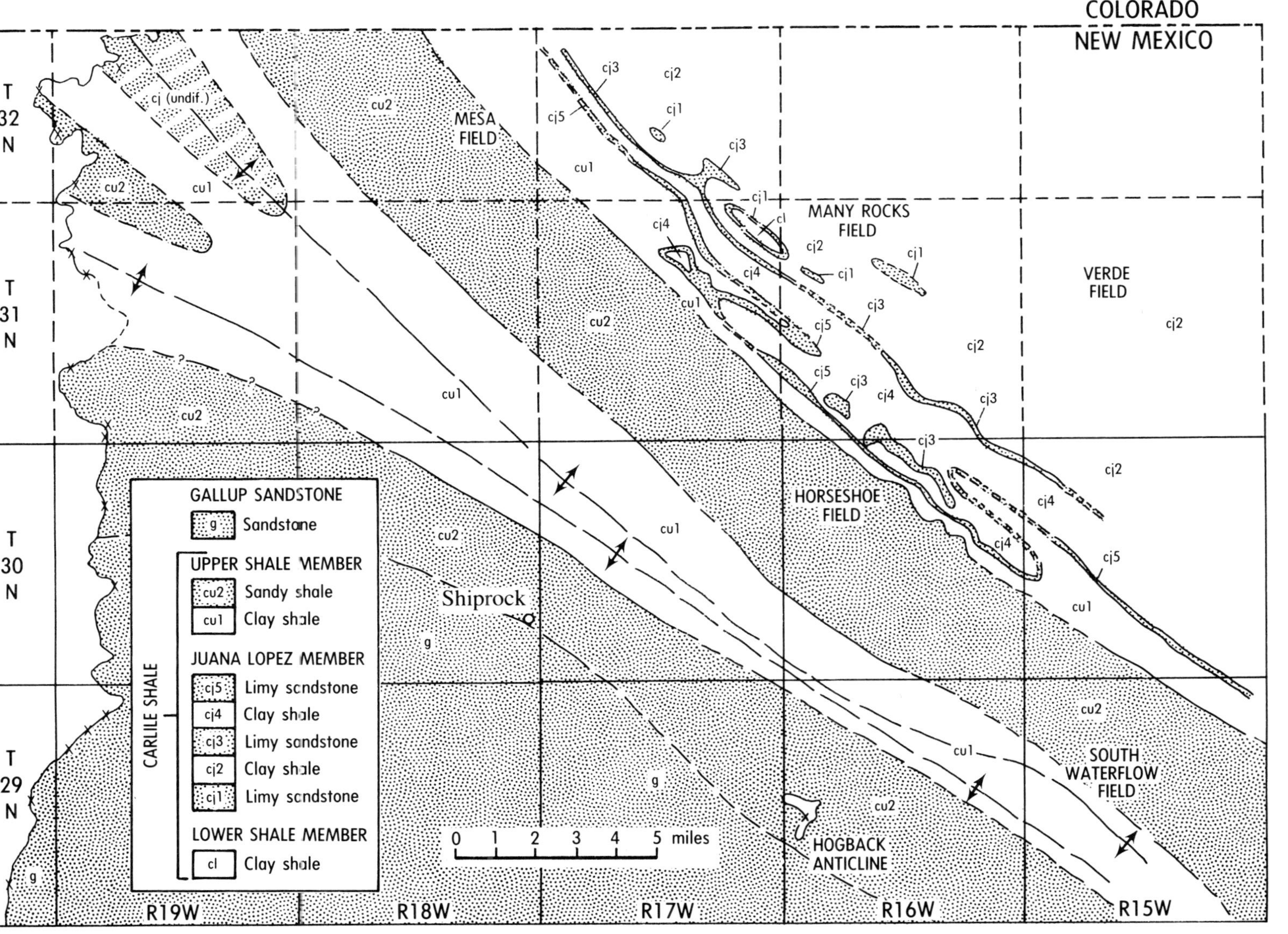

Figure 7-6. Subcrop map beneath pre-Tocito unconformity. Map outlines location of anticline. From McCubbin (1969).

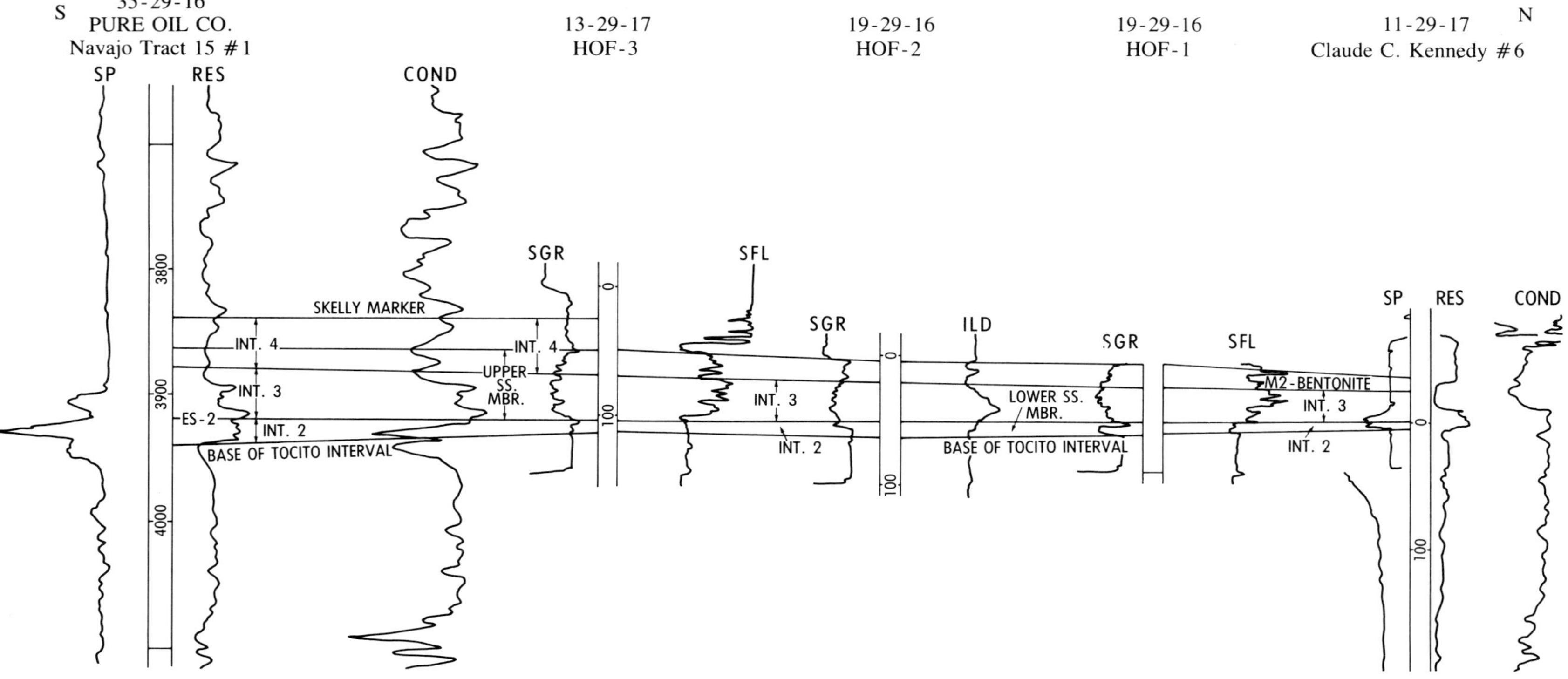

Figure 7-7. Correlation of shallow core holes at Hogback Oil Field to regional subsurface sections (Figs. Q and R). M2 bentonite is believed to be present within upper sandstone member at this location.

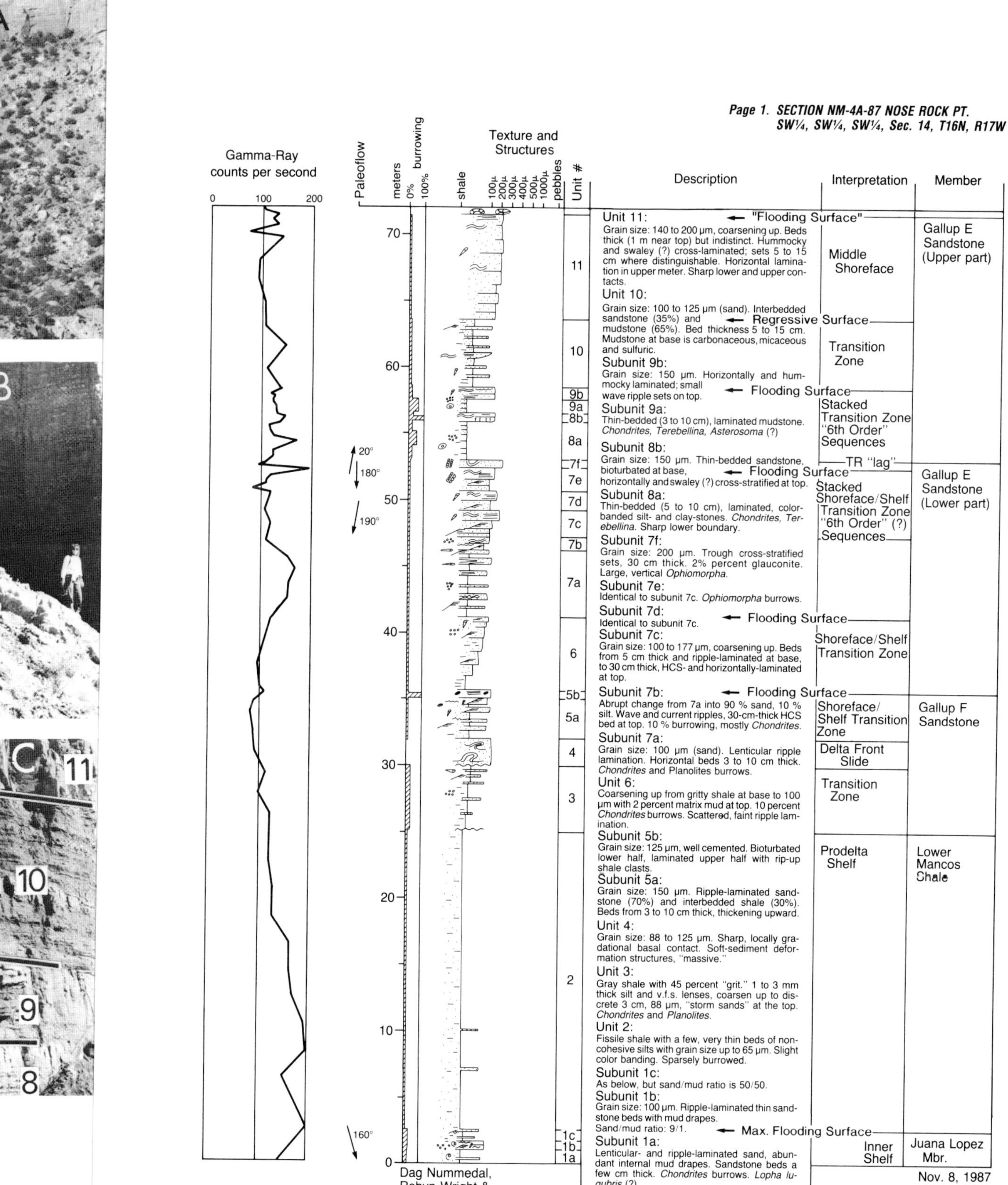

Figure 7-8. Measured and interpreted section of the Mancos Shale and the lower part of the Gallup Sandstone at Nose Rock Point. Note that the Gallup here consists of a forestepping set of shoreface parasequences.

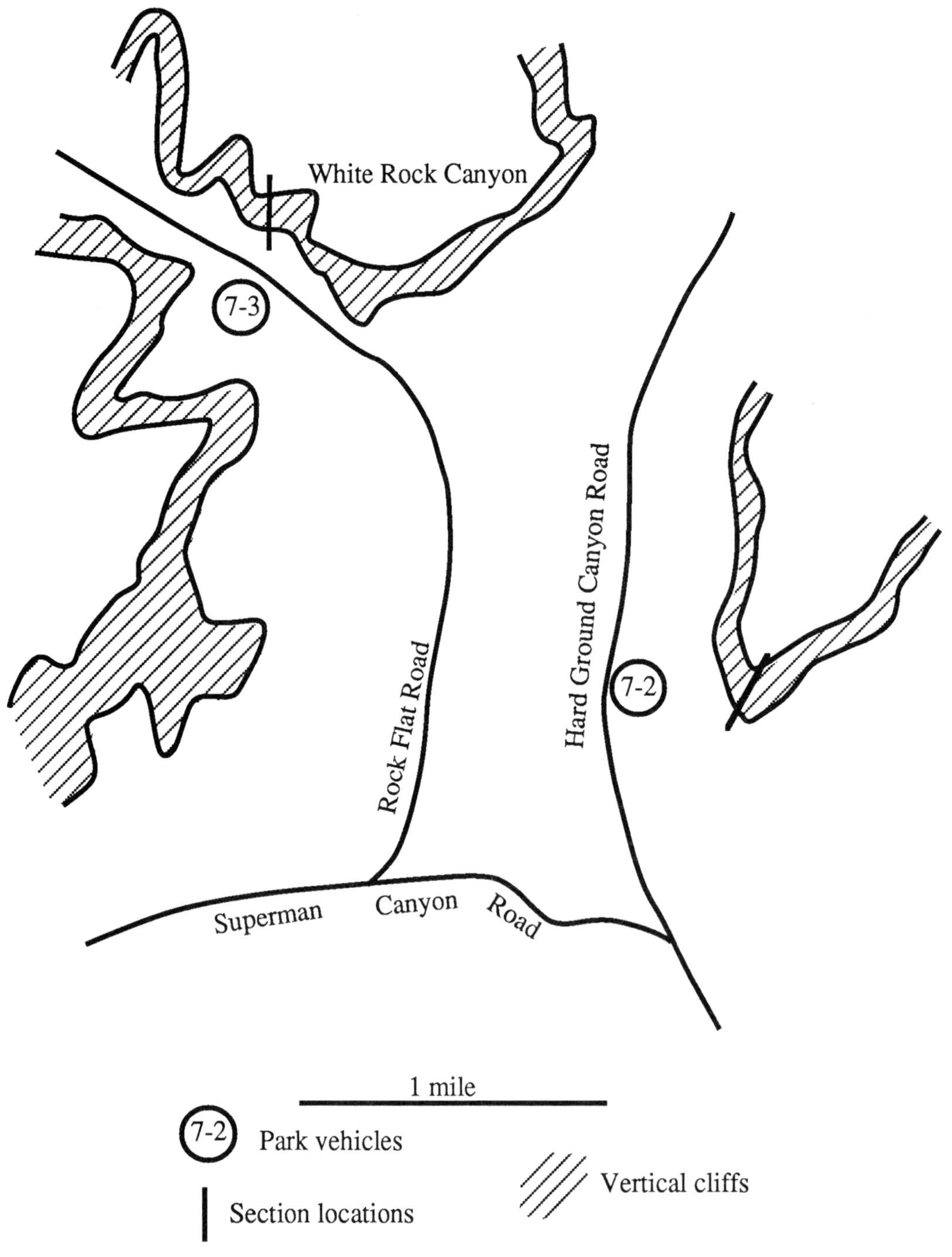

Figure 7-11. Location map for Nose Rock Point and White Rock Canyon stops.

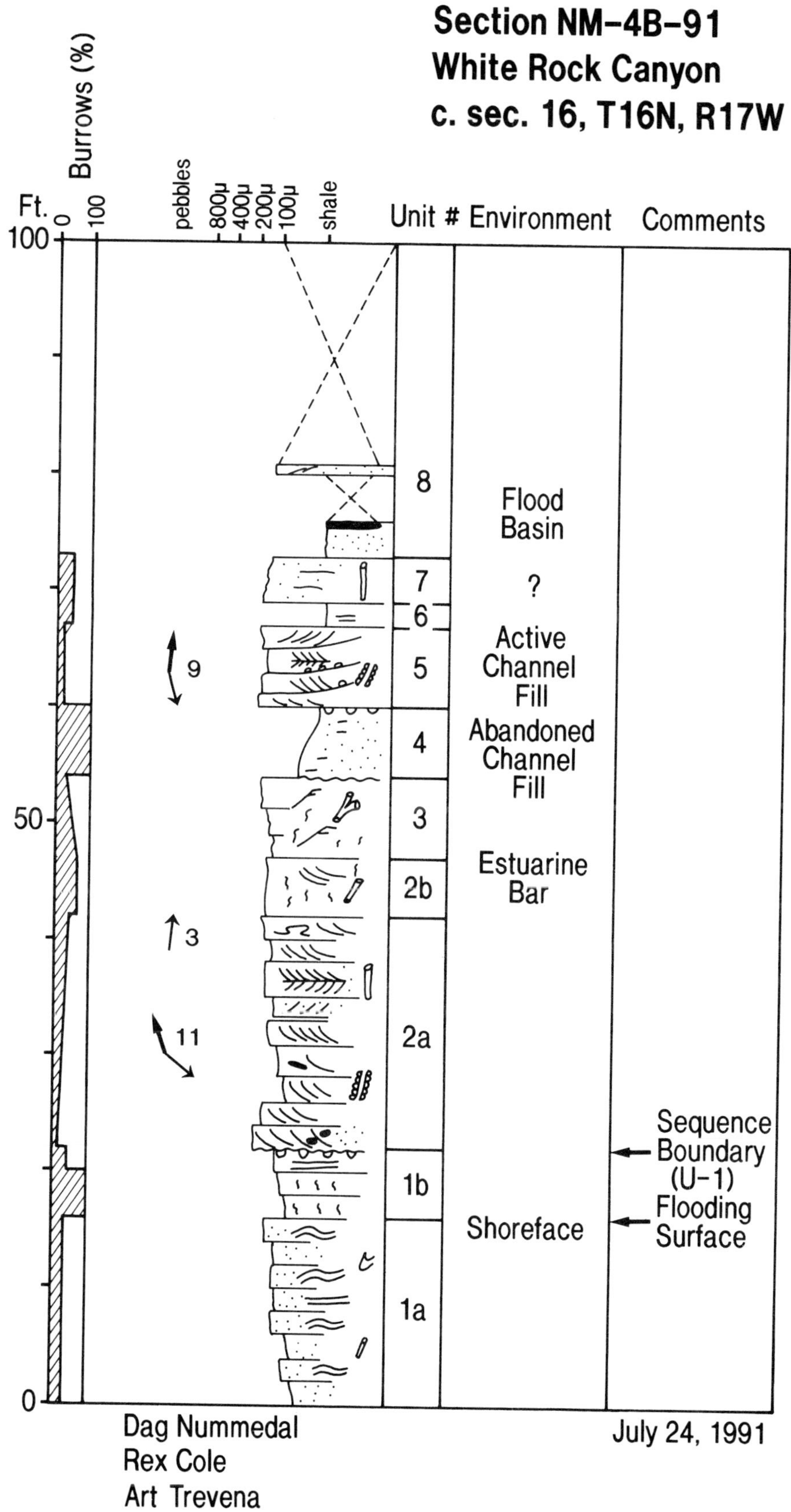

Figure 7-12. Measured section at north side of White Rock Canyon. A distinct sequence boundary occurs at the base of unit 2a, where estuarine facies directly overlie truncated lower shoreface facies of the Gallup Sandstone. The sequence boundary separates Gallup tongues D (below) and C (above) of Molenaar (1983).

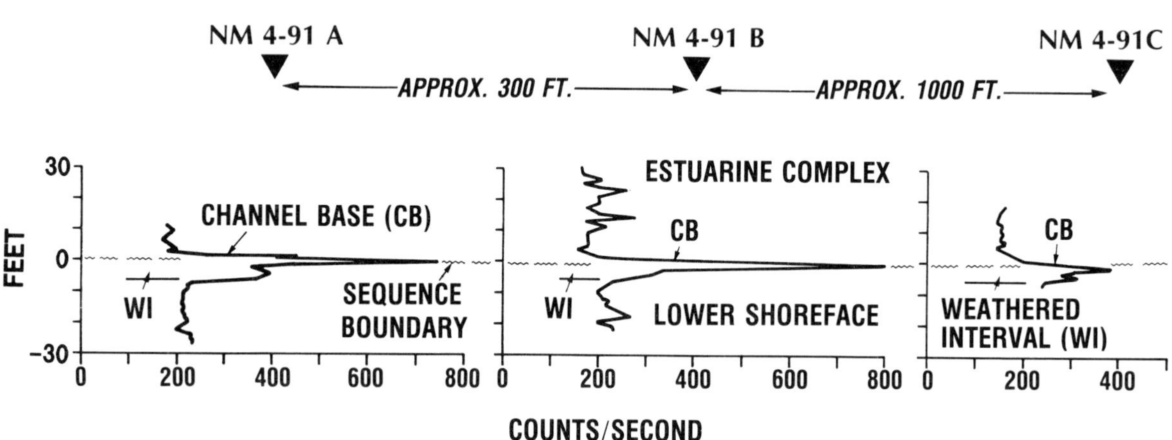

Figure 7-13. Gamma-ray profiles across sequence boundary at three closely-spaced traverses. Data are total gamma-ray counts. Data courtesy of Rex Cole and Art Trevena, Unocal.

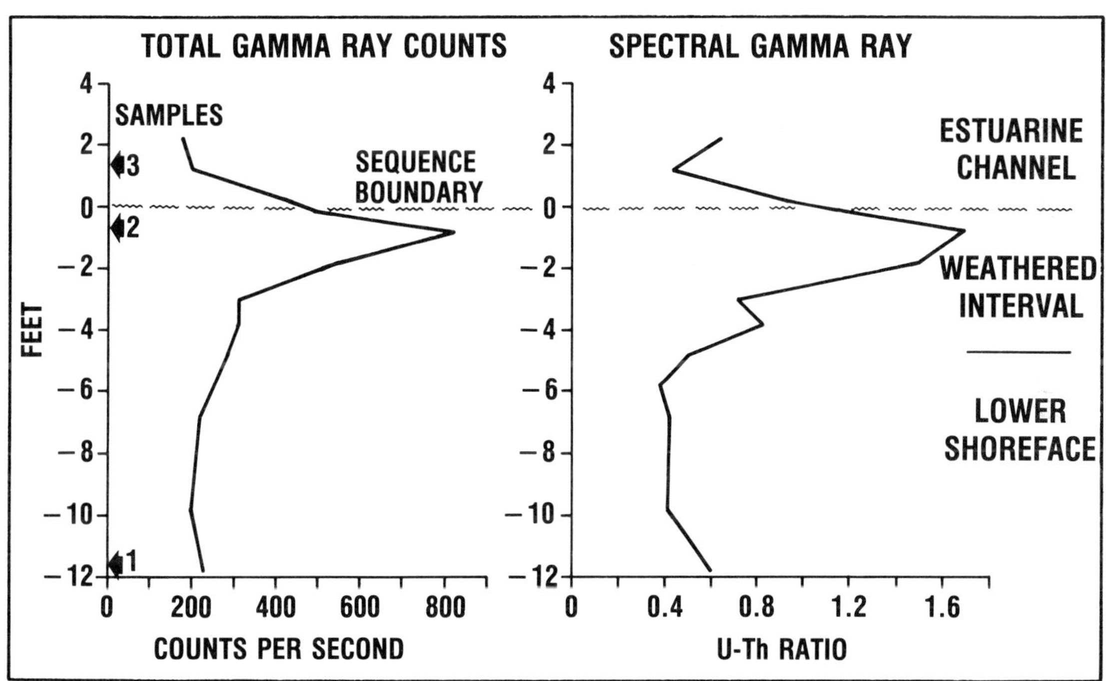

Figure 7-14. Total and spectral gamma-ray profiles across sequence boundary. The U-Th ratio demonstrates that the high gamma-ray count at the sequence boundary is due to U enrichment. Data courtesy of Rex Cole and Art Trevena, Unocal.

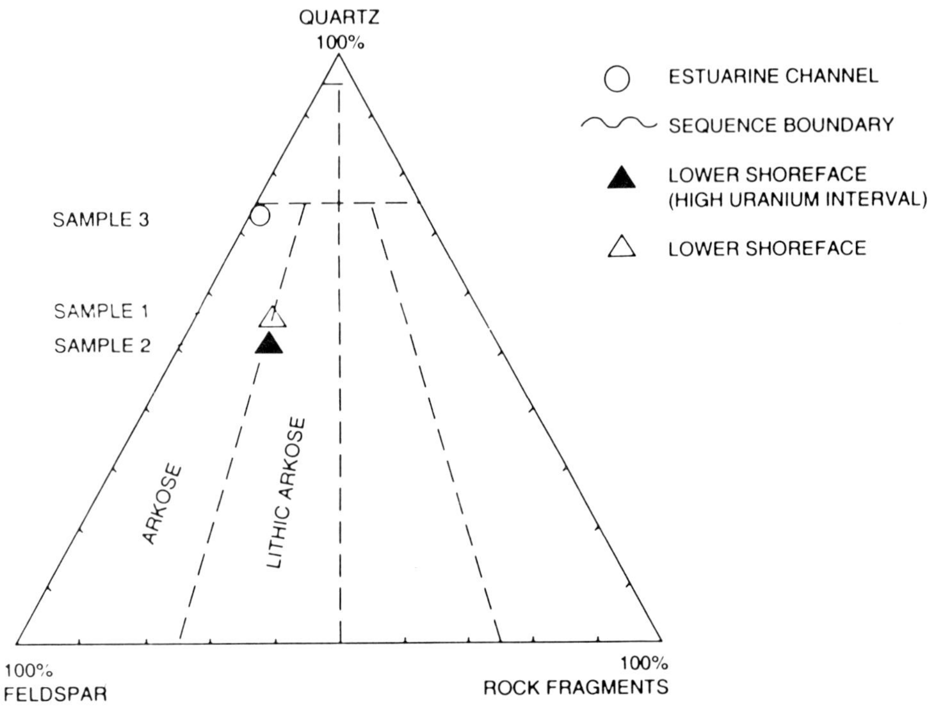

Figure 7-15. QFL ternary diagram for three sandstone samples from profile NM-4-91B. Note the more mature composition of the estuarine sandstone compared to the underlying shoreface samples.

Fig. 1--Ma
Upper Cret

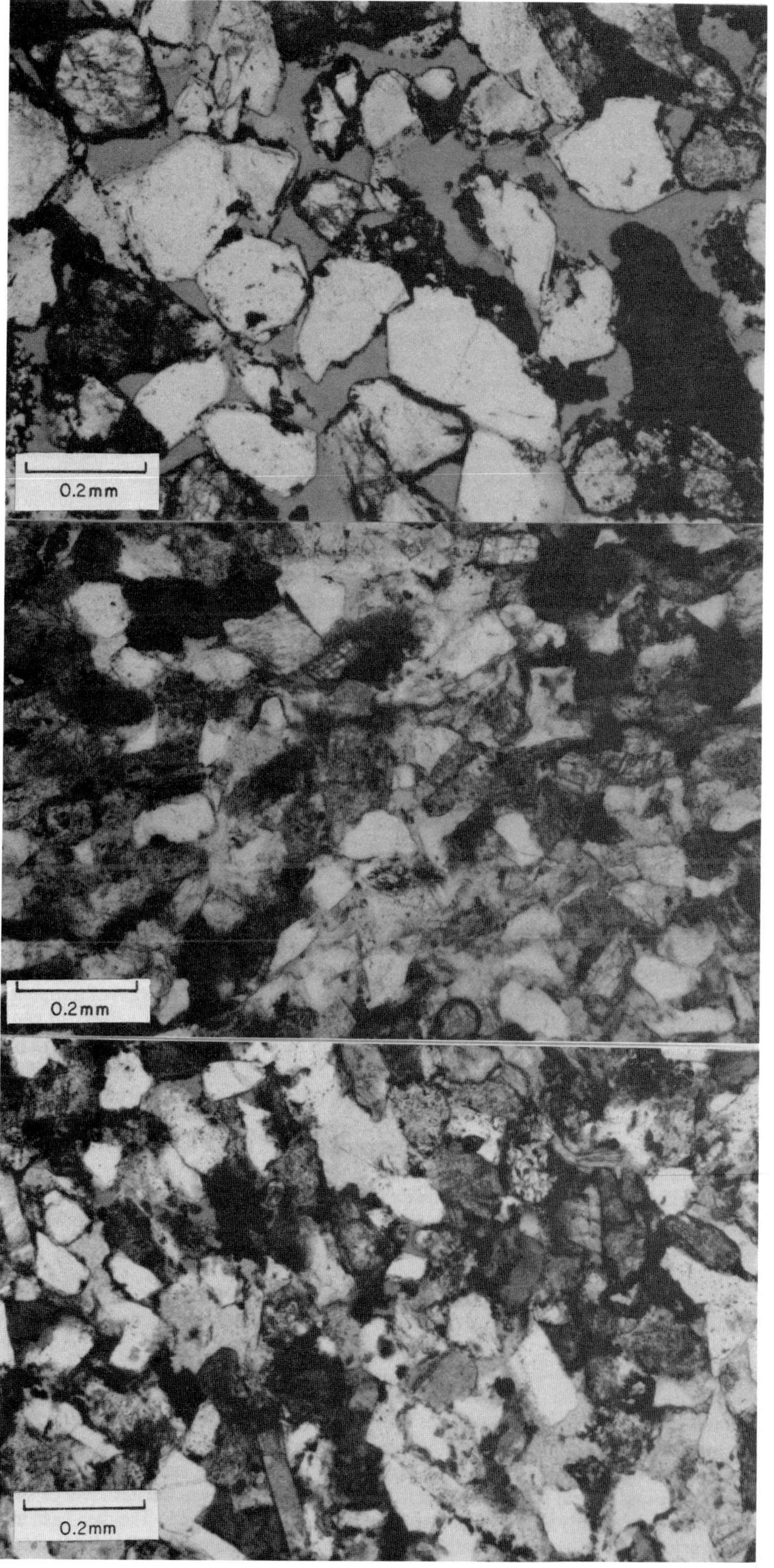

Figure 7-16. Photomicrograph of sandstone samples of the Gallup Sandstone at White Rock Canyon, profile NM-4-91B. Note the large pores and mature composition of sample 3 (estuarine fill) as compared to sample 2 (shoreface).

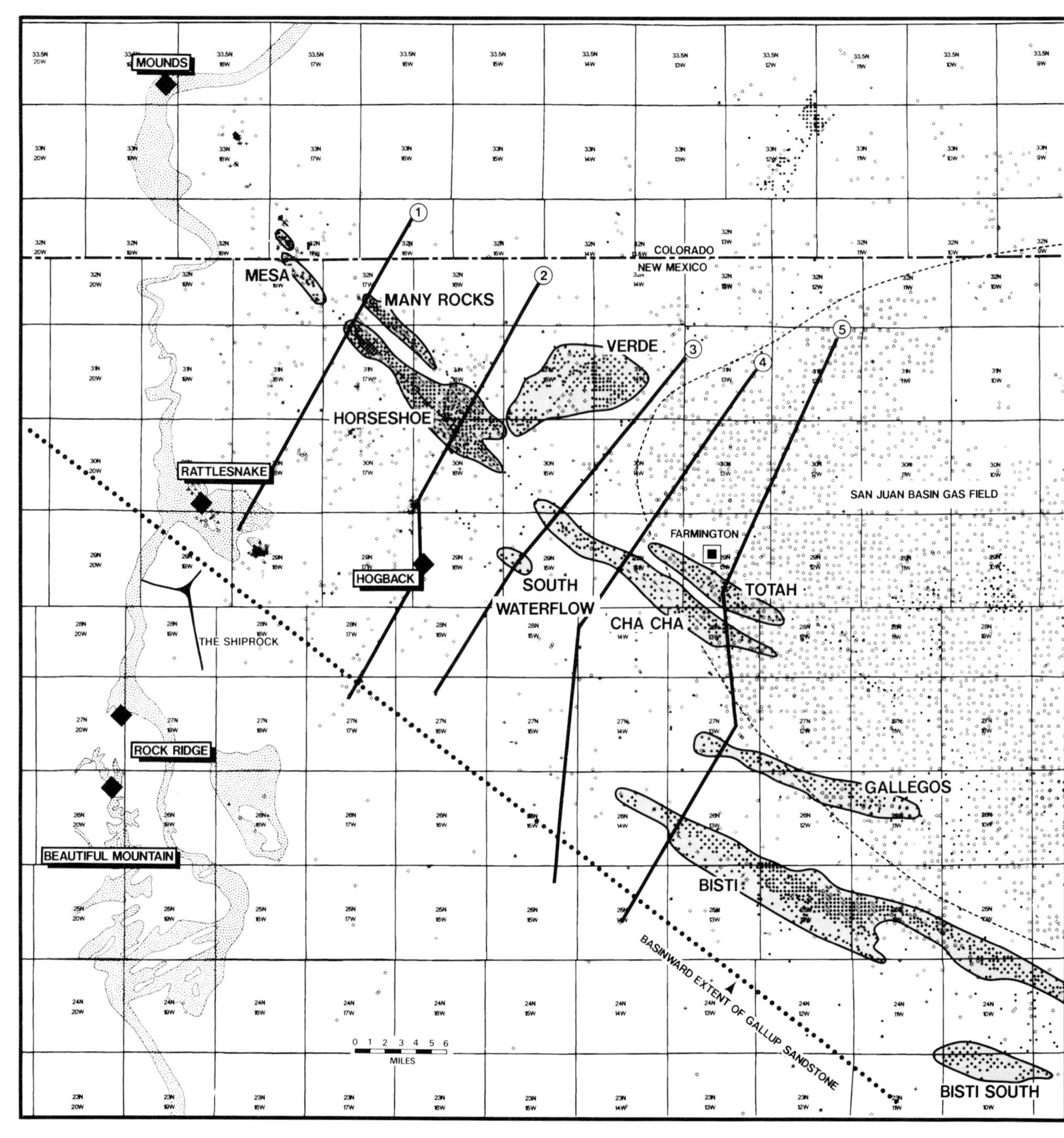

Fig. 2--Map of northwestern New Mexico and the western margin of the San Juan basin. Northwest-southeast-trending Tocito oil fields are outlined and labeled. Seaward limit of Gallup shoreface strata is also shown (from Molenaar, 1983a). The San Juan Basin gas field (chiefly Dakota production) forms the eastern boundary of the study area. The western outcrop belt of the Gallup and Tocito sandstones and key localities are shown. The heavy northeast-southwest lines labeled 1 through 5 mark the position of regional cross sections shown in Figure 5.

sequence boundary across the northern part of the study area. Two linear erosional lows or paleovalleys are cut into the Mancos shale and are named Horseshoe (southern valley) and Many Rocks (northern valley) after formal field names. The Horseshoe valley is the deeper of the two with up to 40 ft of erosional relief. The Many Rocks valley has around 20 ft of incision into underlying strata. Both have asymmetric cross sections, with steeper slopes on the southern edges and more gently sloping northern edges. The sequence boundary is commonly coincident with the Juana Lopez Member of the Mancos Shale, particularly along the Many Rocks valley (Fig. 7). McCubbin (1969) inferred that the calcareous sandstone beds of the Juana Lopez were more resistant to erosion in comparison to the overlying Mancos shale, which forms the steep, southern wall of the Horseshoe valley.

The Tocito-1 sequence boundary is traceable northeast of the Many Rocks field. The surface gently rises in this direction (Fig. 5, Fig. 6). Lamb (1968) reported a similar stratigraphic rise away from the Juana Lopez where the Tocito crops out far to the north at Mesa Verde, Colorado. South of the Horseshoe valley, the sequence boundary rises much more rapidly and is eventually truncated by the overlying Tocito-2 sequence boundary. Southwest of this line of truncation, no record of the Tocito-1 sequence is preserved.

Fill Style. Variable log patterns typify Tocito-1 lowstand strata. Horseshoe valley is characterized by both fining- and coarsening-upward SP patterns, but generally blocky gamma-rays suggest the absence of pure shale interbeds. Figure 8 summarizes facies data from the Solar 151-F core from the Horseshoe valley. A sharp basal contact separates the open-marine rippled siltstones and black shale of the Mancos from the poorly sorted, medium- to coarse-grained glauconitic sandstones that typify the Tocito. This basal contact is considered by previous workers to represent a significant unconformity (Penttila, 1964; McCubbin, 1969; Tillman, 1985a). The oil-producing facies from the cored interval consists of beds of trough- to sigmoidal- cross bedded, medium- to coarse- grained sandstone. Current-rippled beds with clay drapes, clay rip-up clasts and bioturbated fine-grained sandstones are interbedded with the crossbeds. Beds are typically thin (<1 ft thick), with sharp bases. Tangential cross lamination grades rapidly upward into highly burrowed tops. Glauconite is abundant thoughout with minor phosphatic and quartz pebbles, sharks teeth and *Inoceramus* fragments. Thin, delicate leaf imprints resembling pine needles occur along shale partings. The small-scale (< 1 ft thick) variations in cross bedding with clay-drapes and bioturbated intervals indicates deposition under rapidly fluctuating energies. We observed no wave-generated structures such as wave ripples or hummocky stratification in this or any of the Tocito lowstand sandstones. The overall fining-upward profile culminates at the top in a burrow-churned shaley sandstone. Trace fossils are dominated by horizontal *Paleophycus* burrows with lesser *Skolithos*, *Thalassinoides* and *Ophiomorpha* burrows. This relatively low diversity, high abundance assemblage suggests a stressed environment (G. Pemberton, pers. comm, 1990) and is consistent with an estuarine setting. We infer stress may have been induced by fluctuations in salinity, temperature or rates of sedimentation. Observations from nearby cored wells are consistent with the Solar F-151 well (i.e. Humble Oil and Refining Co. wells: F-32, sec 3-31N-17W; F-33, sec 9-31N-17W; F-39, sec 10-31N-17W; G4, sec 11-31N-17W). Cores in this area suggest that bioturbation significantly increases toward the valley margins.

Strata from the Tocito-1 sequence crop out at the Mounds locality in Colorado (Fig. 6). This outcrop locality is on trend with the Many Rocks valley and is interpreted to represent an updip facies of the Many Rocks valley fill. At the Mounds locality, a 10 ft thick, medium- to coarse-grained sandstone rests sharply on distal marine shale and siltstones of the Lower Mancos (Fig. 9). Phosphatic pebbles and caliche nodules are concentrated along the base of the sandstone and sharks teeth and glauconite are dispersed in the lower portion of the unit. Tabular to tangential crossbeds are common and display a uniform, southeasterly paleocurrent direction (120° average vector). Lamb (1968) and McCubbin (1969) observed similar paleocurrent distribution at this locality. We interpret these coarse-grained, cross-bedded sandstones to mark a pronounced basinward shift in facies associated with the Tocito-1 sequence boundary. Tocito strata are likewise interpreted as cross-bedded estuarine deposits.

Shales within the Tocito-1 valley fill contain a diagnostic late Coniacian palynology assemblage (see section on Biostratigraphy). This information is critical in the chronostratigraphic understanding of the Tocito-1 unconformity at the base of the Tocito. At the Mounds locality, the Tocito-1 sequence boundary overlies the Juana Lopez Member of the Lower Mancos Shale which is of middle late Turonian age. Thus, a significant thickness (approximately 400 ft) which is present in Beautiful Mountain outcrops is missing below the Horseshoe and Many Rocks valleys (cf. Penttila, 1964, Lamb, 1968, McCubbin, 1969). This clearly demonstrates the unconformable nature of the basal Tocito surface and places the oldest mapped Tocito valley stratigraphically above the late Turonian-early Coniacian strata of the marine Gallup.

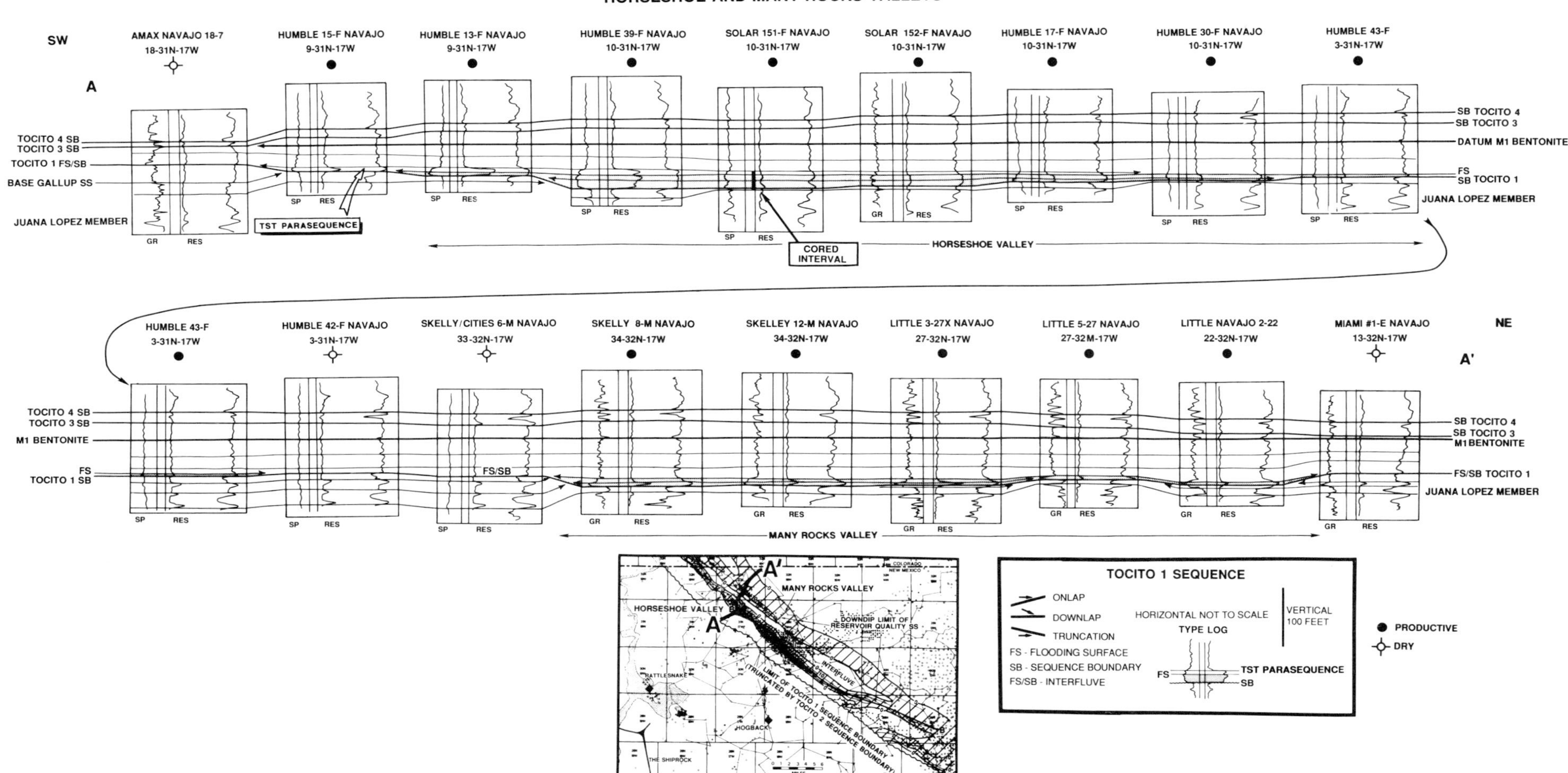

Fig. 7--Transverse cross section across the Tocito-1 incised valleys, Horseshoe and Many Rocks field. The datum is the M1 bentonite of McCubbin (1969). Note the erosional relationship with the underlying Juana Lopez Member of the Mancos Shale. The Horseshoe valley is capped by a thin parasequence that was deposited over the valley fill during the initial transgressive flooding across the valley. This unit downlaps to the northeast (basinward) and onlaps the Tocito-1 sequence boundary to the southwest. The core for the Solar 151-F well is shown in Figure 8.

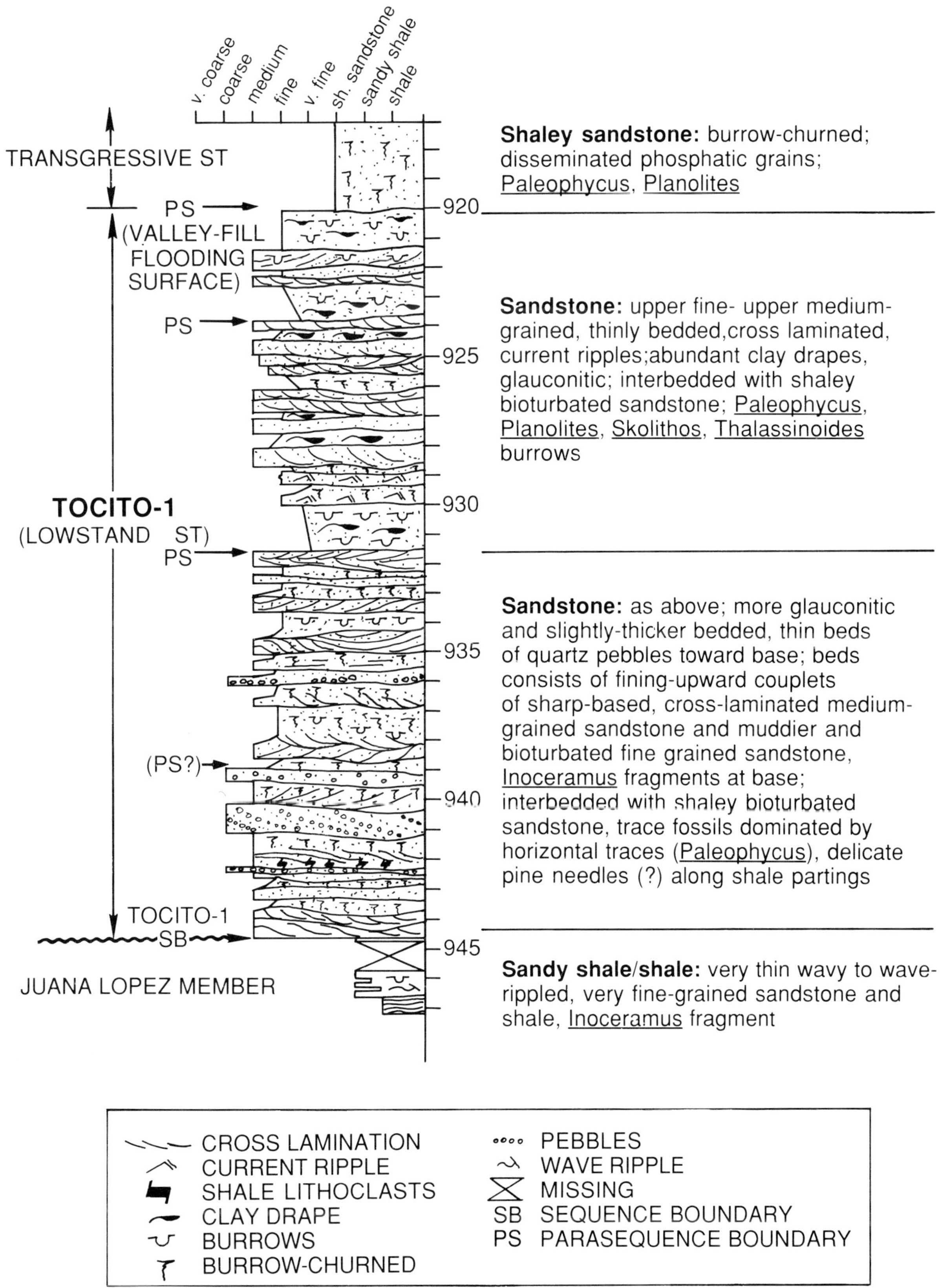

Fig. 8--Lithofacies and sequence-stratigraphic summary of the core from the Solar 151-F well. Sandstones of the Tocito-1 sequence sit unconformably on shale and fossiliferous, fine-grained sandstone beds of the Juana Lopez Member.

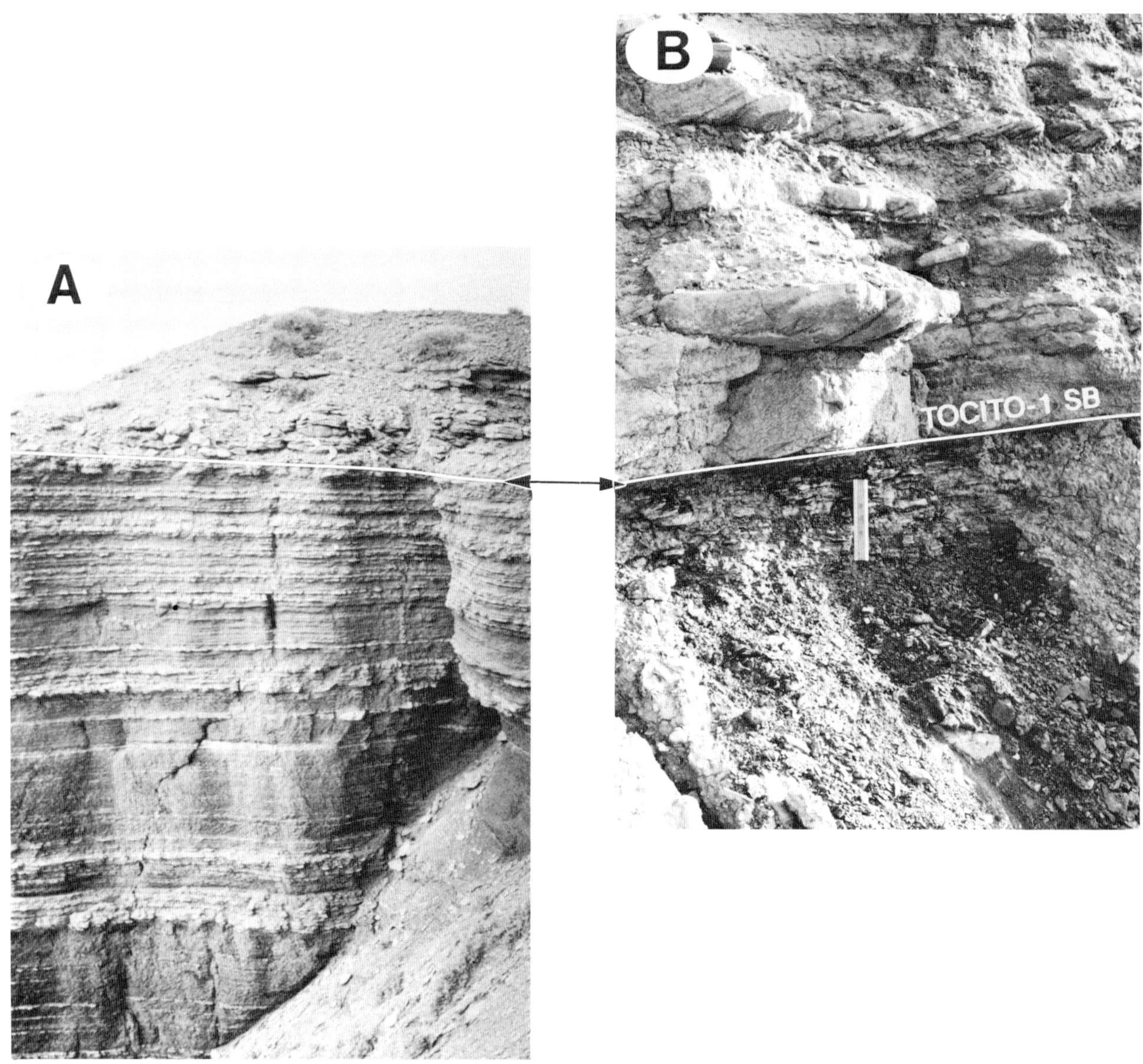

Fig. 9--Outcrop photograph of the Mounds locality (SE NE sec 26-33.5N-19W). **(A)** Distant view show thin sandstone and shale beds of the Juana Lopez Member and the Tocito-1 sequence. **(B)** Detailed of the sequence boundary where coarse-grained sandstones sit sharply on the distal shales of the Mancos Shale. Tabular cross bedding shows a prevailing flow direction to the southeast (120°).

Transgressive Flooding Surface. A thin (10 ft maximum thickness) parasequence caps the main sand body of the Horseshoe valley (Fig. 7). The parasequence is present along the entire length of the Horseshoe valley and gradually thins and downlaps to the northeast. Figure 7 shows the thin parasequence at the top of the sandy valley fill. The unit also terminates to the west by onlap onto the stratigraphically rising Tocito-1 sequence boundary. This parasequence is interpreted as a shoreline deposit that formed during the immediately after the transgressive flooding of the valley. The parasequence is by definition part of the transgressive system tract and not part of the lowstand valley fill. The transgressive flooding surface in the Tocito-1 sequence is placed at the top the valley-fill sandstone to the north (Many Rocks valley), and at the base of the parasequence when present (Horseshoe valley). This parasequence causes the top of the sand to appear to rise stratigraphically relative to the M1 marker (cf. McCubbin, 1969).

Shale-rich marine strata overlie the parasequence and valley-fill flooding surface. The marine shale also contains regionally widespread resistivity markers (M1 marker of McCubbin, 1969; N1, this study) which show apparent onlap onto the parasequence and valley-fill flooding surface (Figs. 5 and 7). We interpret this apparent onlap as downlap onto a progressively northward-tilted valley-fill flooding surface. Tilting occurred prior to and possibly during deposition of the muddy highstand strata. Thus the relative rise in sea level that ended lowstand valley fill sedimentation and initiated the relative transgression may have been induced by a pulse of subsidence. All of the marker beds (M1 and N1) in the muddy highstand strata either lap out to the southwest onto the sequence boundary or are truncated by the overlying Tocito-2 sequence boundary. The pronounced northeast dip on these markers and the Tocito-1 valley-fill flooding surface seen in Figure 5 is related to this post-depositional tilt. The influence of syndepositional tectonism on Tocito deposition is discussed below.

Map Distribution. Distribution of the Tocito-1 lowstand system tract is characteristic of incised valley-fill patterns (Van Wagoner et al., 1990). Gently sinuous fairways with corresponding axial thicks become less sinuous in the downdip direction. Valleys also exhibit a tributary-like merging in this direction. Asymmetric and commonly steep-sided profiles suggest valley incision by entrenched fluvial systems. Sinuous axial thicks may be the product of thalweg-generated deeps. Relief on valley edges decreases to the south as valleys become broader and shallower and sandstone thickness correspondingly decreases in this direction for each valley. Figure 10 illustrates a longitudinal profile down a valley axis. In the Horseshoe valley, resistivity markers downlap to the southeast toward the inferred mouth of the estuary. Reservoir quality of valley-fill sandstones diminishes in the downdip direction. The limit of reservoir facies is shown in Figure 7. Horseshoe valley sandstones extend further down to the southeast than sandstones from the Many Rocks valley.

Tocito-2 Sequence

Sequence boundary. The irregular erosional surface of the Tocito-2 sequence boundary is much like the Tocito-1 sequence boundary but the Tocito-2 surface is more regionally extensive. Four southeast-trending valleys are mapped and include three relatively narrow valleys called Cha Cha, Totah and Bisti and a broad valley named Waterflow (Fig. 11). Well-defined stratal truncation of Tocito-1 sequence and Gallup strata occurs below the Tocito-2 sequence boundary. Along the northern and southern limits of the study area, the Tocito-2 sequence is truncated by the overlying Tocito-3 sequence boundary.

Fill style. Tocito-2 valley fills are generally sand-prone. In the extreme southeastern portion of Cha Cha, Totah and Waterflow valleys, sandstone is restricted entirely to the lowstand systems tracts. Low-resistivity marine shale of the transgressive and highstand systems tracts overlie theses sandy units. When traced updip to the northwest, these shales change facies into sandy parasequences. This facies relationship in the Tocito-2 sequence is shown in Figure 10.

Tocito-2 valley-fill sandstones are similar to those described from the Tocito-1 valley fill. Cross-bedded, medium- to coarse-grained sandstones are common in the Humble L-9 and L-10 cores from the Cha Cha valley (sec. 36-29N-14W). Glauconite, phosphate, quartz pebbles, *Inoceramus* and oyster shell fragments are abundant.

The Tocito exposure along the Hogback anticline is situated along the southern margin of the Waterflow valley (Fig. 11). The sandstone at the base of this measured section is correlated as a tributary to the main valley. This channel sandstone displays sigmoidal-cross bedding with occasional clay drapes which suggests a tidal origin for the Tocito-2 valley-fill deposits.

Valley-Fill Flooding Surface. The Tocito-2 sequence has a well-developed flooding surface across Cha Cha and Totah valleys (Fig. 12). Similar to the Tocito-1 sequence, marine shales with interbedded bentonites downlap onto the top of the Tocito-2 valley-fill sandstones in this area. The Tocito-2 flooding surface may have been likewise affected by structural tilting to the north because overlying strata appear to progressively onlap in a southerly direction (Fig. 12).

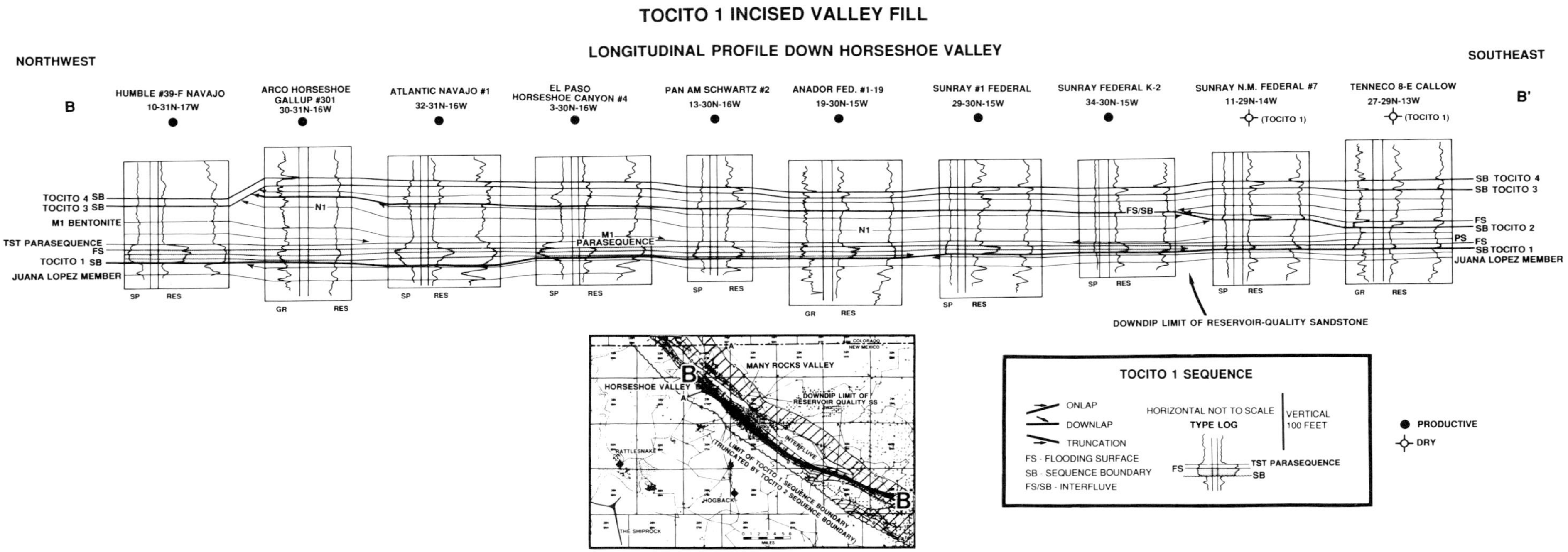

Fig. 10--Longitudinal cross section down the Horseshoe valley. Sandstone thickness decreases to the southeast by progressive downlap. Note lateral connectivity of sandstones down the valley.

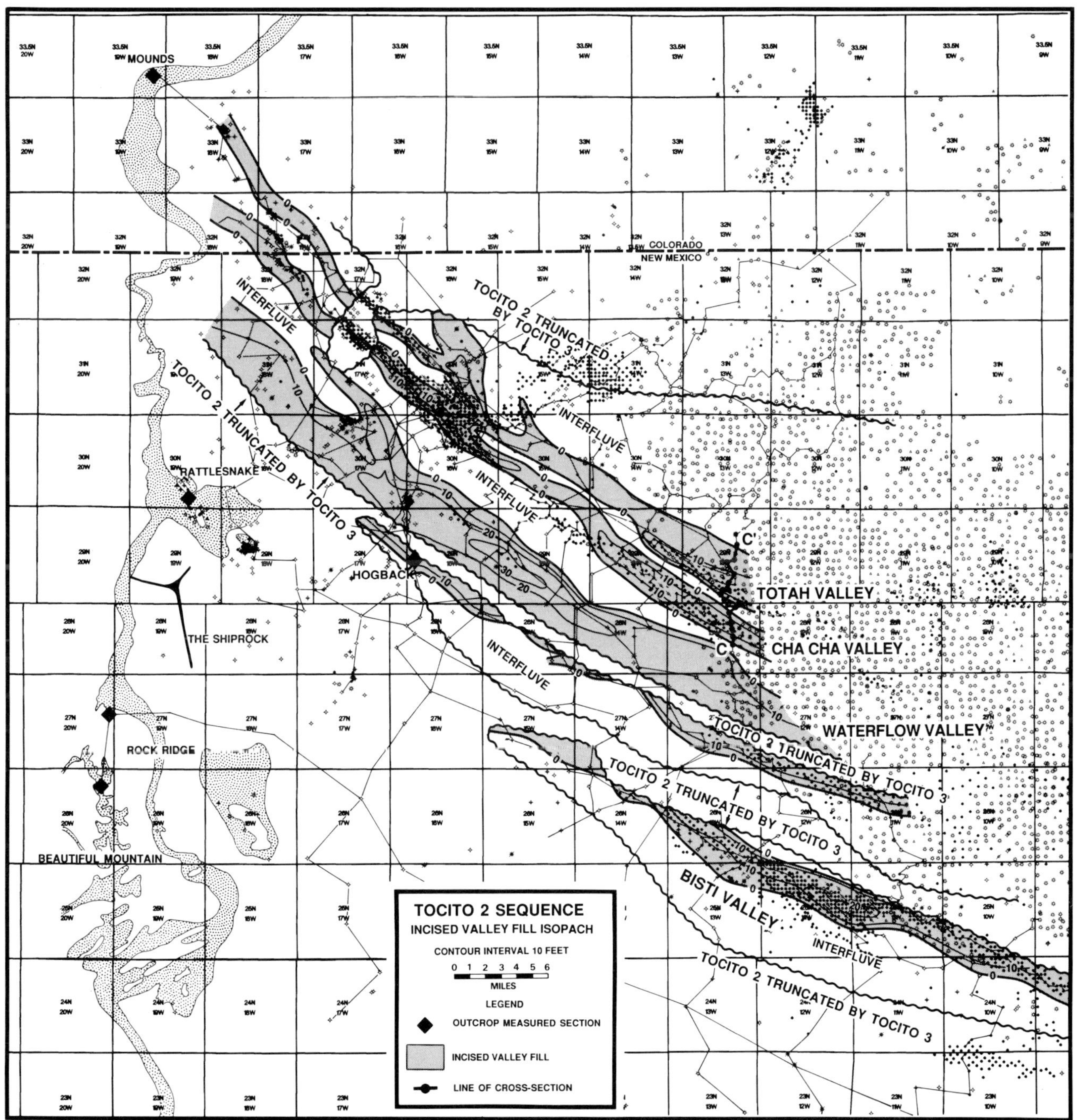

Fig. 11-- Isopach contour of the incised-valley fill for the Tocito-2 sequence. Four parallel valleys separated by interfluves are mapped. The Waterflow valley contains the thickest sandstones. Note the distribution of valley-fill sandstones is more widespread than the Tocito-1 sequence. The Tocito-3 sequence boundary locally incises into and removes the Tocito-2 sequence. This is particularly evident along the southern margin of the Waterflow valley and northern margin of the Bisti valley. These narrow bands of truncation correspond to axial thicks in the Tocito-3 sequence. This erosional relationship has led to a number of hydrocarbon traps in this vicinity. A thin interval of the Tocito-2 sandstone crops out at the Hogback anticline locality.

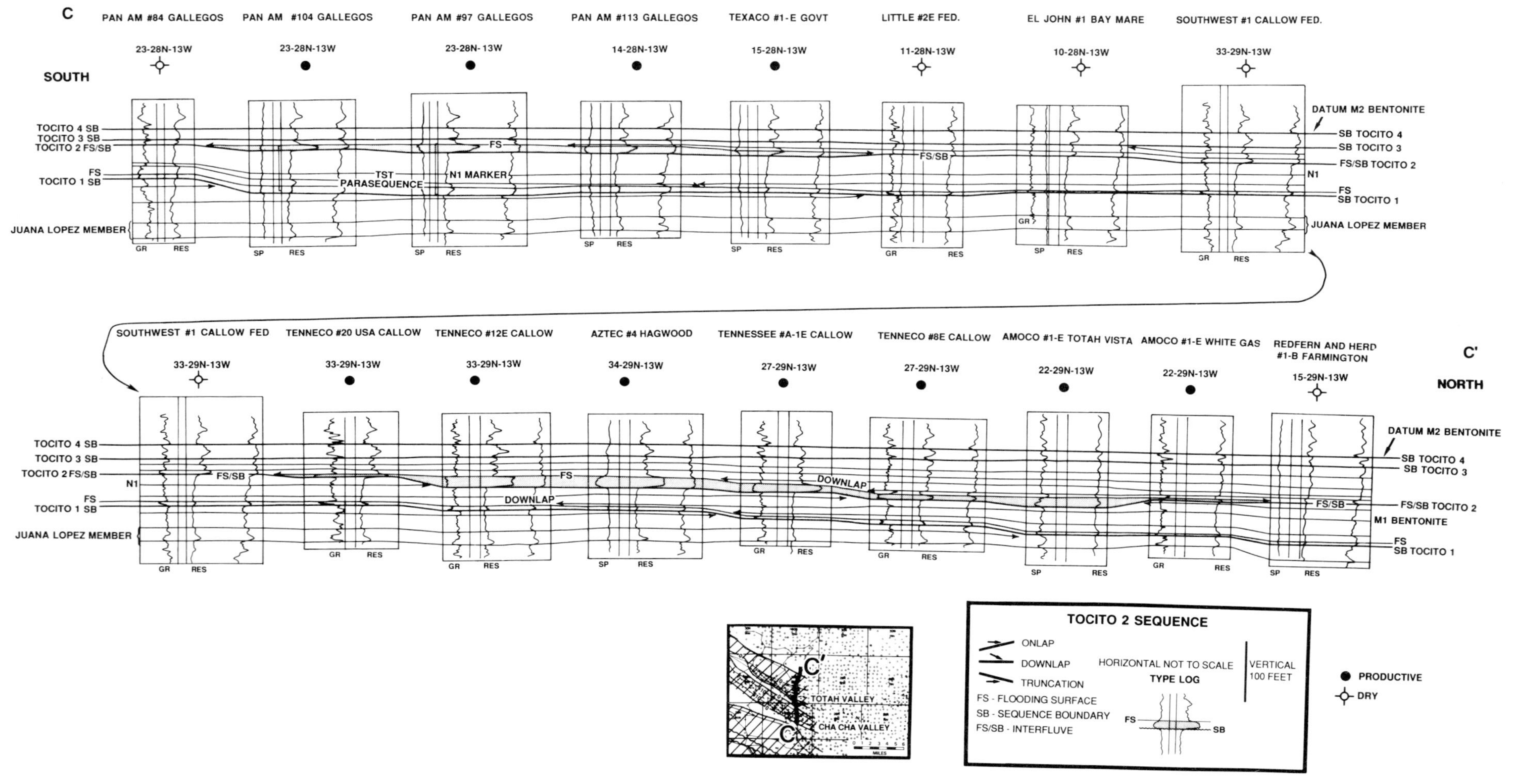

Fig. 12--Transverse cross section across the Totah and Cha Cha valleys, Tocito-2 sequence. This section forms a traverse across the Totah and Cha Cha oil fields. The M2 bentonite of McCubbin (1969) serves as the datum.

As in the Tocito-1 transgressive flooding surface, this stratal relationship was caused by minor tilting to the north.

Map Distribution. The branching array of valleys arose through the entrenchment of a relatively flat surface by rivers. Tocito-2 valleys become slightly wider and less sinuous downdip to the southeast. Reservoir-prone facies within Tocito-2 valleys extend farther seaward relative to those in the Tocito-1 sequence.

The Tocito-2 sequence has a much wider areal extent than the underlying Tocito-1 sequence. In addition, a wide age-range of strata is truncated by the Tocito-2 sequence boundary. The Tocito-2 sequence boundary cuts into Tocito-1 strata in the northern part of the mapped area, into lower Mancos shale and lower Gallup strata in the central portion and into the Torrivio Sandstone in the southern part of the study area. The lowstand sands of the Tocito-2 sequence are the same as the "upper sandstone beds" of McCubbin (1969), who documented an unconformable basal contact along the Waterflow valley. This study supports the northern continuation of this erosional surface below these sands and can be followed far to the northeast until it is truncated by the Tocito-3 sequence boundary.

Tocito-3 Sequence

Sequence Boundary. The Tocito-3 sequence boundary is the most irregular of the Tocito sequence boundaries. Erosional lows range from broad and relatively shallow (Verde valley, 12 miles in width) to very deep and V-shaped (Waterflow Valley, 3 miles in width) (Fig. 13). The Waterflow valley is also limited in the updip direction. It thins from 70 ft to 10 ft over a distance of eight miles.

The erosional incision by the Tocito-3 sequence boundary has profound effects on the distribution of the Tocito-2 sequence. Erosion at the base of the Tocito-3 sequence is responsible for completely truncating the Tocito-2 sequence in the far northern and southern portions of the mapped area. The Tocito-3 sequence boundary also incises through Tocito-2 valley fill along a narrow band at the southern edges of the Waterflow and Bisti valleys (Figs. 11, 13). This erosional pattern has created many of the hydrocarbon traps associated with the Tocito-2 valley system.

Fill Style. Tocito-3 valleys have highly variable fill styles. Verde and Waterflow valleys are dominated by marine shale and siltstones with interbedded bentonites which onlap onto the valley edges (Fig. 14). Sandstones are poorly developed to absent in these valleys. The Bisti valley trend contains thin sandstones updip but becomes shaley in the downdip direction to the southeast, particularly where the valley fill thickens to 50 ft (Fig. 13). The medium- to coarse-grained sandstone contains low-amplitude trough cross beds, disseminated glauconite and abundant *Inoceramus* fragments.

The Hogback outcrop contains a thin interval of the Tocito-3 sequence. At this locality, the Tocito-3 sequence is a bioturbated sandy mudstone with scattered glauconite, phosphate pebbles and *Inoceramus* fragments. In addition, the Tocito-3 sequence has been correlated into the outcrop at Beautiful Mountain where it consists of a fine-grained coastal plain facies. The Tocito-3 sequence maintains its erosional basal boundary in this area. The Tocito-3 sequence boundary locally truncates the Torrivio Sandstone.

Valley-Fill Flooding Surface. Over most of its extent, the transgressive flooding surface that overlies the Tocito-3 valley-fill has been truncated by the overlying sequence boundary (Tocito-4). Consequently, the map of the Tocito-3 sequence is different from the other Tocito sequences. Rather than a valley-fill isopach map, the map of the Tocito-3 sequence is a total-thickness isopach map of the interval from the the Tocito-3 sequence boundary to the overlying Tocito-4 sequence boundary.

Map Distribution. Thicknesses greater than 10 ft are highlighted on the Tocito-3 sequence map. A wide range of valley shapes is evident (Fig. 13). The V-shaped Waterflow valley may be an example of a valley which formed by headward erosion. The sequence boundary cut deeply into the underlying Tocito-2 sequence and completely erodes through the Tocito-2 sequence along a relatively straight and narrow band (Fig. 11). Bisti valley exhibits similar truncation along a linear trend. The straight, near vertical edges of both these Tocito-3 valleys suggest control by syndepositional normal faults which is discussed later. In addition, the Tocito-3 sequence is very thin to absent along a northwest-southeast-oriented ridge that defines the southern edge of the Waterflow valley (Figs. 12, 13). McCubbin (1969) recognized this same feature through subsurface correlation and interpreted it as as a paleotopographic rise or cuesta-like ridge.

Tocito-4 Sequence

Sequence Boundary. The Tocito-4 sequence boundary is the most widespread and regionally extensive of the four sequence boundaries in the study area. Three broad erosional lows are mapped on the Tocito-4 sequence boundary. The Tocito-4 sequence boundary is characterized by less truncation of underlying strata than the older Tocito sequences. In fact, clear evidence of truncation is limited to three zones: in the

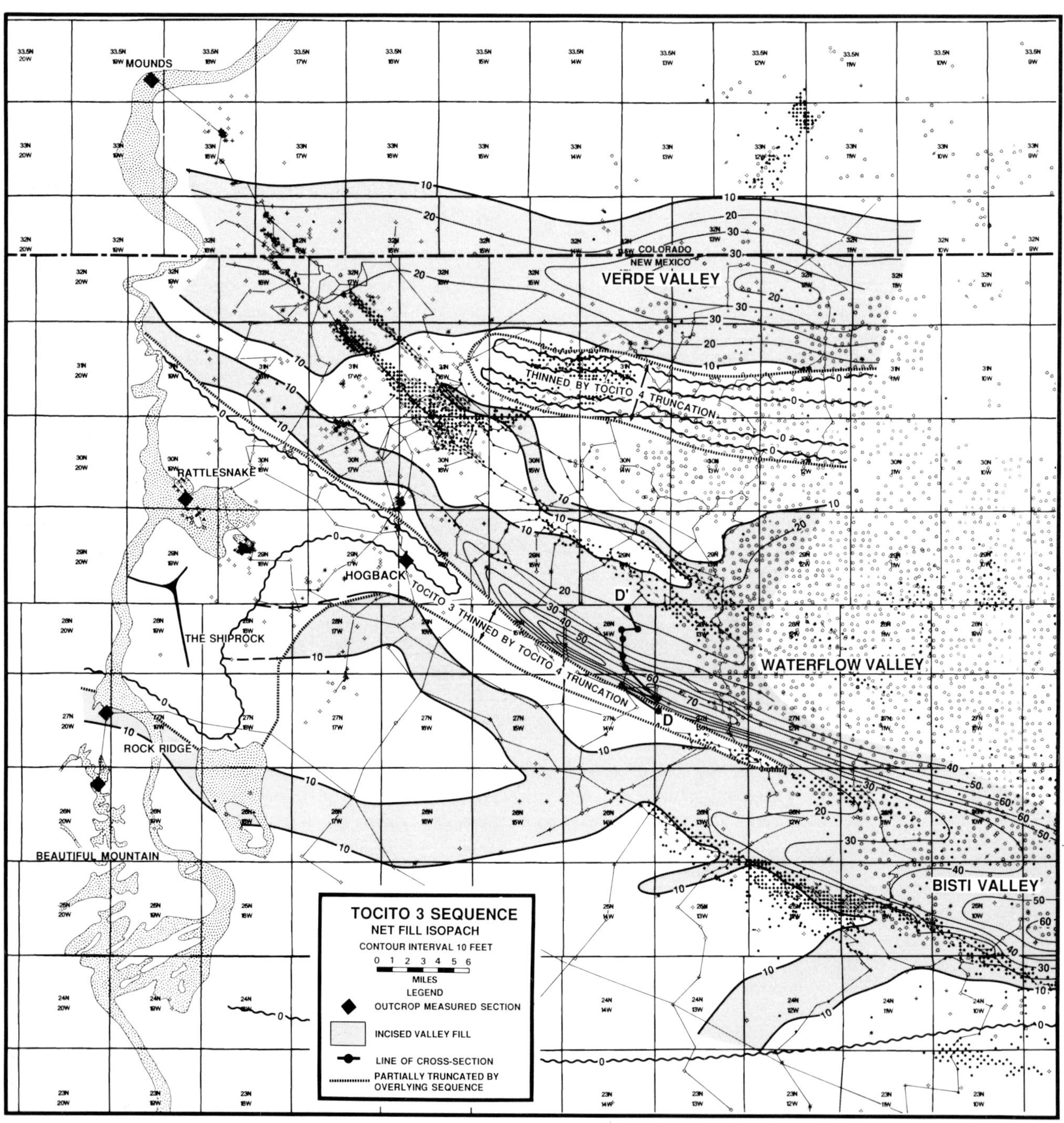

Fig. 13--Isopach map of the Tocito-3 sequence. The interval mapped is from the Tocito-3 sequence boundary to the Tocito-4 sequence boundary. Interval thickness of 10 ft and greater is shaded. A different array of valley shapes is evident: the broad Verde valley, the deep, v-shaped Waterflow valley and the asymmetric Bisti Valley. Note areas thinned by truncation by the overlying Tocito-4 sequence boundary, particularly along the southern margin of the Waterflow valley and toward the outcrop area.

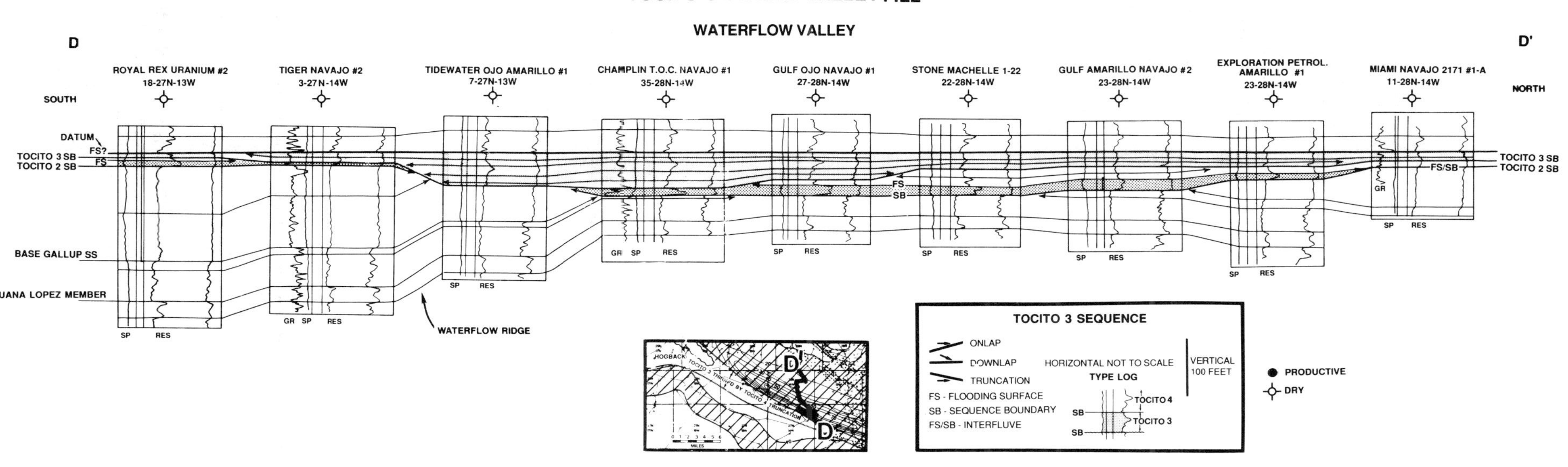

Fig. 14--Transverse cross section across the Waterflow valley. The Tocito-3 sequence boundary truncates into the underlying strata of the Tocito-2 valley-fill. Along this axis of truncation, the Tocito-2 valley fill is removed by erosion and strata of the Tocito-3 sequence rest on the Gallup strata. Toward the south, the Tocito-3 sequence thins rapidly by onlap as the sequence boundary rises. Note this is consistent with the rapid rise in the Juana Lopez and Gallup markers toward the north. This pattern in the Gallup and Juana Lopez beds is persistent along the entire length of the study area and is probably evident along the outcrop north of the Rattlesnake anticline.

north between Verde and Waterflow valleys (Figs. 13, 15); along the southern margin of the Waterflow valley, and to the southeast along the outcrop belt from Beautiful Mountain to the Rattlesnake Anticline. The isopach map of the Tocito-3 sequence shows the location of this truncation (Fig. 13). Given these relationships, the Tocito-4 may be more accurately characterized as a series of large embayments which are tectonically controlled rather than an incised valley. Recognition of the Tocito-4 sequence boundary is therefore defined by gentle truncation of underlying strata and by onlap. North of the Waterflow valley, the sequence boundary is marked by the M2 bentonite.

A strong angular discordance is observed at outcrops north of the Rattlesnake anticline (secs 6- and 31-31N-19W). These outcrop localities project into the subsurface trend associated with the southern margin of the Waterflow valley where a high degree of truncation is observed (Fig. 5).

Fill Style. The broad, deep, erosional lows that were created by the Tocito-4 sequence boundary are filled with parasequences that exhibit an aggradational to progradational stacking pattern (Fig. 16). This vertical arrangement is observed in outcrop across the study area. The base of the unit is marked by a thin (0-6 in), coarse, basal lag which consists of a mixture of granular sand, phosphate and calcareous pebbles, and shell fragments. This is overlain by tan mudstone and thin-bedded, highly bioturbated sandstone. The unit coarsens and thickens upward into rippled and sigmoidally cross-bedded sandstones with abundant phosphate pebbles and *Inoceramus* fragments. In the subsurface, the Tocito-4 valley fill is muddier than the cross-bedded fill found in Tocito-1 or Tocito-2 valleys. The Tocito-4 parasequences are widely correlatable across the study area. Regionally persistent bentonites, such as the M2 bentonite (McCubbin, 1969), also characterize the fill, particularly in lower portions. This stratigraphy suggests more distal but still tidally-influenced sedimentation within much wider and broader valley systems. As valley fills thin toward interfluves, the number of parasequences decreases through onlap. This is observed in both the subsurface and outcrop where only one or two tidal parasequences occur within the Tocito-4 valley fill.

Valley-Fill Flooding Surface. The flooding surface is defined by the marine shale at the top of the valley-fill parasequence set (Fig. 16). This surface onlaps and merges with the sequence boundary to mark the edges of interfluve areas. The flooding surface is locally overlain by thin parasequences that exhibit a retrogradational or backstepping pattern.

Map Distribution. The broad valleys are oriented northwest-southeast. Three erosional axes are mapped and range in width from about 7 miles (Rattlesnake) to 25 miles (Waterflow and Verde valleys). The relatively straight and wide Tocito-4 valleys contrast with the narrow and sinuous valleys found in the Tocito-1 and Tocito-2 sequences.

The greatest thickness of Tocito-4 valley fill is located along the southern margin of the Waterflow valley. The Tocito-3 isopach displays a similar relationship along the same axis. The superposition of valley-fill thicks through time suggests a persistent, syndepositional structural control on the erosion of these two features. The linear and asymmetric geometry of the Tocito-4 fill along the Waterflow valley suggests the presence of a down-to-the north fault. Active growth along faults could act as a catchment for river systems in the vicinity. Thus, the Waterflow valley may be an example of an asymmetric valley in which erosion and deposition may have been influenced by increased accommodation along the down-dropped segment of an active fault.

BIOSTRATIGRAPHIC SUMMARY

Lower Mancos Shale/ Gallup Sandstone

The Juana Lopez Member of the Lower Mancos Shale is the oldest unit correlated in this study. The unit is widespread across the San Juan Basin and serves as an excellent stratigraphic marker in both outcrop and subsurface correlations. Dane et al. (1966) assigned an early late Turonian age to the Juana Lopez based on ammonites. This age is corroborated by the collection of the subzonal ammonite *Scaphites whitfieldi* during outcrop studies.

The youngest strata of the marine Gallup are assigned a late Turonian to early Coniacian age by Molenaar (1973; 1983a). In-house palynology on marine Gallup samples from outcrop indicate a late Turonian age (Y.Y. Chen, EPR, pers. comm; Fig. 17).

The nonmarine Gallup did not yield palynomorphs. We postulate an early Coniacian age for this unit because of its stratigraphic position in the outcrop (cf. Molenaar, 1973).

Tocito Sandstone

The oldest Tocito sequence, Tocito-1, yielded an early late Coniacian palynomorph assemblage from cores in the Horseshoe Field (e.g. F-32, sec 3-31N-17W). Tocito-1 valley-fill sandstones are exposed at the Mounds outcrop in Colorado, where Dane (1960) identified *Inoceramus deformis* and interpreted a mid-Coniacian age for these rocks. These age data were instrumental in the correlation of

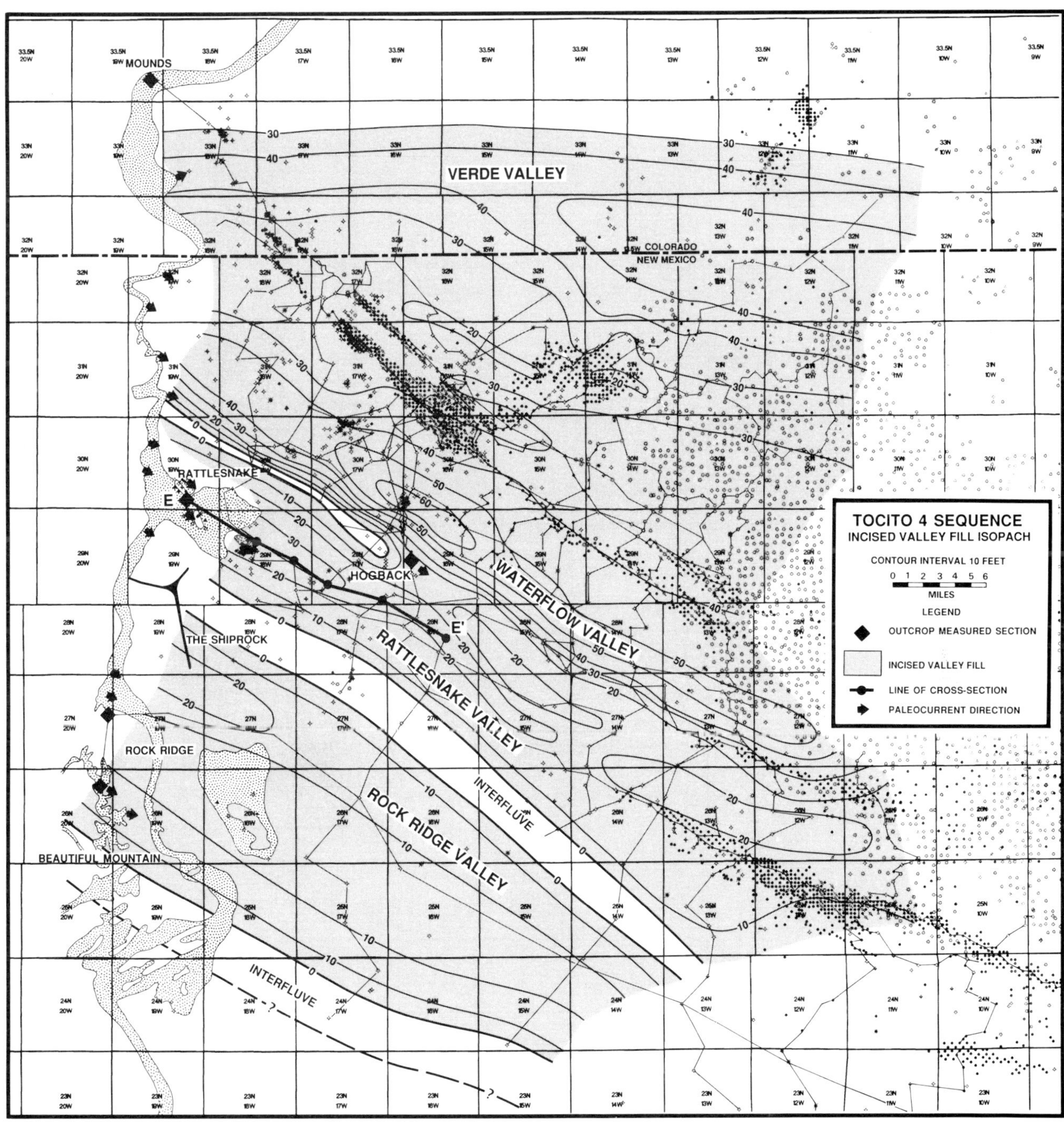

Fig. 15--Isopach map of the Tocito-4 incised valley fill sequence. Most of the map is represented by valleys with only minor, narrow interfluvial areas. Thins and thicks mapped in the outcrop correspond well with measured thicknesses of the Tocito at outcrop. Most of the Tocito cropping out along Rock Ridge and Beautiful Mountain are strata from the Tocito-4 sequence.

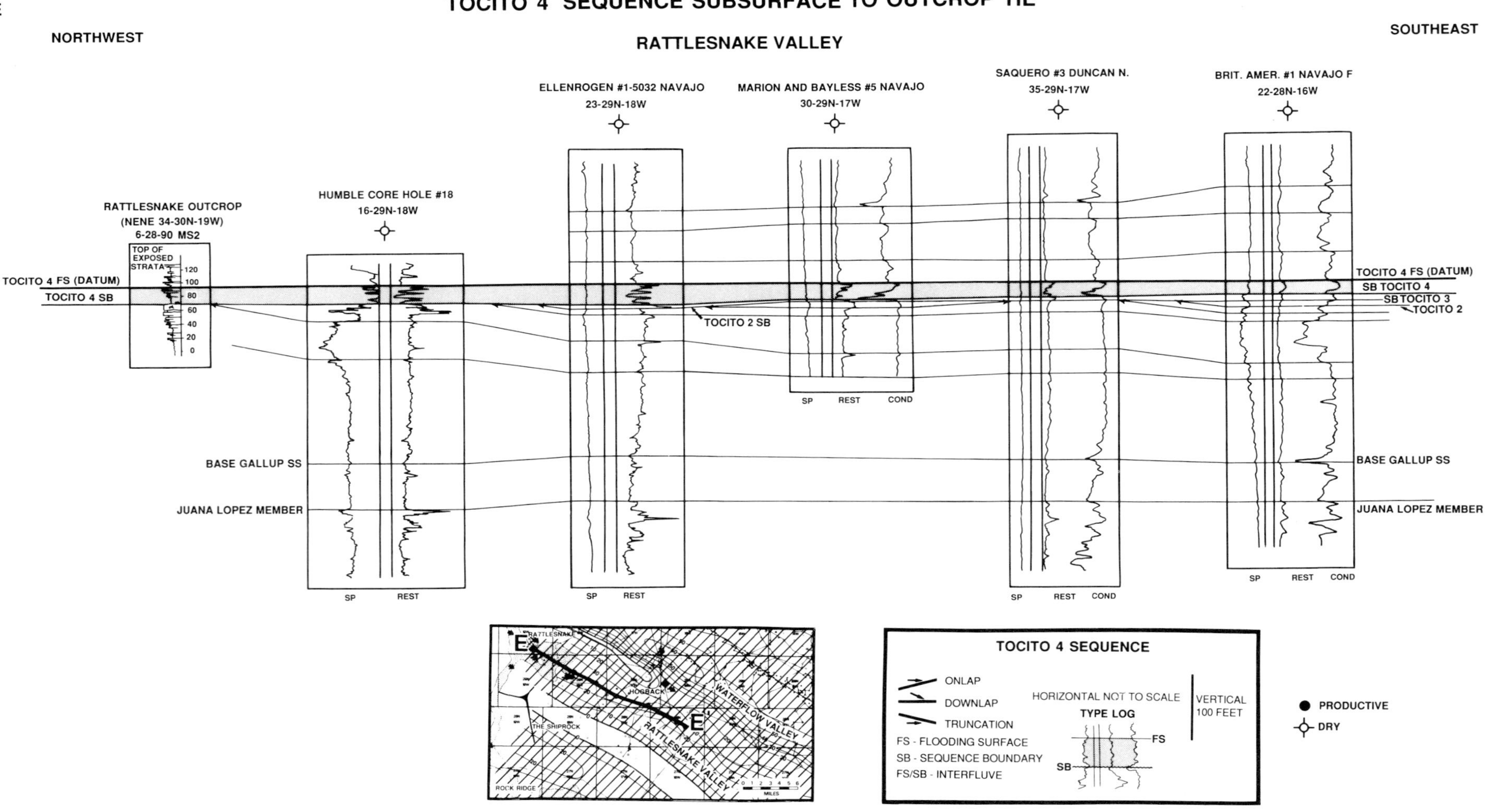

Fig. 16--Longitudinal cross section down the axis of the Tocito-4 Rattlesnake valley. Gradual truncation of underlying Tocito and Gallup strata is evident. The subsurface cross section is correlated into the Rattlesnake outcrop section. At this locality, a number of stacked tidal-bar parasequences comprise the Tocito and this pattern is readily correlated into the subsurface. Location of section shown on Figure 15.

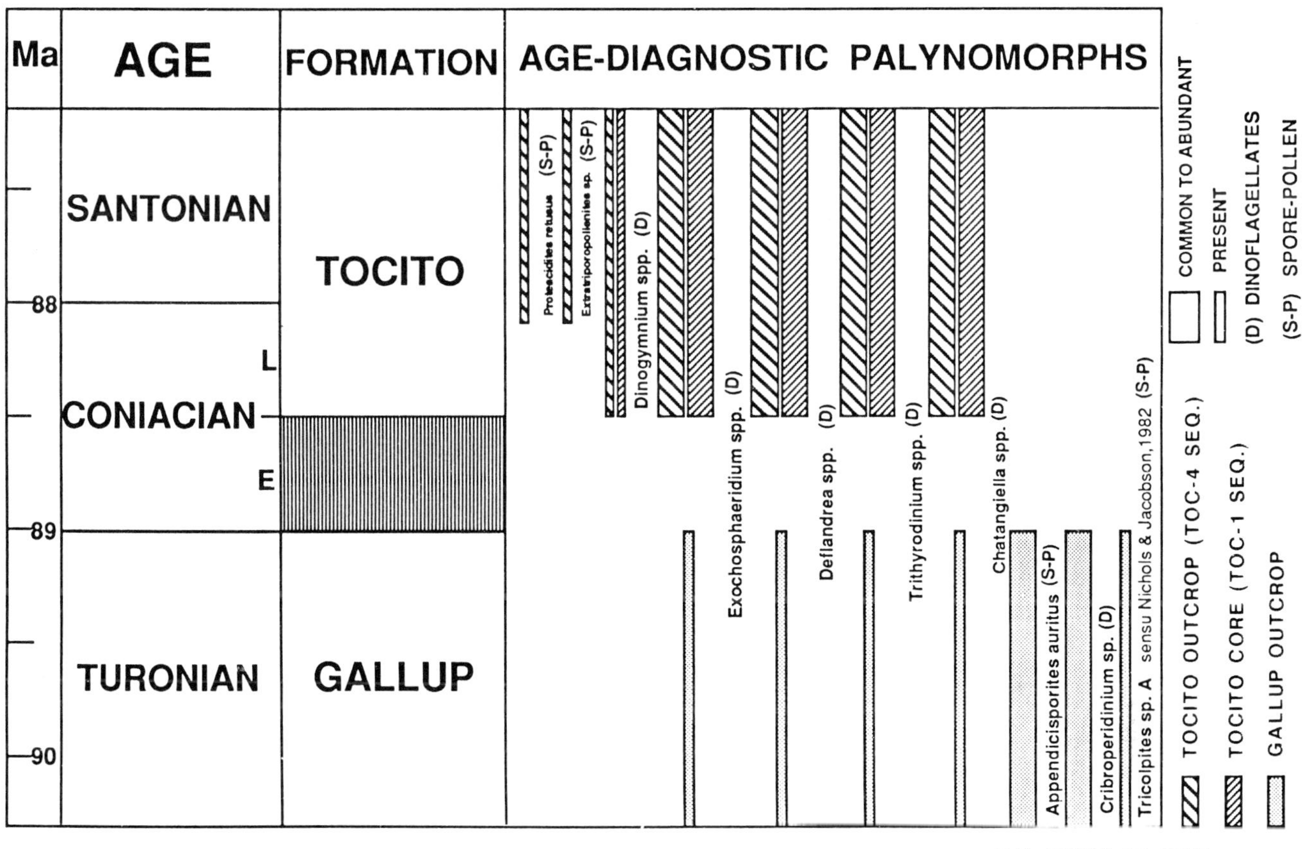

Fig. 17--Preliminary biostratigraphic interpretation of the Tocito using selected samples from the palynological data set. Samples were collected from both the outcrop (Tocito-4 sequence, Rock Ridge and Beautiful Mountain and subsurface (Tocito-1 sequence, Horseshoe Field). Most assemblages indicate a brackish, marginal marine environment with a terrigenous influx. Samples from the Tocito-1 sequence yield a late Coniacian age and the Tocito-4 samples are latest Coniacian to possibly earliest Santonian in age. Samples from the Gallup record a late Turonian age. An hiatus is found across every Gallup and Tocito contact. This preliminary work is based on interpretations by palynologist Y.Y. Chen (Exxon Production Research).

the Tocito-1 sequence in the subsurface because they precluded the correlation of any Tocito markers into the Turonian-age strata of the marine Gallup. Previous workers have correlated Tocito sandstones into the marine Gallup (eg. Campbell, 1979; Nummedal et al., 1989).

The Tocito-4 sequence yields a similar palynomorph assemblage to the Tocito-1 sequence. However, two pollen species are present in the Tocito-4 strata which have their earliest range in the latest Coniacian to early Santonian. Thus, the four Tocito sequences record deposition from the earliest late Coniacian possibly into the early Santonian (Fig. 17).

DISCUSSION

Incised Valley and Sequence Development

Erosional Phase

In this study, we interpret sinuous fairways of incision common at each sequence boundary to erosional valleys cut by fluvial action. This interpretation requires at least four significant withdrawals of relative sea level and widespread exposure. No direct evidence for subaerial exposure such as rooted beds and mature soil horizons is observed in either the subsurface or outcrop. However, two lines of evidence strongly suggest formation under subaerial processes. The first is paleogeomorphology of the Tocito erosional surfaces and the second is geochemical analyses of nodular carbonate. Geomorphic features such as sinuous valley axes, steep to near vertical valley walls, and tributary-like junctures of valleys imply fluvial action. Axial sinuosity decreases in the southeasterly direction in the Tocito-1 and Tocito-2 valley systems, suggesting that rivers became straighter and possibly less erosive as they lost power in this direction. The narrow and deeply incised valleys of the Tocito-3 sequence along Waterflow and Bisti valleys alternatively suggests formation by headward erosion. In these valleys, erosion propagated in the updip direction as streams entered high-gradient, V-shaped drainages. McCubbin (1969) similarly reasoned that submarine erosion could not be responsible for the nearly 400 ft of missing sub-Tocito strata under the Horseshoe field. He also noted the lack of direct evidence of subaerial exposure in cores and outcrop but postulated that any evidence of subaerial exposure was probably removed or planed off during subsequent transgression. A similar explanation is given by Van Wagoner et al. (1990) and Walker and Eyles (1991) to explain the lack of subaerial evidence such as roots and soils in a similar paleogeographic setting.

The second line of evidence involves the recognition of pedogenic or lacustrine calcareous nodules from the Tocito sandstones. Isotopic analysis of both detrital and in-situ carbonate nodules collected from both core and outcrop indicate the calcite formed under shallow, meteoric conditions. The subsurface sample is from the basal Tocito-1 lag deposits from the Humble F-39 in Horseshoe oil field (917.5 ft, sec 10-31N-17W). Outcrop samples come from the Hogback anticline (basal lag, Tocito-4 sequence; sec 19-29N-16W) and from in-place caliche nodules at Lichii Wash (Tocito-4 sequence; NE-NE sec. 8-30N-19W). Assuming the samples had not been deeply buried, a meteoric origin is implied by the consistent ratio of $\partial^{18}O/\partial^{16}O$. These ration cannot be obtained with normal-marine fluid chemistry in a shallow burial setting (1-1.5 km burial depth) since temperatures greater than $60^{o}C$ are required (Fig. 18 a). In addition, $\partial^{13}C$ values ($^{o}/oo$ PDB, range -5 to -1) plotted against $\partial^{18}O$ values ($^{o}/oo$ PDB, range -14 to -10) also suggest a non-marine origin for the calcareous nodules. Figure 18b shows that the Tocito samples have carbon-oxygen cross plot values similar to those obtained from recent and fossil calcretes, soils and lacustrine carbonates (Salomons et al, 1978, Cerling 1984, Talbot 1990, Lander 1991). We infer from these results that the nodules may represent pre-existing pedogenic or lacustrine carbonates deposited in the vicinity of the Tocito lowstand valley systems. Overbank lows along interfluvial areas may have been sites for pedogenic or lacustrine carbonate deposition and subsequent erosion of these carbonates probably caused the nodules to be incorporated as lags within the incised valley fills.

Valley-Filling Phase

Core and outcrop data indicate that tidally influence sedimentation prevailed in the valley systems. As relative sea level rose, marine waters gradually flooded into the valleys and estuaries developed. Given that no fluvial strata are preserved with the valleys, it is likely that submarine erosion modified valley shape during initial flooding. Bedding relationships found in core and outcrop indicate highly fluctuating energy conditions characteristic of estuarine setting. The increase in bioturbation toward the valley edges in the Horseshoe field may be related to dampened tidal energies along valley edges. Log patterns calibrated to cores indicate that Tocito-1 and Tocito-2 valley fills are composed of a stacked succession of tidally influenced bars which prograded down the valleys. These units commonly have thickening- and coarsening-upward profiles separated by surfaces of relative deepening and are considered parasequences. This pattern is illustrated where the Tocito-4 sequence crops out along Rock Ridge and Beautiful Mountain (Fig. 16). Sigmoidal cross-bed geometries locally with double clay drapes and lamina set bundling are prevalent are these localities. Figure 19 details tidal bedding of the

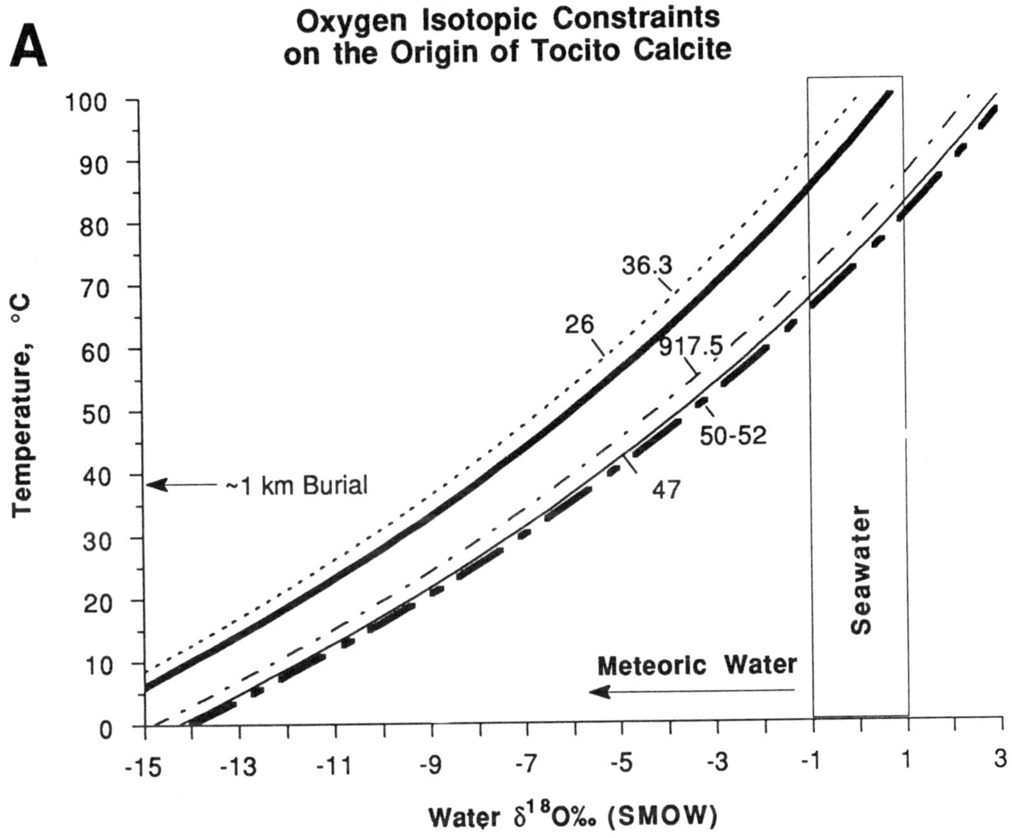

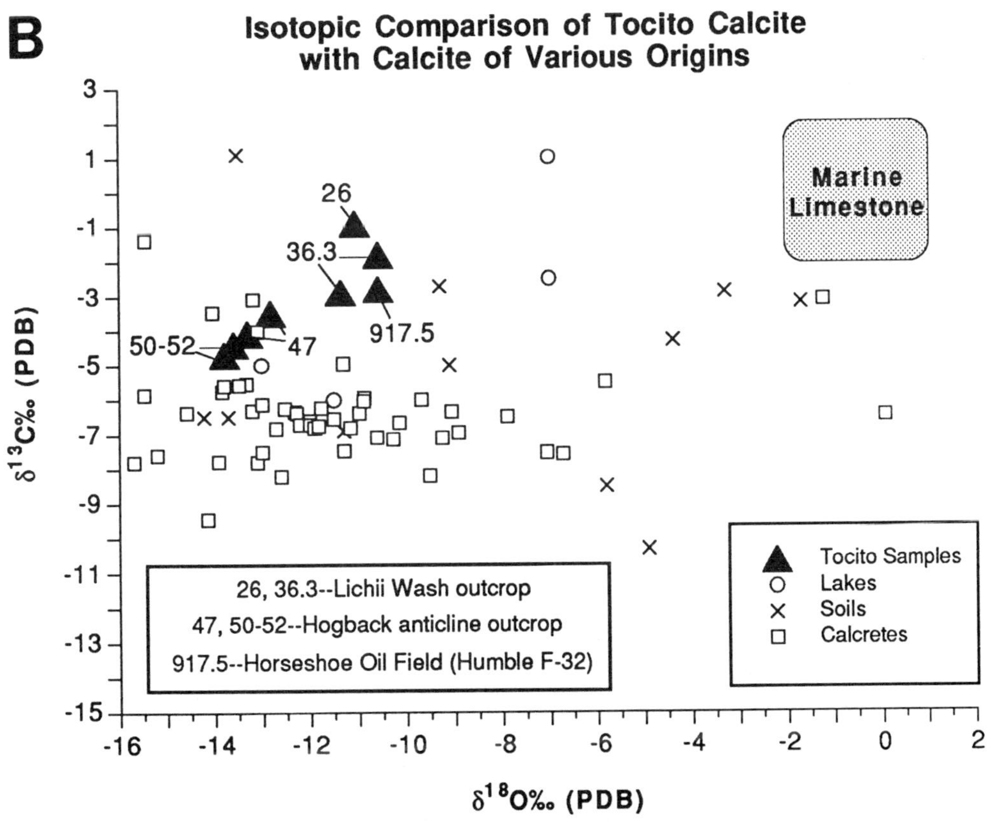

Fig. 18--Isotopic analyses of carbonates nodules from the Tocito Sandstone: **(A)** Plot of $\partial^{18}O$ ‰ (SMOW) versus temperature of calcite precipitation. **(B)** Plot of $\partial^{13}C$ ‰ (PDB) versus, $\partial^{18}O$ ‰ (PDB).

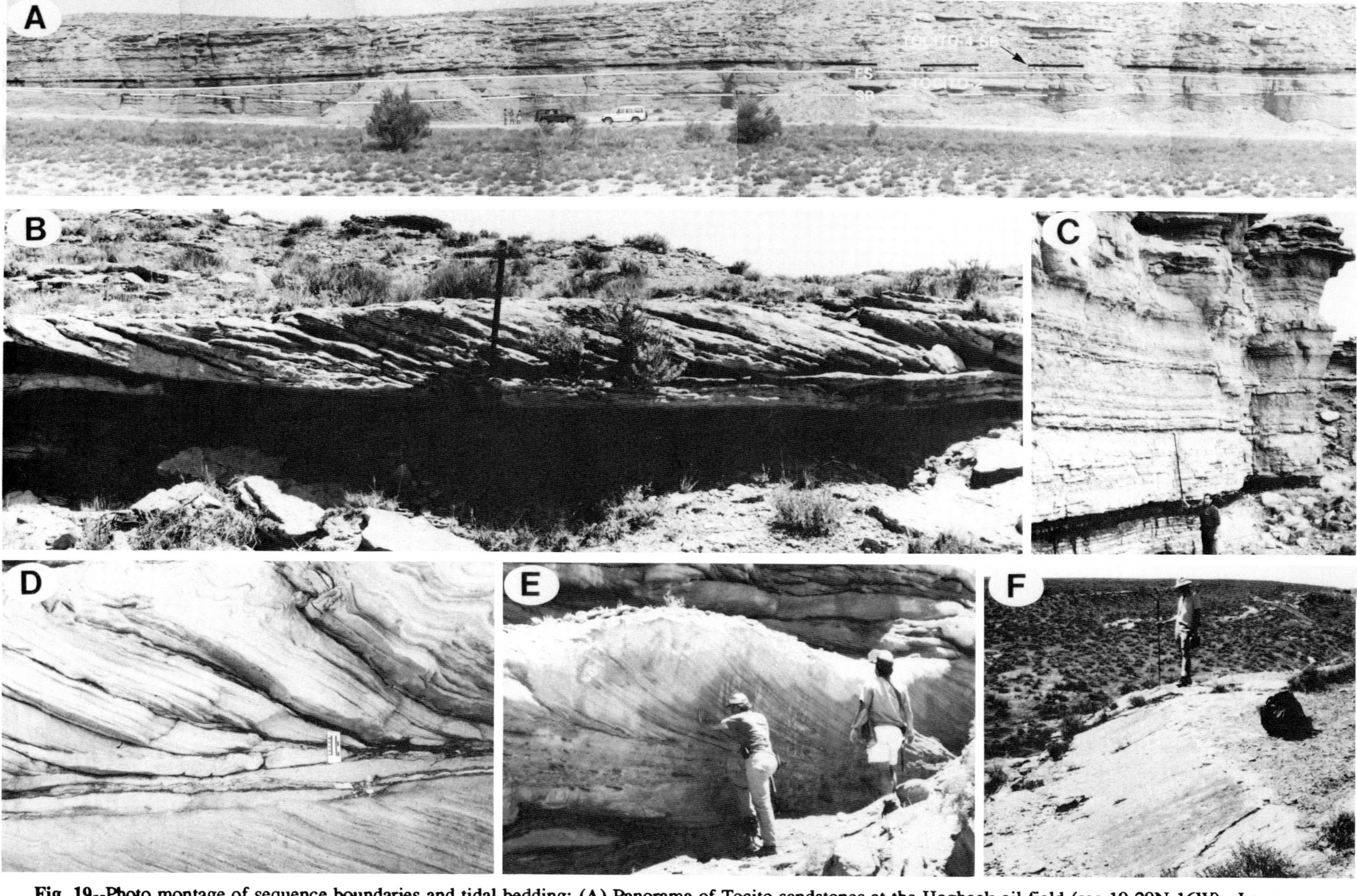

Fig. 19--Photo montage of sequence boundaries and tidal bedding: **(A)** Panorama of Tocito sandstones at the Hogback oil field (sec 19-29N-16W). Lower channelized sandstone is part of the Tocito-2 sequence. A thin (2-3 ft) interval of bioturbated sandy mudstone separates the Tocito-2 valley-fill from the overlying Tocito-4 sequence (black dashes). **(B)** Detail of sigmoidally cross-stratified sandstone bed from the Tocito-4 sequence in the Lichii Wash area (NE NE sec 8-30N-19W). Along this bed, tangential lower toes grade laterally into rippled toe sets and systematic variations in laminae dip and reactivation surfaces are evident. Current direction is also to the southeast at this locality. **(C)** Typical vertical profile of the Tocito-4 sequence along Beautiful Mountain. Top of staff is at the Tocito-4 sequence boundary. The Tocito-4 sequence is relatively thick (30 ft) but notably poor in cross-bedded sandstone at this locality. **(D)** and **(E)** Detail of sigmoidal cross-bed geometries from exposures along Chaco River (sec 13-29N-17W. Note size of cross beds and occurence of regularly spaced muddy

TIDAL-BAR STRATIGRAPHY AND FACIES: TOCITO SANDSTONE

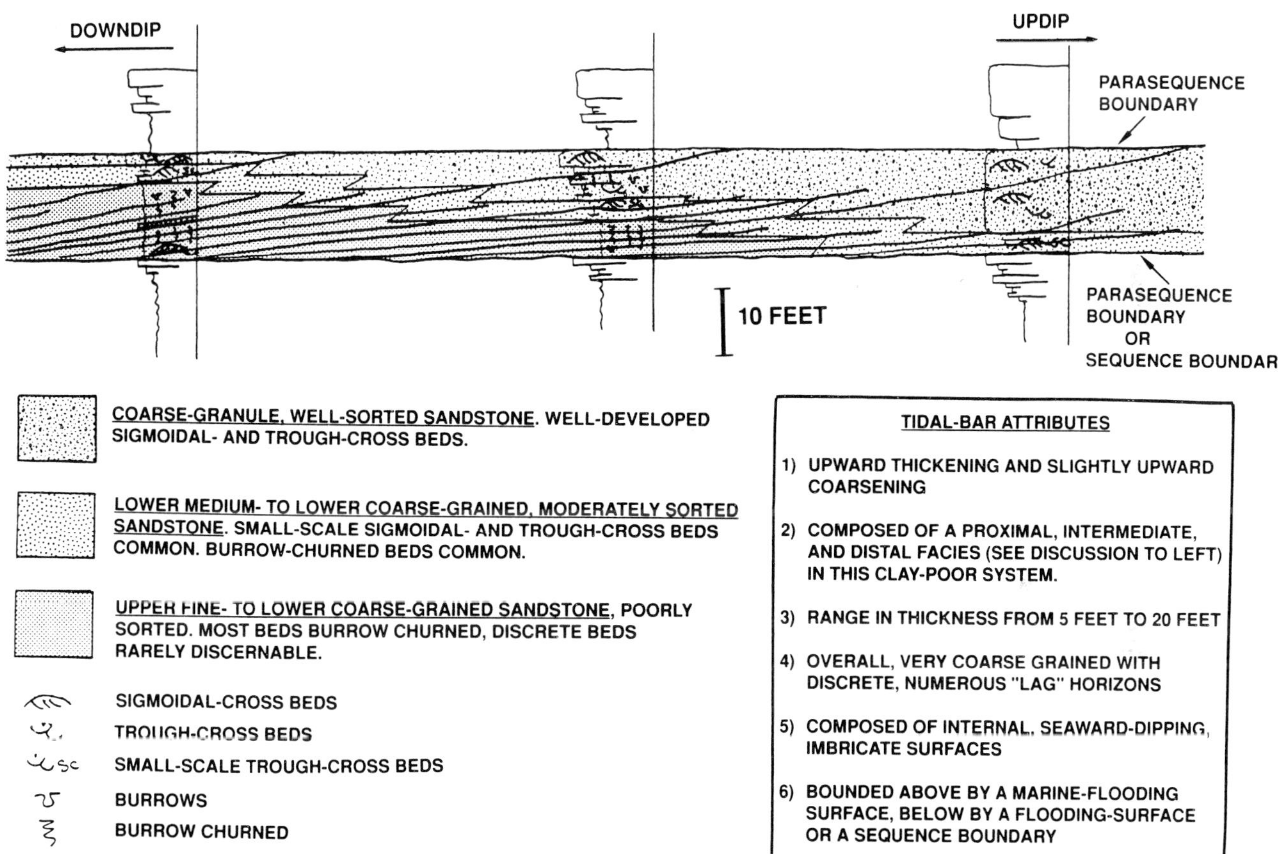

Fig. 20--Depositional model of a tidal-bar parasequence from the Tocito Sandstone.

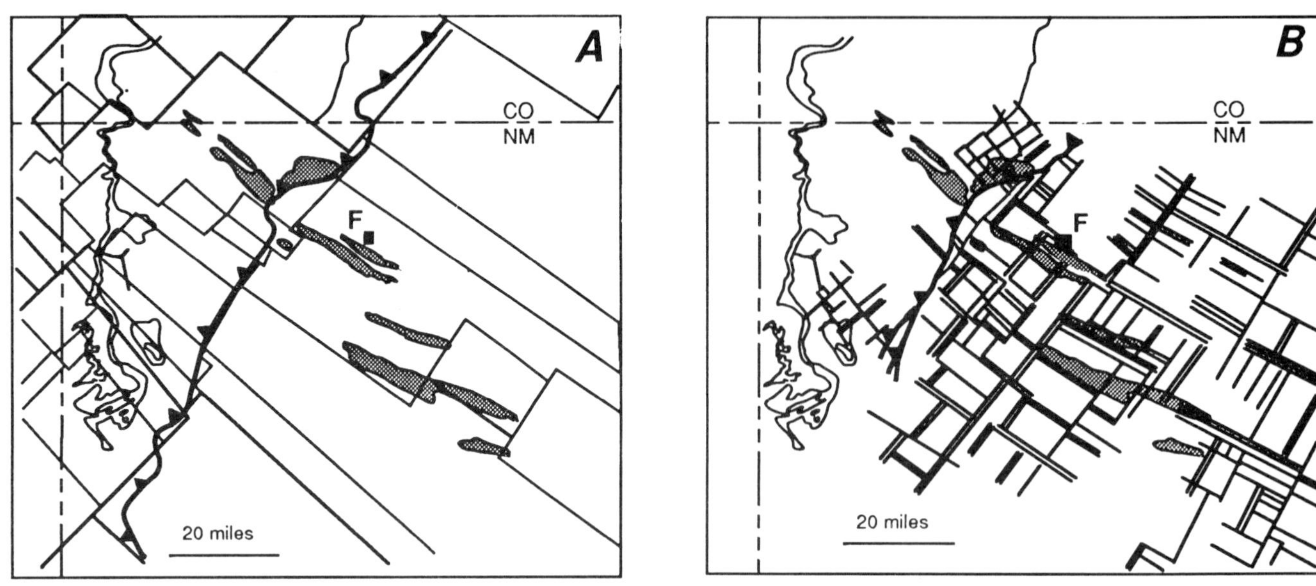

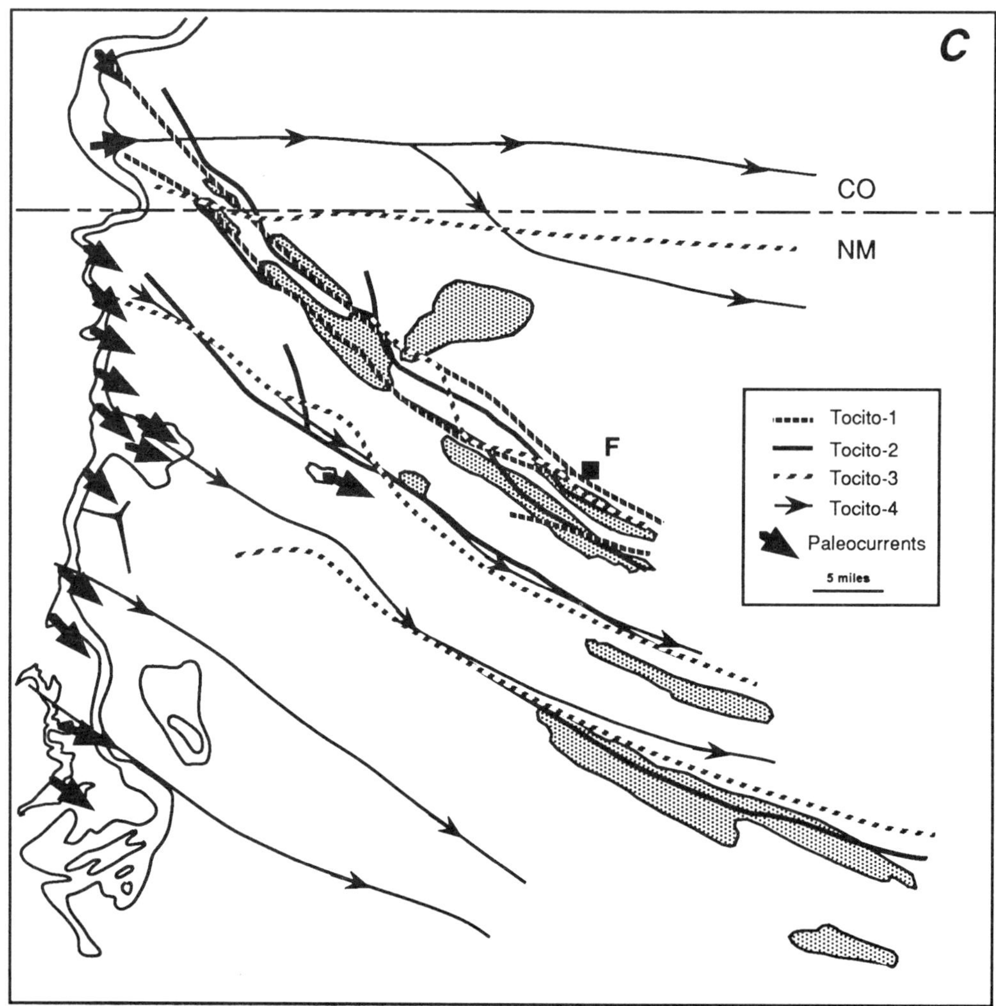

Fig. 21--Synthesis of Tocito oil fields, basement structural elements and incised-valley axes: **(A)** Basement fault map from Stevenson and Baars (1986). Block-fault patterns are based on differential thickness of Paleozoic strata measured from well-log data; **(B)** Structural element map of Huffman and Taylor (1989). The Tertiary Hogback basement thrust and the city of Farmington, N.M. (F) , is shown on both **A** and **B**; **(C)** Summary diagram of all Tocito valley axes. Valley axes are strongly parallel and stack vertically over one another. This relationship indicates the repeated involvement of the faults described in A and B. Paleocurrent data show a persistent southeast drainage system throughout Tocito deposition (modified from McCubbin,1969).

Tocito and Figure 20 illustrates the stratigraphy and formation of the tidal parasequences seen in the Tocito.

In the subsurface, the Tocito-3 and Tocito-4 sequences represent a departure from the sand-prone style of valley fill observed in the underlying Tocito-1 and Tocito-2 sequences. Lowstand deposits in the Tocito-3 and Tocito-4 sequences tend to be relatively sand-poor. Based on evidence discussed above, fluvial/estuarine processes were responsible for the erosional incision of the Tocito-3 and Tocito-4 valley systems. However, coarse sediments were not widely deposited in the valleys. This may be attributed to two reasons: (1) rivers in the vicinity carried less sand, or (2) the relative sea-level rise during the late lowstand was rapid and caused sand-prone facies tracts to retreat too quickly for appreciable sand to be deposited. In addition, valley formation by headward erosion may have occurred during Tocito-3 sequence boundary formation. In this case, sand-poor valleys are expected. The overall result in these downdip areas are valleys filled with distal estuarine deposits that approach open-marine deposits.

Valley-Flooding Phase

Continued relative sea-level rise led to the transgressive flooding of both the valleys and interfluve areas. Estuaries were replaced with and fully open marine, shelfal conditions and the blanketing marine muds of the transgressive systems tract were laid down. If this rise was interrupted by a relative still-stand, parasequences, like the one found above the Tocito-1 Horseshoe valley fill, developed as part of the transgressive systems tract. Valley-flooding surfaces were frequently downlapped by the muddy toes of prograding parasequences of the highstand systems tract. This relationship is best developed in Tocito-1 and Tocito-2 sequences.

Incised Valleys versus Lowstand Shorelines

Linear asymmetric erosional features with the steep limb facing toward the basin have been described in the Turonian-Coniacian Cardium Formation by Plint (1988) and Walker and Eyles (1991), among others. These workers interpreted a stepped erosional surface with asymmetric profiles as the product of erosion by a transgressing shoreline system across a tilted depositional plane. Shoreface entrenchment, coupled with subaerial stream modification, created a branching array of linear erosional lows which were commonly filled with sandstone and conglomerates. These coarser grained beds changed facies into marine shale. These workers envisaged a fully open-marine, wave-dominated setting throughout deposition of the lowstand facies. Importantly, no tidally influence deposits were observed in the Cardium.

In view of the northeast-prograding Gallup Sandstone complex and the northwest-southeast-trending erosional axes from the overlying Tocito sandstone, a similar transgressing shoreline model may be applied. However, several aspects of the Tocito are in conflict with this model. Chief among them is the tide-dominated nature of the Tocito deposits, the stressed ichnofauna assemblage, the sinuous and branching distribution erosional lows and the deeply intercutting nature of the sequence boundaries. The lack of wave generated bedding indicates a restricted access to open-marine processes. In this study, we relate the gradual northward thinning of sandstones with certain valleys not to gradual facies change but to onlap onto a gently dipping valley flank. Post-depositional tilting of this surface toward the northeast created the apparent onlap relationship with the overlying strata.

Structural Control on Incised-Valley Orientation

We propose the dominant influence on paleodrainage patterns during Tocito deposition was folds and faults associated with reactivation of a northwest-trending basement fabric. The San Juan basin lies along a prominent basement lineament called the Wichita-Olympia Lineament. In northern New Mexico, this feature is expressed as a series of northwest-trending faults which periodically reactivated to form a series of horsts and grabens. Several studies have documented these basement-related features using a variety of subsurface data sets. Stevenson and Baars (1977, 1981, 1986) mapped a fan-like array of horst-and-graben basement faults based on thickness of Paleozoic strata from subsurface well control. Huffman and Taylor (1989, 1991) identified a similar trend of basement elements using a conventional CDP seismic data. Figure 21 shows the strong relationship between these basement-fault trends, hydrocarbon accumulations and valley axes mapped in this study. Valley axes are clearly parallel to the basement grain. In addition, valley axes stack through time. This point is illustrated along the Waterflow valley where both the Tocito-3 and Tocito-4 sequences have asymmetric thicks (Fig. 5). These thickness relationships suggest that a near vertical fault with a down-to-the-north displacement influenced these valley patterns. This inferred fault is named the Waterflow fault. A corresponding thin Tocito interval occurs along the southern margin of the Waterflow valley which extends toward the outcrop. The Tocito-3 and Tocito-4 sequences are thin to absent along this ridge (Figs. 13, 15).

This fault trend takes on additional significance when comparing sub-Tocito thicknesses across the fault. Figure 5 and Figure 14 show dramatic thinning of the Gallup interval on the north side of the inferred Waterflow fault. Abrupt stratal termination of Gallup strata below the Tocito sequence boundaries implies truncation by erosion. This relationship requires that the north side of the fault was uplifted and eroded relative to the south side. This movement is opposite to that documented in the Tocito-3 and Tocito-4 sequences. Thus, we infer that Waterflow fault had initially a reverse sense of movement when erosion of the Gallup took place and then reactivated in the opposite sense as a normal fault during deposition of the Tocito sequences. These observations are consistent with Stevenson and Baars (1986) who documented sets of steeply dipping faults with vertical displacement that periodically changed from normal to reverse. They inferred these relationships implied episodic relaxation of compressional forces and the rejuvenation of tensional forces due to wrench-style tectonics. Huffman and Taylor (1991) documented a similar pattern of high-angle faults which had variable fault histories and sense of movements. They documented three phases of rejuvenation: Pennsylvanian-Permian, Jurassic-Cretaceous and early Tertiary. Above the Permian strata, most offsets are detected by drape, different thickness and lithologic changes across faults. In this study, no faults were observed to cut the Gallup or Tocito intervals so we similarly support the occurrence of drape across faulted blocks.

Other studies have emphasized structural control on paleovalley formation. Many of the early Cretaceous paleovalleys of the Powder River Basin in Wyoming and Montana were governed by rejuvenated basement blocks (Slack, 1981; Weimer et al., 1982). Similar to the Tocito, these workers observed that incised fluvial drainages repeatedly focused along graben-like lows. Horsts served as drainage divides. Greb (1989) interpreted the sub-Absoroka unconformity in eastern United States to represent entrenched paleovalleys whose distribution was related to syndepositional faulting. Evidence included valley parallelism with faults, diversion of paleovalley trends corresponding to fault intersection, and locally rectangular paleodrainage patterns. Greb (1989) also noted streams flowed parallel to paleostrike until they intersected a cross fault where it would turn 90°. Overall drainage patterns were thus controlled by faults regardless of the orientation of the regional paleogradient.

Lamb (1968) and McCubbin (1969) attributed Tocito sedimentation patterns to structural features. They felt that the primary control on erosional patterns was variable erosion though folds and faults in the underlying Gallup and Mancos strata. Following a drop in relative sea level, a ridge-and-valley or questa-like paleolandscape arose as erosion across resistant units formed near vertical escarpments and non-resistant shales formed the valleys. Lamb (1968) related the apparent folding to syndepositional uplift. He also noted small-scale faults locally acted as escarpments which led to accumulation of Tocito sandstone on the downthrown side.

The stratigraphic evidence presented in this and previous studies indicate that a phase of pronounced structural deformation followed deposition of the Gallup sandstone. The passive, northeast-dipping ramp present during Gallup deposition tectonically reorganized into a fault-segmented platform with a northwest-southeast structural grain. Basement normal faults reactivated during the relative sea-level lowstand and created linear depressions which governed drainage patterns. Streams flowed across depositional strike until they met the sea to form linear estuaries. A persistent southeast-directed paleodrainage system is evident by uniform paleocurrent data observed at all Tocito outcrops (Fig. 21). The step-wise evolution of the Tocito Sandstone is illustrated in Figure 22.

HYDROCARBON SYSTEM

Production Statistics

Early drilling in the San Juan Basin was primarily for shallow gas objectives. This changed in the mid-1950s when a number of oil discoveries were made from lenticular sandstone reservoirs which are now known as the Tocito Sandstone. Eighty percent of the oil produced in the San Juan Basin comes from these sandstones (Ross, 1980). Other major producing intervals in the basin are the Upper Cretaceous Dakota Sandstone and the Fruitland Formation, both of which produce gas. The Hospah field, located 40 miles south of the Bisti field, is the only field which produces from the Gallup Sandstone.

Bisti, Horseshoe, Many Rocks, Totah, Cha Cha, Gallegos and Verde are among the significant Tocito fields and are officially classified as oil fields with minor associated gas (Fassett, 1983). Data through 1987 indicate that these seven fields have produced over 100 MBO and 112 GCF. Field sizes vary considerably. Bisti and Horseshoe fields are the largest with EURs of 40 million barrels and 54 million barrels, respectively. All fields are considered mature, with over 90% of primary hydrocarbons recovered (Table 1). Bisti is the most active with current production rates of 1,145 BOPD and 1036 KCFD.

Tocito fields in the San Juan Basin are dominantly stratigraphic traps. Individual fields are dissolved-gas-drive (pressure depletion) reservoirs with typical primary recovery factors of 15% of original oil in place. During initial development, production histories typified a pressure-depletion reservoir. Wells had very high initial potentials, with 500 to 1000 BOPD with no water production. Oil production

Table 1. Tocito Oil and Gas Fields, San Juan Basin

(as of 1987)

FIELD NAME	# PRODUCING WELLS	DISCOVERY DATE	RESERVOIR UNIT	DAILY PRODUCTION	CUM. PRODUCTION	EUR MBO	EUR GCF	EUR TOTAL
Horseshoe	114	Sep-56	TOCITO-1, TOCITO-2	458 BO 11 KCF	38.2 MBO 7.8 GCF	38.8 MBO	7.9 GCF	40.2 MOEB
Bisti	167	Oct-55	TOCITO-2	1145 BO 1036 KCF	36.4 MBO 76.2 GCF	40.0 MBO	79.5 GCF	54.1 MOEB
Cha Cha	41	Sep-59	TOCITO-2	189 BO 337 KCF	9.9 MBO 19.0 GCF	10.2 MBO	19.8 GCF	13.7 MOEB
Verde*	28	Sep-55	TOCITO-3, TOCITO-4	49 BO 16 KCF	7.9 MBO 3.2 GCF	8.0 MBO	3.3 GCF	8.6 MOEB
Many Rocks	36	Nov-62	TOCITO-1	30 BO 25 KCF	2.9 MBO 1.1 GCF	3.0 MBO	1.2 GCF	3.2 MOEB
Gallegos	52	Sep-54	TOCITO-2	115 BO 1397 KCF	2.2 MBO 38.3 GCF	2.4 MBO	40.8 GCF	9.6 MOEB
S. Waterflow	0	Nov-63	TOCITO-2	abandoned	0.2 MBO 0.3 GCF	0.2 MBO	0.3 GCF	0.25 MOEB
Totah	1	Sep-59	TOCITO-2	0 BO 3 KCF	3.4 MBO 6.7 GCF	3.3 MBO	6.8 KCF	4.5 MOEB

* production from fractures in shale and low matrix porosity sandstone

134 TOTAL MOEB

rates declined rapidly, and gas:oil ratios showed a corresponding increase through time.

Secondary recovery waterflood projects were initiated at several fields in the early 1960s to maintain reservoir pressure. Waterfloods are still active to varying degrees at Bisti, Cha Cha, Horseshoe and Totah fields (Table 1). The Bisti Field waterflood is quite large with 167 producers and 66 injectors. Recent drill wells and daily well production indicate that waterflooding may be reaching its effective limit. Reservoir pressures are currently very low. In 1981, an offset to very high producers had an initial potential of only 8 BOPD and 40 BWPD (Navajo 225-G, 31N-17W-section 1).

Bisti South is the most recent new field and was discovered in 1984. No cumulative production data is available for this field, but the initial potential at time of discovery was 36 BOPD and 24 KCFD (Dugan Mary Lou #1, sec 32-24N-10W).

Reservoir Characteristics

Tocito reservoir sandstones are generally medium- to coarse-grained, glauconitic sandstones with thin clay interbeds. Porosity values in producing intervals range from 4 to 20% and average about 15%. Permeability values range from 0.5 to 150 millidarcies and average between 50 to 100 millidarcies. Reservoir thickness varies from 5 to 50 feet, and producing interval depths range from 400 to 7000 feet across the basin.

Hydrocarbon Source Rock

The major source facies for Tocito oil is the marine Mancos Shale. The unit is over 2000 ft thick in the basin center and constitutes the major part of the distal, open-marine siltstones and shales in the San Juan Basin.

Oils analyzed from six producing Tocito fields correlate with bitumen extracts from Mancos shales sampled at outcrop (Ross, 1980). Associated and non-associated gases in the Tocito were probably generated from more thermally mature intervals of the Mancos Shale (Rice, 1983).

In-house studies of a well from the basin center, the Schumacher 1-E (sec 8-30N-12W), supports published data that the Mancos is the primary source rock. Total organic carbon values are as high as 5% in places. A Mancos shale sample taken below the Gallup Sandstone at Shiprock Wash near Shiprock, N.M., yielded a TOC value of 9.5%. The abundance of downlapping organic-rich shales above the valley-fill flooding surfaces suggests that source facies are juxtaposed against Tocito reservoir facies.

The Mancos generally produces a sweet, low-sulfur, paraffin-base oil that ranges from 38 to 43° API gravity in the Tocito fields and from 24 to 32° API gravity in the marine Gallup Hospah field to the south. Associated and non-associated gases are wet with C_2-C_5 ranging approximately 15-20% by volume (Rice, 1983).

Hydrocarbon Maturation

Maximum oil generation occurred in the Oligocene in association with maximum burial (Rice, 1983). However, Rice (1983) observed that the high rank of coal in the overlying Fruitland Formation cannot be accounted for by burial alone. He postulated that the emplacement of Oligocene batholiths just to the north of the basin provided the additional heat source for the advanced maturation levels found in these coals.

Trapping Styles

Hydrocarbon traps are formed by a combination of stratigraphic trapping elements. Structure plays a minor role in most accumulations. Principal trapping elements are (1) truncation (e.g. top seal provided by a shaley, overlying sequence), (2) valley edges (lateral seal is dependent on valley incision into sealing, shale-dominated lithologies), (3) change in valley orientation (e.g. trapping against sinuous bends in the valley system), and (4) or structural culminations/noses. Trapping as a function of facies changing into marine shale (i.e. offshore bars in the Mancos sea) was not observed in this study. Figure 23 illustrates the major trap types, and the attributes of each are summarized below.

Bends in Valley Axis

Hydrocarbons are trapped against arcuate bends in valley edges. Regional structural dip to the northeast causes hydrocarbons to trap against the southern margin of the valleys in the study area. Horseshoe, Many Rocks, Totah and Cha Cha fields have hydrocarbons reservoired along bends in the valley edges (Fig. 24 a). McCubbin (1969) recognized an small anticline along the northwest limit of the Horseshoe and Many Rocks field. This feature likely contributes to the trapping of hydrocarbons.

Truncation by Younger Sequence Boundary

Lateral and top seal are provided by a younger, shale-prone valley fill that overlies the sequence boundary. Hydrocarbons are trapped against valley edges against updip area. Bisti and Gallegos fields are examples of traps formed by truncation of the Tocito-2 valley by a shale-filled Tocito-3 valley (Figs. 23, 24b).

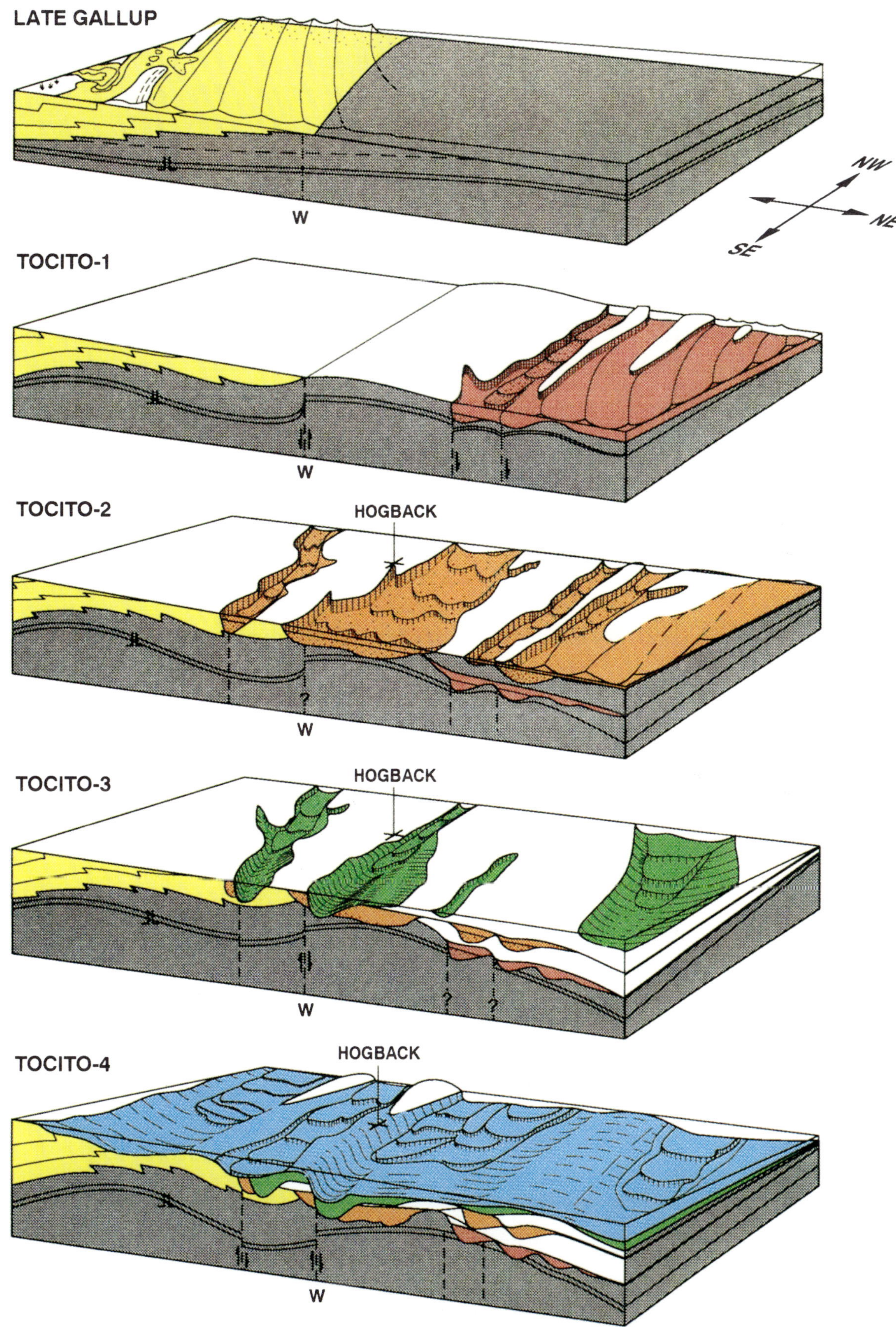

Fig. 22--Sequential evolution of Tocito paleovalleys. Inferred faults are shown as vertical dashed lines with arrows noting sense of movement. Note the Juana Lopez Member (JL) and the Waterflow fault (W). Stratigraphic evidence suggests a reversing sense of movement through time.

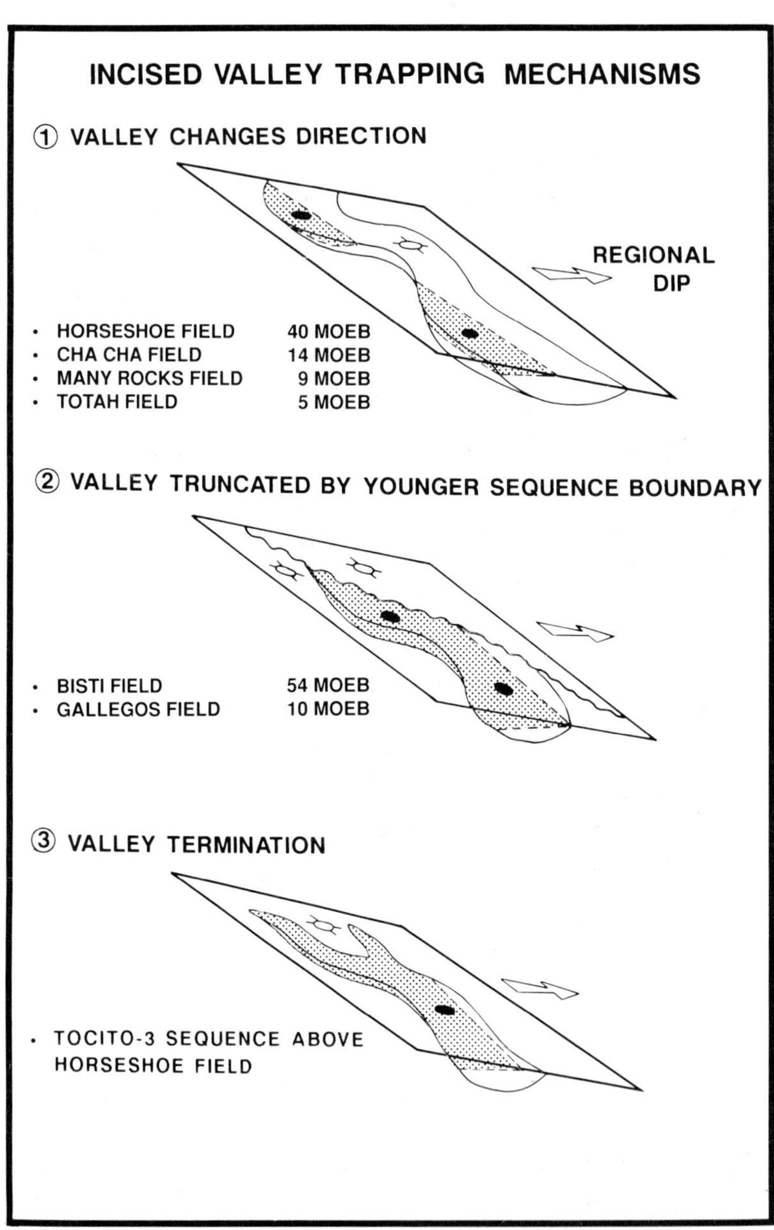

Fig. 23--Summary of trap styles found in the Tocito.

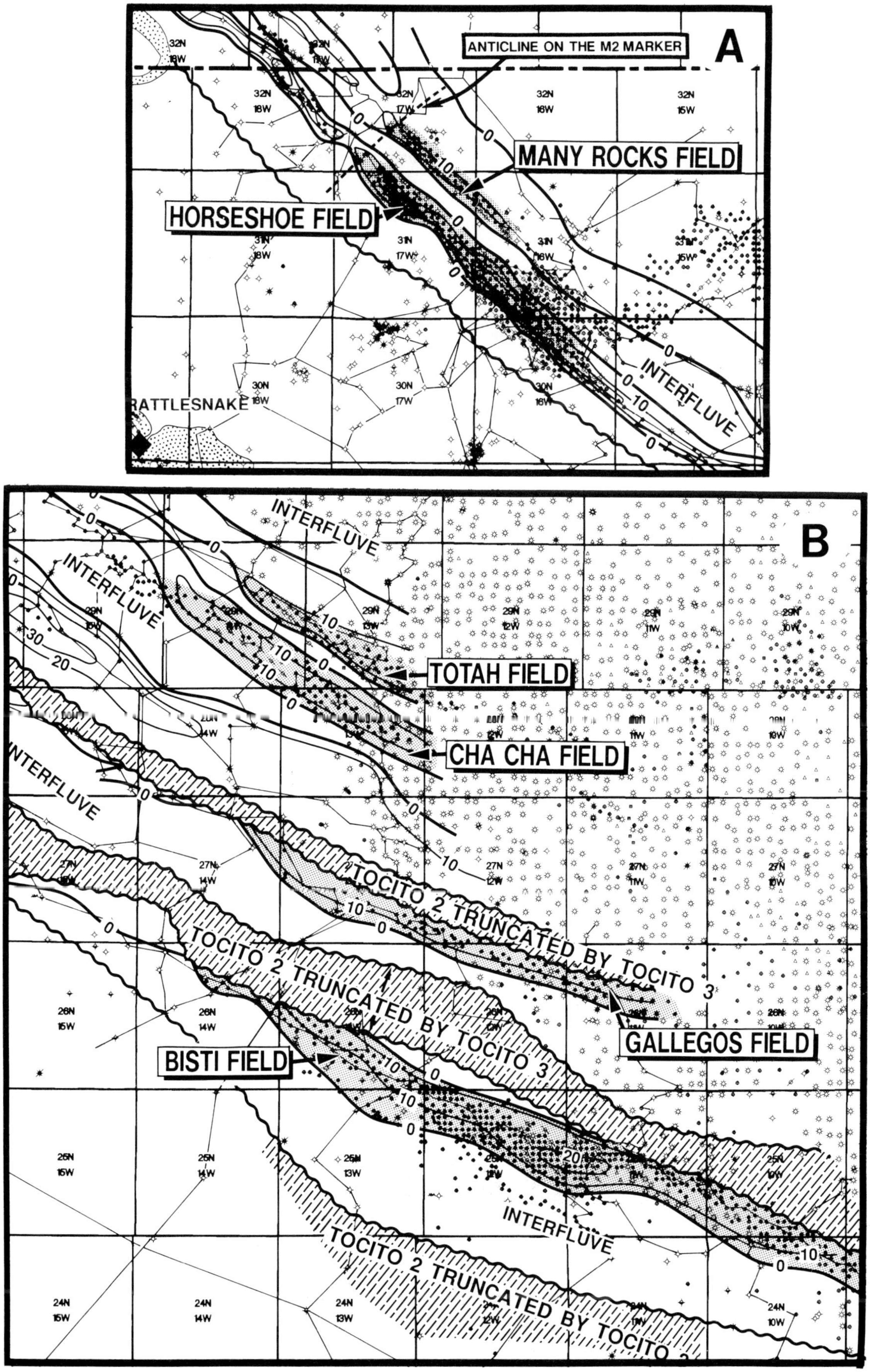

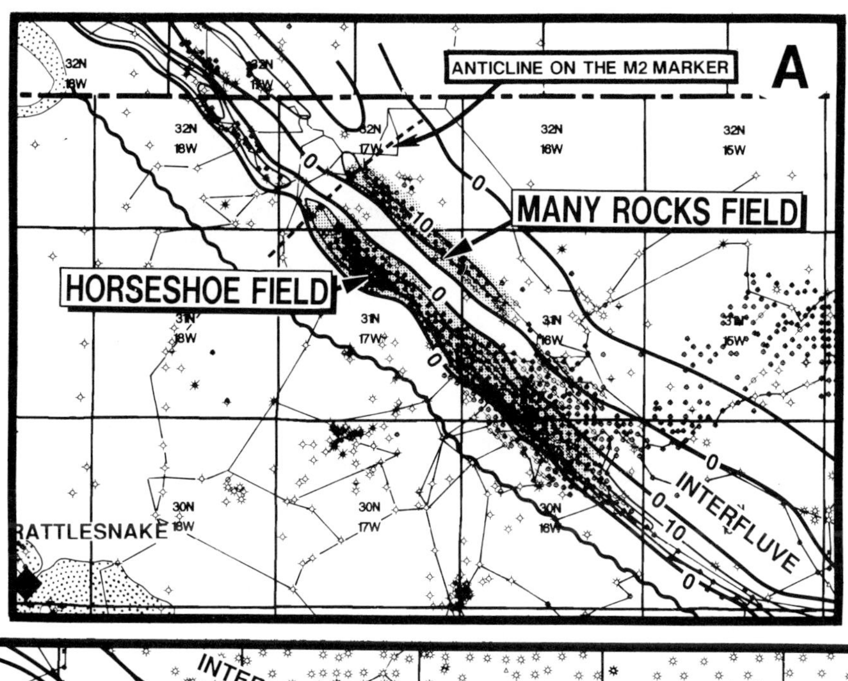

Fig. 24--Detail of selected hydrocarbon accumulations (stippled patterns). Structural dip is down to the northeast in both examples: (A) Horseshoe and Many Rocks fields. Hydrocarbons are trapped in the Tocito-1 sequence along of updip bends in the valleys (Type 1, Figure 23). A small anticline along the northern boundary of each field likely contibutes to the trap (modified from McCubbin, 1969). (B) Bends in the Tocito-2 valley axes form Totah and Cha Cha field (Type 1 trap). Hydrocarbons at Bisti and Gallegos field are trapped updip by the valley edge and along strike by the shale-filled Tocito-3 valley which has truncated the Tocito-2 sequence.

Updip Valley Termination

Valleys with sandy fills, presumably cut by headward erosion, have blind terminations up depositional dip. Trapping is dependent on the sealing capacity of underlying and overlying lithologies. An example is a narrow, oil-producing Tocito-3 valley fill which has its updip termination at Horseshoe field (Fig. 13).

Structural Closure

Accumulations due solely to structural closure was not observed in the study area. However, a small structural culmination exists north of the Horseshoe and Many Rocks fields and the South Waterflow field and may contribute to the trapping elements in this area.

CONCLUSIONS

1. The surface that separates Tocito-age strata from the underlying Gallup is not a single unconformity. Along the western boundary of the study area, the basal Tocito unconformity is a composite erosional surface produced by the complex erosional interaction of at least four relative sea-level falls. Much of the stratigraphic detail of the Tocito is therefore found largely in the subsurface.

During periods of low sea level, rivers incised to form a parallel and branching network of valleys which subsequently filled in with tidally influenced estuarine strata. Given that no fluvial strata are preserved with the valleys, it is likely that submarine erosion modified valley shape during initial valley flooding.

2. This study demonstrates the interplay between sedimentation, structure and basin evolution. We infer that reactivated structural grain controlled the distribution of valleys. Older Tocito valleys (Tocito-1) are preserved only in the north, where basin subsidence rates remained relatively high. To the south, valleys cut into one another, causing erosional segmentation of underlying valley fills. Structural influence is more pronounced along fault zones and consequently the stratal relationships are more complex along these boundaries.

3. Sequences appear to stack in a retrogradational pattern. The narrow, entrenched valleys of older sequences give way to broader, more gently sloping valley profiles in younger sequences. Reservoir quality in the valley fills decreases through time, and the infilling deposits become more marine in character. The Tocito sequences appear to represent a transgressive-sequence set. Overall, the Tocito (Tocito-1, Tocito-2, Tocito-3, Tocito-4) represent a phase of lowstand deposition that followed the marine Gallup highstand.

4. Faults, differential uplift, subsidence and eustacy interacted to create complex stratal geometries. However, stratigraphic complexities as well as hydrocarbon trapping mechanisms can be resolved when a high-resolution, sequence-stratigraphic approach is applied. The result is a more accurate understanding of the complex time-stratigraphic character and the hydrocarbon-trapping mechanisms of the Gallup and Tocito sandstones.

ACKNOWLEDGEMENTS

The authors thank Rob Lander for isotopic analyses and Y.Y. Chen for shedding light on palynology of the Tocito. Jo Ann Baker assembled the manuscipt in preparation for the field conference. An earlier version of this text was reviewed by Wendy Burgis and Gordon Moir. We also gratefully acknowledge Exxon Production Research Company for the resources and time to conduct the study and prepare this paper.

REFERENCES

Campbell, C. V., 1979, Model for beach shoreline of Gallup Sandstone (Upper Cretaceous) of northwestern New Mexico: New Mexico Bureau of Mines and Minerals Resource Circular 164, 32 pp.

Cerling, T.E., 1984, The stable isotope composition of modern soil in carbonate and its relation to climate: Earth and Planetary Science Letters, v. 71, p. 229-240.

Dane, C. H., 1960, The boundary between rocks of Carlile and Niobrara age, San Juan Basin, New Mexico and Colorado: American Journal of Science, v. 258A, p. 46-56.

Dane, C. H., W. A. Cobban and E. G. Kaufmann, 1966, Stratigraphy and regional relationships of a reference section for the Juana Lopez member, Mancos Shale, in the San Juan Basin, New Mexico: USGS Bulletin, 1224-H, 15 pp.

Fassett, J. E., 1983, Stratigraphy and oil and gas production of northwest New Mexico updated through 1983, *in* Oil and Gas Fields of the Four Corners Area: Four Corners Geological Society, p. 849-854.

Flores, R.M., J.C. Hohman, and F.G. Ethridge, 1991, Heterogeneity of Upper Cretaceous Gallup sandstone regressive facies, Gallup Sag, New Mexico, *in* Nations, J.D., and Eaton,J.G., eds., Stratigraphy, depositional environments, and sedimentary tectonics of the western margin of the Cretaceous Western Interior Seaway: GSA Special Paper 260, p. 189-209.

Greb, S.F., 1989, Structural controls on the formation of the sub-Absaroka unconformity in the U.S. Eastern Interior basin: Geology, v. 17, p. 889-892.

Huffman, A.C., Jr., and D.J. Taylor, 1989, San Juan Basin faulting-more than meets the eye (abs): AAPG Bulletin, v.73, no. 9, p. 1161.

Huffman, A.C., Jr., and D.J. Taylor, 1991, Basement fault control on the occurrence and development of San Juan basin energy resources: GSA Bulletin, abstracts with programs, vol. 23, no. 4, p.34.

Lamb, G. M., 1968, Stratigraphy of the Lower Mancos Shales in the San Juan Basin: GSA Bulletin, v. 79, p. 827-854.

Lander, R.H., 1991, Whiter River Group diagenesis, Ph.D. Thesis, University of Illinois, 143 pp.

McCubbin, D. G. 1969, Cretaceous strike valley sandstone reservoirs, northwestern New Mexico: AAPG Bulletin, v. 53, p. 2114-2140.

Molenaar, C.M., 1973, Sedimentary Facies and correlation of the Gallup Sandstone and associated formations, northwestern New Mexico, *in* J.E. Fassett, ed., Cretaceous and Tertiary rocks of the southern Colorado Plateau: Four Corners Geological Society Memoir, p. 85-110.

Molenaar, C.M., 1983a, Major depositional cycles and regional correlations of Upper Cretaceous rocks, southern Colorado Plateau, *in* N. W. Reynolds and E. D. Dolly (eds.), Mesozoic Paleogeography of West Central United States: Rocky Mountain Section of SEPM Symposium, No. 2, p. 201-224.

Molenaar, C.M., 1983b, Principal reference section and correlation of Gallup Sandstone, northwestern New Mexico, *in* Contributions to Mid-Cretaceous Paleontology and Stratigraphy of New Mexico, Part II: New Mexico Bureau of Mines and Mineral Resources Circular, v. 185, p. 29-40.

Nummedal, D., D.J.P. Swift, and R. Wright, 1986, Depositional sequences and shelfal sandstones in Cretaceous strata of the San Juan Basin, New Mexico: Field Guide for the Seventh Annual Research Conference Gulf Coast Section SEPM, p. 1-162.

Nummedal, D., R. Wright, D.J.P. Swift, R.W. Tillman, and R. W. Wolter, 1989, Depositional systems architecture of shallow marine sequences, *in* D. Nummedal and R. Wright (eds.), Cretaceous Shelf Sandstones and Shelf Depositional Sequences, 28th International Geological Congress Field Trip, p. 35-73.p

Penttila, W. C., 1964, Evidence for the pre-Niobrara unconformity in the northwestern part of the San Juan Basin: Mountain Geologist, v. 1, p. 3-14.

Plint, A.G., 1988, Sharp-based shoreface sequences and "offshore bars" in the Cardium Formation, Alberta: their relationship to relative changes in sea level, *in* W. Wilgus, B. Hastings, C.A. Ross, H. Posamentier, J.C. Van Wagoner, and C.G. St. C Kendall (eds.), Sea-Level Changes: An Integrated Approach, SEPM Special Publication no. 42, p. 357-370.

Rice, D. D., 1983, Relation of natural gas composition to thermal maturity and source rock type in San Juan Basin, northwestern New Mexico and southwestern Colorado: AAPG Bulletin, v. 67, No. 8., p. 1199-1218.

Ross, L. M., 1980, Geochemical correlation of San Juan Basin oils-a study: Oil and Gas Journal, v. 78, p. 102-110.

Sabins, F. F., Jr., 1963, Anatomy of a stratigraphic trap, Bisti field, New Mexico: AAPG Bulletin, v. 47, p. 193-228.

Sabins, F. F., Jr., 1972, Comparison of Bisti and Horseshoe Canyon stratigraphic traps, San Juan Basin, New Mexico: AAPG Memoir 16, p. 610-622.

Salomons, W., A. Goudie, and W.G. Mook, 1978, Isotopic composition of calcrete deposits from Europe, Africa, and India: Earth Surface Processes, v. 3, p. 43-57.

Slack, P.B., 1981, Paleotectonics and hydrocarbon accumulation, Powder River Basin, Wyoming: AAPG Bulletin, v. 65, p. 730-743.

Stevenson, G. M. and D. L. Baars, 1977, Pre-Carboniferous paleotectonics of the San Juan Basin, *in* J.E. Fassett, and H.L. James, (eds.), San Juan Basin III: New Mexico Geological Society 28th Field Conference Guidebook, p. 99-110.

Stevenson, G. M. and D. L. Baars, 1981, Pre-Carboniferous paleotectonics of the San Juan Basin, *in* D. W. O'Leary and J. L. Earle (eds.), Proceedings of the Third International Conference on Basement Tectonics: p. 331-346.

Stevenson, G. M. and D. L. Baars, 1986, The Paradox: a pull-apart basin of Pennsylvanian Age: *in* J.A. Peterson (ed.), Paleotectonics and Sedimentation in the Rocky Mountain Region, United States, AAPG Bulletin, Memoir 41, p. 513-539.

Talbot, M.R., 1990, A review of the paloehydrological interpretation of carbon and oxygen isotopic ratios in primary lacustrine carbonates: Chemical Geology (Isotope Geoscience Section), v. 80, p. 261-279.

Tillman, R. W., 1985a, The Tocito and Gallup sandstones, New Mexico, a comparison, *in* R. W. Tillman, D. W. J. Swift and R. G. Walker (eds.), Shelf sands and sandstone reservoirs: SEPM Short Course, Notes No. 13, p. 403-464.

Tillman, R. W., 1985b, Tocito Sandstone core, Horseshoe field, San Juan County, New Mexico, *in* R. W. Tillman, D. W. J. Swift and R. G. Walker (eds.), Shelf sands and sandstone reservoirs: SEPM Short Course Notes No. 13, p. 559-576.

Tillman, R. W. and R. S. Martinsen, 1984, The Shannon shelf-ridge sandstone complex, Salt Creek anticline area, Powder River Basin, Wyoming, *in* R. W. Tillman and C. T. Siemers (eds.), Siliciclastic Shelf Sedimentation: SEPM Special Publication 34, p. 85-142.

Van Wagoner, J.C., H.W. Posamentier, R.M. Mitchum, P.R. Vail, J.F. Sarg, T.S. Loutit, J. Hardenbol, 1988, An overview of the fundamentals of sequence stratigraphy and key definitions, *in* W. Wilgus, B. Hastings, C.A. Ross, H. Posamentier, J.C. Van Wagoner, and C.G. St. C Kendall (eds.), Sea-Level Changes: An Integrated Approach, SEPM Special Publication no. 42, p. 357-370.

Van Wagoner, J. C., R.M. Mitchum, K. M., Campion, and V. D. Rahmanian, 1990, Siliciclastic Sequence Stratigraphy in Well Logs, Cores and Outcrops: Concepts for High-Resolution Correlation of Time and Facies: AAPG Methods in Exploration Series, No. 7, 52 pp.

Walker, R. G., and C.H. Eyles, 1991, Topography and significance of a basinwide sequence bounding erosion surface in the Cretaceous Cardium, Alberta, Canada: Journal Sedimentary Petrology, v. 61, p. 473-496.

Weimer, R.J., J.J. Emme, C.L. Farmer, L.O. Anna, T.L. Davis and R.L. Kidney, 1982, Tectonic influence on sedimentation, Early Cretaceous, east flank, Powder River basin, Wyoming and South Dakota: Colorado School of Mines Quarterly, v. 77, 61 p.

ORIGIN OF LATE TURONIAN AND CONIACIAN UNCONFORMITIES IN THE SAN JUAN BASIN

DAG NUMMEDAL
GREGORY W. RILEY

Department of Geology and Geophysics, Louisiana State University, Baton Rouge, Louisiana

INTRODUCTION

Objectives

The objective of this field research conference is to address basic issues in shallow marine sequence stratigraphy. One central issue is the origin of the many discontinuities (unconformities and diastems) we observe: Three possible mechanisms may produce such discontinuities: (1) eustatic sea level fall, (2) tectonic uplift or tilting of the basin, and (3) lateral shifts in depositional environments (the origin of most diastems). Late Turonian and Coniacian rocks of the western San Juan Basin provide an excellent section of rocks in which to address these questions, because we believe that the three major discontinuities present within this part of the stratigraphic column are examples of all these listed modes of origin.

The analysis of the origin of these unconformities is based on: (1) a synthesis of relevant biostratigraphic data (including much not previously published) in order to constrain their ages with the highest possible precision, (2) documentation of parasequence stacking patterns and depositional systems to relate unconformities to their correlative deposits, and (3) subsurface data that demonstrate local tectonic movement.

Sequence boundaries and "source diastems"

Erosional unconformities reflect episodes of shifting base level in sedimentary basins. Wherever erosional surfaces (unconformities or diastems) are formed, there must be a corresponding depositional system, for which the erosional surface represents the (removed) source (Frazier, 1974; Swift et al., in press). This basic principle works on any scale. The spatial separation of a depositional system and its "source diastem" may be large, as when deep-sea fans are sourced by upland erosion during episodes of major eustatic lowstands. The separation is less, however, in settings such as foreland basins where both thrust-belt, forebulge (and basement blocks ?) may source central basin fills (Jordan and Flemings, 1991).

Relative rates of eustatic and tectonic changes

Rates of base level change are very poorly constrained. There is no a priori reason to assume that eustatic changes are faster (or slower) than tectonic ones. Currently active tectonic regions commonly subside (or rise) at rates of several mm/year. Several recent studies on Mesozoic compression in the western U. S. have inferred horizontal rates in the range of 2 to 6 mm/yr (see summary in Elison, 1991) and probable foredeep subsidence rates of similar orders of magnitude. Geohistory analysis of Late Cretaceous strata in the San Juan Basin (Pang, in prep.) document subsidence rates on the order of 0.1 mm/yr.

These observations of the spatial and temporal scales of base level fluctuations imply that stratigraphic analysis of sedimentary basins must evaluate both rates of tectonic change and eustasy in the explanation for unconformities, and relate each depositional system, parasquence and sequence to its corresponding source discontinuity. Based on the rates presented, time-scales of thrusting, and flexural models it is clear that tectonic subsidence in the Cretaceous basin of the western interior cannot be treated as linear on the time-scale of many depositional sequences (1-3 My).

San Juan Basin structure

The driving mechanism for subsidence in the San Juan basin during the Late Cretaceous remains highly conjectural, but it probably includes a component of flexure in response to loading on the Sevier orogenic belt and younger, possibly "Laramide style" compression. Regional basement fault systems in the Colorado Plateau province are well known to have been reactivated several times (Baars, 1966). Evidence exists for both dip- and strike-slip movment on these northwest-trending faults in the San Juan Basin (Stevenson and Baars, 1977, 1986; Fig. 1). Huffman and Taylor (1991) recently documented significant movement on these faults during the Pennsylvanian to Permian (ancesteral Rocky Mountains orogeny), Jurassic to Early Cretaceous ("Nevadan") and Late Cretaceous to early Tertiary (Sevier and Laramide orogenies). They observe that few faults show actual offset above Permian strata but their movment affected sedimentation patterns in much younger rocks. Our interpretation of Late Turonian and Coniacian unconformites support their view.

Biostratigraphic overview

Recognition and interpretation of the unconformities require integration of biostratigraphic, lithostratigraphic and sedimentologic data. In some areas the unconformities are based on purely physical criteria either because their lacunas are below biostratigraphic resolution or the adjacent rocks are devoid of age-diagnostic fossils.

Ammonites, inoceramids and oysters provide the key data base for the dating of regional unconformities and their correlative conformities. Site-specific fossil data from three areas to be visited on the field trip are presented in Figure 2. Our fossil range chart (Fig. 3) is developed from data in Kauffman et al. (1991). Only those fossils used to constrain age ranges in this paper are included in these simplified charts. Fossil data for individual stratigraphic sections come in part from the published literature but most is based on recent collections by Cobban, Molenaar, Nummedal, Riley, Tillman and Valasek.

A composite chronostratigraphic chart for Late Turonian and Coniacian rocks in the San Juan Basin has been constructed from these range charts and all pertinent megafossil occurrences (Fig. 4). Unconformities are cross-hatched, and queried where biostratigraphic resolution is inadequate. As expected, poor chronostratigraphic control exists within the non-marine Gallup, Torrivio Sandstone and the Dilco Coal Member. All marine units, however, are now quite well constrained biostratigraphically.

UNCONFORMITY NO 1 (LATE TURONIAN)

Physical character

In the San Juan Basin this unconformity is best developed in depositionally updip regions, near Gallup, NM. The unconformity is entirely contained within the Gallup Sandstone and is expressed throughout the region as a sharp contact between truncated shoreface facies below and tidally-influenced estuarine (locally fluvial ?) facies above (see Stop 7-3). The abrupt upward shoaling inferred from this facies contrast imply base-level lowering. Immediately below the unconformity there is a highly uranium-enriched zone, suggestive of weathering. Also, a unique suite of burrows of the *Glossifungites* ichnofacies indicate a fairly long lacuna associated with this surface.

Traced downdip, unconformity 1 forms the base of several incised channels truncating the Gallup shoreface around Sanostee. McCubbin (1982) inferred that these channels fed a "distributary plain" farther to the northeast (Fig. 5). Fluvial deposits at the base of some of the incised Gallup channels at Sanostee (McCubbin, pers. comm., 1991) imply that the incision was caused by subaerial erosion during a relative sea level fall. A sharp base of the Gallup shoreface in the region around Sanostee also supports the inferred sea-level lowering.

Sequence stratigraphic implications

We cannot trace unconformity no. 1 farther seaward and infer, therefore, that the distal Gallup Sandstone shoreface (on the Four Corners' Platform) is the conformable lowstand deposit correlative with this updip bypass surface (Fig. 6), or "source unconformity", a concept discussed in the introduction . This inference is also driven by the biostratigraphic data. The marine Gallup Sandstone updip of Sanostee contains *Inoceramus perplexus, Prionocyclus novimexicanus* and older megafossils; implying that none of it is younger than late T10 (Fig. 4). In contrast, the Gallup Sandstone on the Four Corners Platform farther north contains an Early Coniacian megafossil assemblage. This includes *Bacculites yokoyamai* and *Lopha sannionis.* We infer, but cannot prove, that this distal, lowstand Gallup is time-equivalent to the estuarine fill above the updip sequence boundary.

The presence of an internal sequence boundary within the Gallup Sandstone has been suspected for some time (e. g. Nummedal, 1990) because of the stacking pattern of the Gallup (para?)sequences. As well illustrated in Molenaar (1983, his Fig. 15) his Gallup C-tongue separates an aggradational to progradational stacking pattern below the C from one which is more aggradational again above. The implied change in rate of generation of accomodation space is consistent with a relative sea level fall between D and C-tongue progradations. The subsequent lowstand rise in sea level caused aggradation of the most distal Gallup strandplains (tongues B? and A), backfilling of updip valleys with estuarine and lagoonal deposits (parts of Gallup C; Fig. 6), and the start of deposition of Fort Hays Limestone at Pueblo (Fig. 4).

Regional extent of Late Turonian sequence boundary

Biostratigraphic data in Fig. 4 imply that this unconformity is correlative with the one below the Fort Hays Limestone at Pueblo. That unconformity, the Carlile/Niobrara boundary, has been recognized across wide areas of the western interior (Fisher et al., 1985). Its wide extent and the absence of evidence for structural movement in the western San Juan Basin at this time suggest that this unconformity is of eustatic origin. This surface is a sequence boundary and we infer that it corresponds to the boundary between UZA-2 and UZA-3 "supersequences" in Haq et al. (1988). It formed in the latest part of the Late Turonian, probably in the *Scaphites whitfieldi* assemblage zone. Using the time scale in Kauffman et al. (1991) its age is about 89.2 Ma. Because of different absolute time scales this is 0.8 My younger than implied by Haq et al. (1988).

UNCONFORMITY NO 2 (EARLY CONIACIAN)

Physical character

We infer the existence of an unconformity at the base of the "main body" of the Torrivio Member (of the Gallup

Sandstone) based on its regional erosional base and gradual "convergence" with the underlying Gallup Sandstone. The Torrivio Member, however, presents several complex stratigraphic issues not yet fully resolved. The unit is named for a coarse fluvial sandstone at Torrivio Mesa, west of Gallup (Molenaar, 1983). Field observations by the authors and Molenaar suggest that the Torrivio, in part, represents fluvial feeder systems to the Gallup strandplains. One major Torrivio unit, which overlies the older feeder systems can be traced in several outcrop exposures northward from Gallup. At Nose Rock point (Stop 7-2) this "main Torrivio" lies about 80 ft above the top of the Gallup shoreface (D tongue), at Sanostee (Stop 6-1) it is typically 10 - 20 ft above the Gallup, in outcrops around Mitten Rock (near Stop 6-5) it is directly on top of, or slightly incised into, the upper shoreface and beach of the Gallup, and at the Roadside Dump (Stop 6-6) it is incised into the lower Gallup shoreface.

This documented convergence is consistent with a progressive increase in (broad regional?) erosion towards the northeast. In this interpretation the base of the Torrivio is regional unconformity no. 2. This surface was initially inferred to be a sequence boundary by Nummedal and Swift (1987). We now recognize that there also are additional sequence boundaries, and that the one at the base of the Torrivio is not the one of greatest regional extent.

The offshore unconformity

Seaward of the Gallup Sandstone there is a major unconformity juxtaposing the marine, coarse-grained Tocito Sandstone with the underlying Mancos Shale and, locally, the truncated top of the Juana Lopez Member (McCubbin, 1969; Molenaar, 1973). Several subsurface studies have traced this unconformity landward above shales equivalent to the distal (A and B tongues) Gallup SS (see Stops 6-6, 6-7, and 7-1). This erosion surface, therefore, is younger than the Late Turonian regional sequence boundary (unconformity 1). This erosional episode occurred in the Coniacian because the distal Gallup falls in assemblage zone CO 1 (Kauffman et al., 1991). The oldest fossils recovered from the Tocito Sandstone above this unconformity include *Cremnoceramus erectus, Cr. deformis* and *Forresteria* sp. of assemblage zone CO 2.

This Early Coniacian unconformity is (probably) traceable throughout the San Juan Basin and in many places it truncates the Late Turonian sequence boundary, such as on structural highs associated with the big lacunas at the Plunge Pool and The Mounds (Fig. 4). Pebbles at the base of the Cooper Arroyo SS at El Vado, and at a shale-on-shale contact in the southeastern San Juan Basin (Highway NM 44 near San Luis turnoff) indicate wide-spread high energy conditions across the basin seafloor at this time. There is, however, no documented evidence of subaerial exposure on this basal unconformity anywhere seaward of the termination of the Torrivio, nor are there any fluvial deposits of this age in the San Juan Basin. We argue, therefore, that the seaward parts of unconformity 2 were cut by marine erosional processes.

Both the biostratigraphic data and the lithostratigraphic correlations are consistent with the inference that this offshore unconformity correlates landward with the basal Torrivio erosion surface.

Timing of structural movement

Formation of the Early Coniacian erosion surface coincides with a period of significant structural movement along the Four Corners lineament (terminology from Stevenson and Baars, 1986). The existence of a regional high just seaward of the Gallup SS pinch-out was well documented by McCubbin (1969) and has been reaffirmed by all subsequent workers. Correlations presented at Stops 6-6 and 7-1 document that this structure ("Waterflow anticline") formed after deposition of the distal Gallup (CO1). The uplift ceased and was locally reversed before deposition of the youngest Tocito in assemblage zone CO5.

The height of this anticlinal feature and the timing constraints imposed above permit calculation of its rate of uplift. Using the time scale of Kauffman et al. (1991) we find that this structure rose at an average rate of 0.10 mm/yr for a period of about 0.5 My. In order to get back to the issue we raised in the introduction about rates of eustatic change compared to tectonic ones it is interesting to compare this to typical Cretaceous "third-order" eustatic cycles. In order for a eustatic million-year (sinusoidal) sea level cycle to produce a maximum rate of rise equal to the calculated rate of structural uplift it would need a height of more than 50 m. This is higher than any currently believed amplitudes for Cretaceous sea level changes (Kauffman, pers. comm., 1991). This leads to the inevitable conclusion that it would be almost impossible **not** to create shoaling, and hence erosion of an unconformity at the Waterflow anticline during its uplift.

Source and transport of Tocito sands

The physical model developed for this unconformity is consistent with the idea that coarse sediments bypassing the coastal plain at the time of formation of unconformity 2 were the source of the Tocito Sandstone. Biostratigraphic data (Fig. 4) support the same inference. Sea level fall, by itself, would not suddenly introduce new coarse sediments into the basin. For example, the inferred sea level fall in the latest Turonian did not produce a significantly coarser Gallup strandplain. We argue that the arrival of coarse sand and gravel along the shoreline in Early Coniacian time is a consequence of tectonic uplift in the source area, essentially contemporaneous with the movement on the Waterflow anticline and other basinal blocks. Because the tectonics of the source area (Mogollon highlands, AZ?) is very poorly known we cannot evaluate the nature of the uplift any further.

One additional argument against a eustatic origin of unconformity no. 2 is the absence of correlatable unconformities farther east. For example, the Early Coniacian at Pueblo was characterized by deposition of relatively deep-water chalks of the Fort Hays Member (of the Niobrara Fm.).

Depositional environment

The coarse Torrivio sand was delivered to a shoreline affected by strong tidal currents, possibly locally enhanced in emabyments, straits (?) or "estuaries" produced by structural movements. Dispersal of "lowstand" coarse sediments away from these river mouths into a tide-dominated delta and shelf sand ridge depositional system was geologically instantaneous. A persistent, dominant current regime from the northwest to the southeast is reflected in crossbedded facies of nearly all outcropping Tocito sandbodies. Since all these outcrops are landward of the Waterflow anticline this strongly unidirectional paleocurrent regime may indicate segregated tidal currents in a strait landward of a sea floor high or island (?) located at the crest of the structure (the mythical highland of "Molenaaria").

UNCONFORMITY NO. 3

Physical character

A sharp erosional surface, commonly associated with lags of coarse sand and gravel, separate the base of the Tocito Sandstone in the Beautiful Mountain area (Stop 6-3) from the underlying Torrivio and Dilco Coal Members. The overlying marine Tocito Sandstone has no megafossils other than *Pseudoperna congesta* which ranges throughout Coniacian and younger rocks (Fig. 3). In some areas this surface contains large wave ripples with pebble lags (Bergsohn, 1988). Based on its physical appearance this surface is interpreted as a ravinement and the associated lacuna is probably small.

Age and duration

Tracing this surface seaward has proven to be a difficult task due to discontinuous Tocito outcrops. Regional outcrop work by Riley (in prep.), however, strongly suggests that this is the same surface which truncates the phosphatic pebbly mudstone at the Hogback oil field (Stop 7-1) and also separates the Tocito at The Mounds (Colorado) into upper and lower members. If this correlation is correct then the age of the surface is constrained, because the upper Tocito at the Hogback oil field contains *Peroniceras westphalicum* which ranges from CO4 (Middle Coniacian; Fig. 3) into the Late Coniacian. The upper Tocito at The Mounds is probably also of late Middle Coniacian age. The correlation of these seaward outcrops provide age constraints which suggest that unconformity no. 3 does, indeed, represent a very short lacuna.

The regional transgression responsible for this ravinement surface reflects local, relative sea level rise. Whether this rise was caused by renewed subsidence across this belt of active tectonism or a eustatic rise is presently unresolved.

Environments of deposition of the "upper" Tocito

The "upper" Tocito Sandstone is everywhere marine. It contains abundant inoceramids, common ammonites (mostly *Placenticeras* sp.), marine authigenic minerals such as glauconite and phosphorite (calciumfluorapatite), and an abundant and diverse assemblage of large trace fossils (*Ophiomorpha; Thalassinoides, Teichichnus, Trichichnus, Asterosoma, Terebulina*). Although all these genera can occur in restricted (brackish) water, they would be expected to be smaller in such settings than those actually observed in the Tocito. We believe that most of the "upper" Tocito Sandstone represents shallow marine ("shelf") sandbodies formed by the transgressive reworking of the Torrivio Sandstone and the "lower" Tocito tide-dominated delta facies during formation of unconformity no. 3.

CONCLUSIONS

There are three distinct breaks in the Late Turonian to Coniacian stratigraphic succession along the western margin of the San Juan Basin.

Unconformity no. 1 is of latest Turonian age and separates the main body of the Gallup Sandstone (updip parts of tongue C; above) from the underlying higstand shoreface deposits (tongue D and downdip parts of tongue C). This surface can be traced seaward to about Sanostee, at which point it changes into a conformable (?) sequence boundary at the base (?) of the lowstand Gallup shoreface (A tongue). We infer that this sequence boundary correlates with the "global" Upper Zuni A2/A3 supersequence boundary (Haq et al., 1988). It was probably caused by a eustatic sea level fall.

Unconformity no. 2 is a regional erosion surface traceable at the base of the fluvial Torrivio Member in updip regions and correlating with the basal Tocito unconformity farther seaward. We infer that this unconformity is due to an episode of tectonic uplift which changed fluvial gradients and loads, thus introducing a new, coarse sand suite into the marine environment. The same tectonic episode was associated with uplift of some reactivated basement blocks trending NW-SE across the Four Corners platform. This tectonic episode occurred during the latter part of Early Coniacian time.

Unconformity no. 3 is a transgressive surface associated with shoreline retreat during a mid-Coniacian relative sea level rise. This transgression further remobilized some of the shallow marine, tide-dominated sandbodies which had formed during the Early Coniacian lowstand.

There is no evidence of subaerial exposure on the basal Tocito unconformity seaward of the pinch-out of the Torrivio Member (about the Gallup SS pinch-out). Erosional processes operating across the San Juan Basin in the Coniacian, therefore, are inferred to have been submarine. Enhanced tidal currents and storm (?) scour during lowstand, combined with local amplification of tides in straits and estuaries, probably account for the extensive, structure-parallel erosional scours. The fact that most of the erosion is located at a broad structural high (Jennette et al., this volume) further suggests erosion by submarine rather than fluvial currents.

REFERENCES

Baars, D. L., 1966, Pre-Pennsylvanian paleotectonics - key to basin evolution and petroleum occurrences in the Paradox Basin, Utah and Colorado: AAPG Bull., 50, p. 2080-2111.

Bergsohn, I., 1988, Lithofacies architecture of the Tocito Sandstone, northwest New Mexico; Thesis, LSU 170 p.

Elison, M. W., 1991, Intracontinental contraction in western North America: continuity and episodicity: GSA Bull., v. 103, p. 1226-1238.

Fisher, C. G., E. G. Kauffman, and L. Van Holdt Wilhelm, 1985, The Niobrara transgressive hemicyclothem in central and eastern Colorado: the anatomy of a multiple disconformity; *in* L. M. Pratt, E. G. Kauffman and F. B. Zelt (eds.); Fine-grained Deposits and Biofacies of the Cretaceous Western Interior Seaway: Evidence of Cyclic Sedimentary Processes; SEPM Midyear Mtg. Field Trip Guidebook No. 9, p. 184-198.

Frazier, D. E., 1974, Depositional episodes: their relationship to the Quaternary stratigraphic framework in the northwestern part of the Gulf Basin: Geol. Circ., 74-1, Bureau of Economic Geology, Univ. Texas, Austin, 28 p.

Haq, B.U., J. Hardenbol and P.R. Vail, 1988, Mesozoic and Cenozoic chronostratigraphy and cycles in sea level change; *in* C.K. Wilgus, B.S. Hastings, C.G.St. C. Kendall, H.W. Posamentier, C.A. Ross, and J.C. Van Wagoner (eds.), Sea-Level Changes: An Integrated Approach, Spec. Publ., Soc. Econ. Paleon. Mineral.,42, 71-108, 1988.

Huffman, A. C. Jr. and D. J. Taylor, 1991, Basement fault control on the occurrence and development of San Juan Basin energy resources: Abstract with Programs, GSA Rocky Mountain section mtg., v. 23, p. 34.

Jordan, T. E., and Flemings, P. B., 1991, Large-scale stratigraphic architecture, eustatic variation, and unsteady tectonism: a theoretical evaluation; J. Geophys. Res., 96, 6681-6699.

Kauffman, E.G., Cobban, W.A., and Obradovich, J.D., 1991, in press, Cretaceous Western Interior time scale, *in* W.G.E., Caldwell, and E.G., Kauffman, (eds.), The evolution of the Western Interior Basin, Geological Association of Canada, Special Paper 39.

McCubbin, D. G., 1969, Cretaceous strike-valley sandstone reservoirs, northwestern New Mexico: AAPG Bull., 53, p. 2114-2140.

McCubbin, D. G., 1982, Barrier island and strandplain facies; *in* Scholle, P. A. and Spearing, D., (eds.), Sandstone Depositional Environments, AAPG Mem. 1982, p. 247-280.

Molenaar, C. M., 1973, Sedimentary facies and correlation of the Gallup Sandstone and associated formations, northwestern New Mexico; *in* J. E. Fassett (ed.); Cretaceous and Tertiary Rocks of the Southern Colorado Plateau, Four Corners Geol. Soc. Mem., p. 85-110.

Molenaar, C. M., 1983, Principal reference section and correlation of Gallup Sandstone, northwestern New Mexico, *in* S. C. Hook (compiler), Contributions to Mid-Cretaceous Paleontology and Stratigraphy of New Mexico; New Mexico Bureau of Mines and Mineral Resources, Circ. 185, 29-40.

Nummedal, D., 1990, Sequence stratigraphic analysis of upper Turonian and Coniacian strata in the San Juan Basin, New Mexico, U.S.A; *in* R. N. Ginsburg and B. Beaudoin (eds.), Cretaceous Resources, Events and Rhythms, 33-46.

Nummedal, D., and D. J. P. Swift, 1987, Transgressive stratigraphy at sequence-bounding unconformities: some principles derived from Holocene and Cretaceous examples; *in* D. Nummedal, O. H. Pilkey and J. D. Howard (eds.), Sea-Level Fluctuation and Coastal Evolution; SEPM Spec. Pub. 41, 241-260.

Stevenson, G. M., and D. L. Baars, 1977, Pre-Carboniferous paleotectonics of the San Juan Basin; *in* J. E. Fassett (ed.), San Juan Basin III, New Mexico Geol. Soc. Guidebook, 28th Field Conference, p. 99-110.

Stevenson, G. M., and D. L. Baars, 1986, The Paradox: a pull-apart basin of Pennsylvanian age; *in* J. A. Peterson (ed.), Paleotectonics and Sedimentation in the Rocky Mountain Region, U. S.; ; AAPG Mem. 41, p. 513-539.

Swift, D. J. P., S. Philips and J. A. Thorne, in press, Sedimentation on continental margins; Part IV: lithofacies and depositional systems; *in* Swift, D. J. P., R. W. Tillman, G. F. Oertel, (eds.), Shelf sands and sandstone bodies: geometry, facies and sequence stratigraphy; Internat. Assoc. Sedimentol., Spec. Pub. 12.

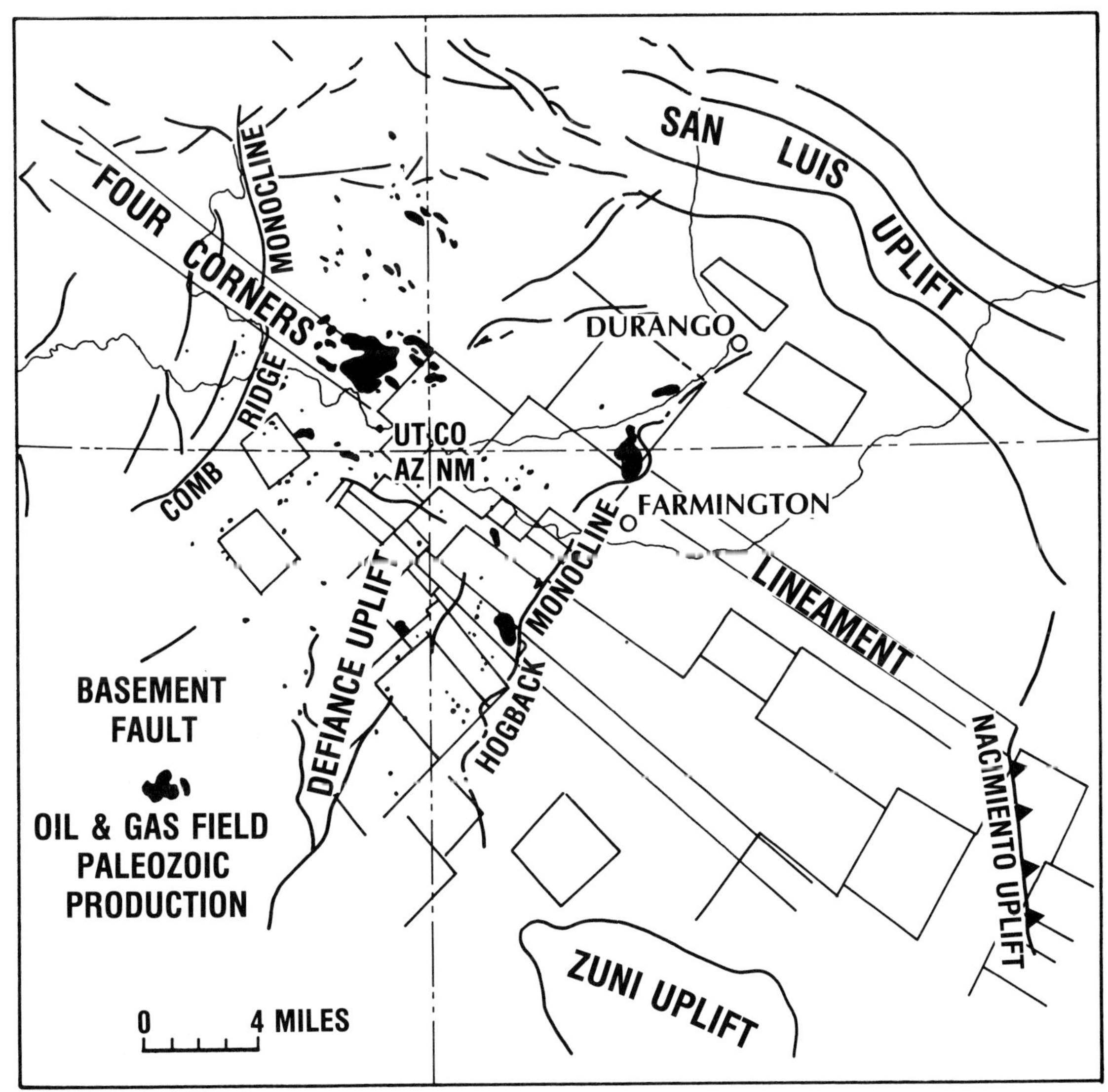

Figure 1. Map showing location of basement lineaments near the Four Corners Platform. From Stevenson and Baars (1986).

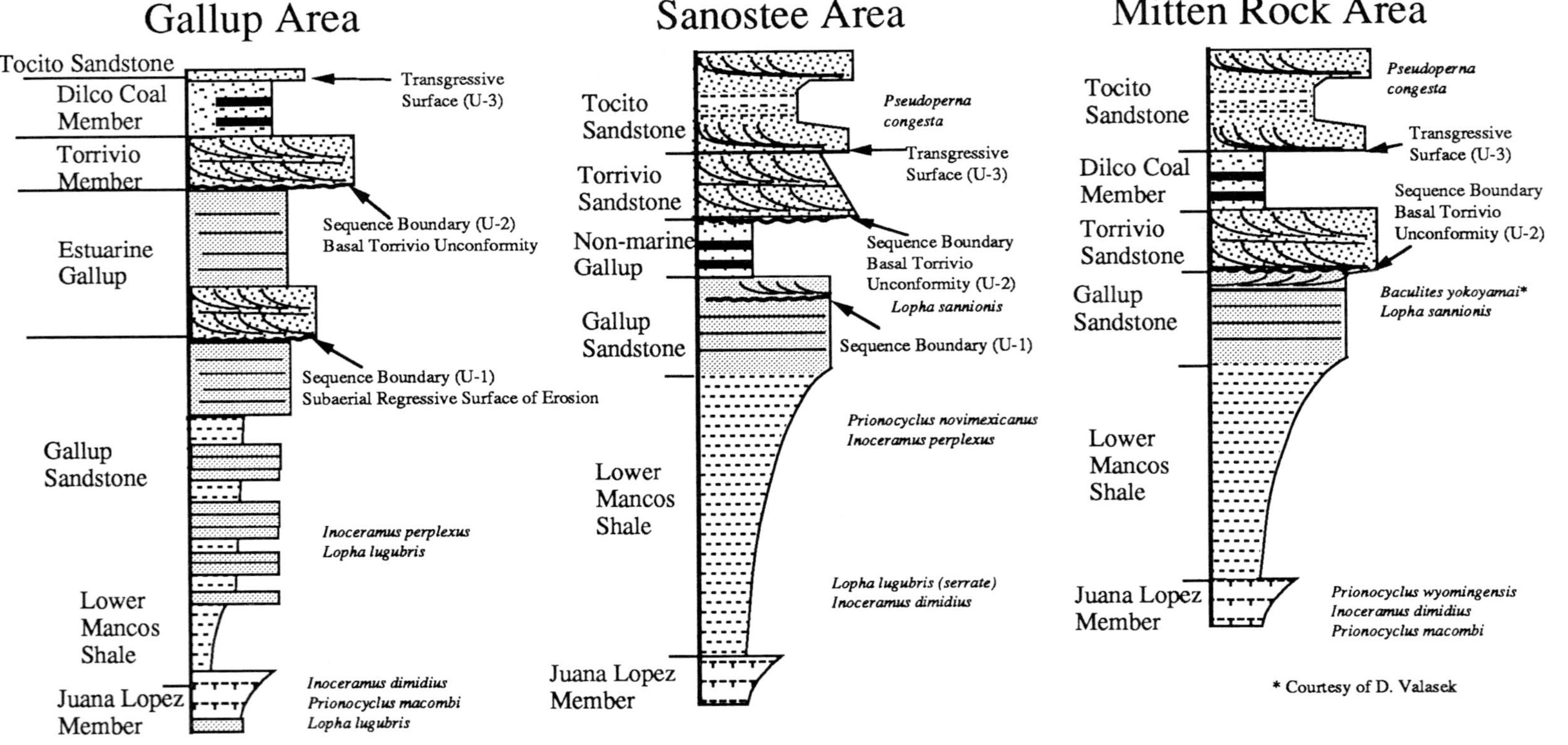

Figure 2. Megafossil control for the Gallup Sandstone and associated strata at Nose Rock Point, Sanostee, and Mitten Rock.

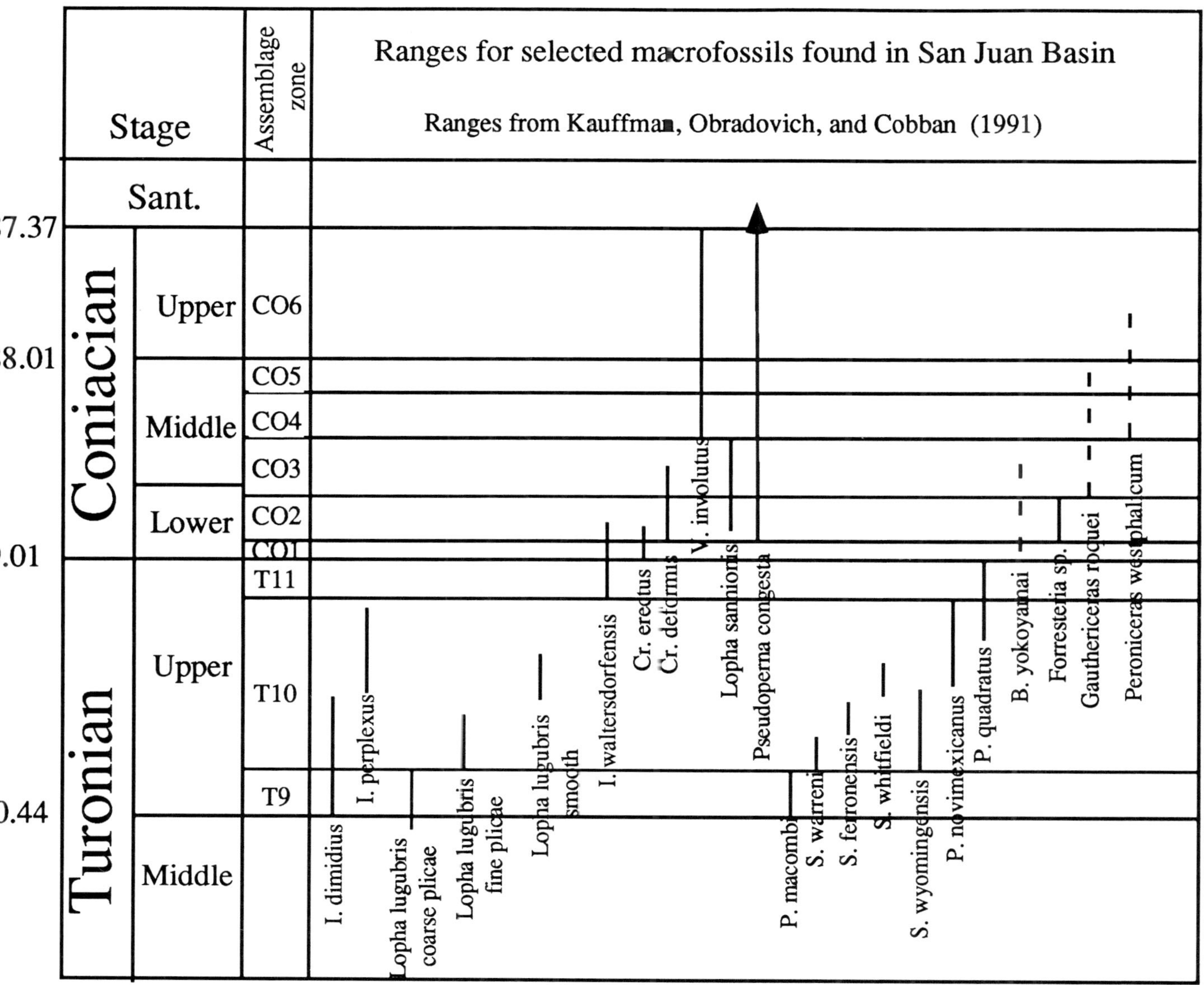

Figure 3. Age ranges for common megafossils within Late Turonian and Coniacian marine deposits of the San Juan Basin. Ranges from Kauffman, Obradovich, and Cobban (1991).

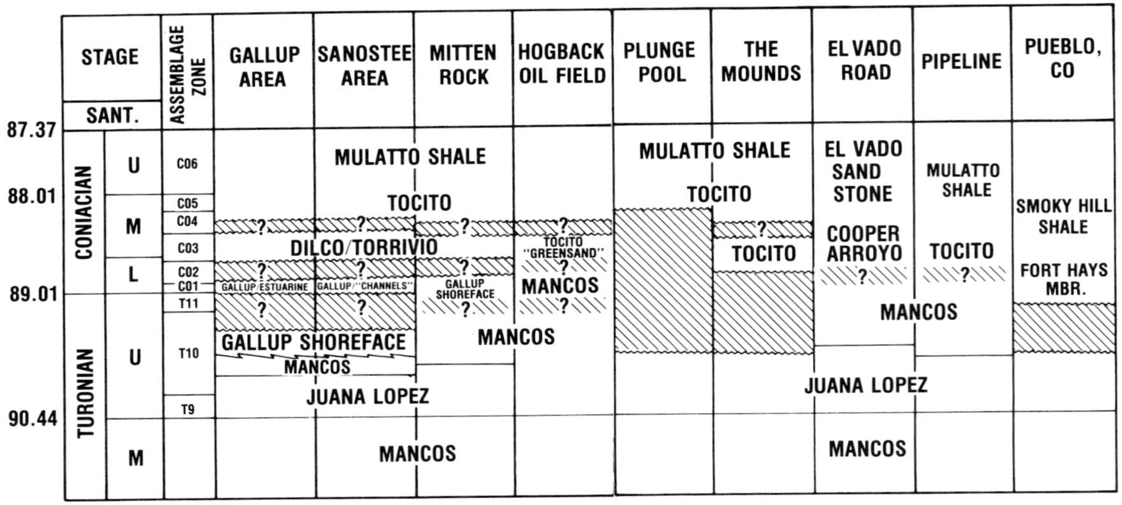

STAGE
SANT.
ASSEMBLAGE ZONE
GALLUP AREA
SANOSTEE AREA
MITTEN ROCK
HOGBACK OIL FIELD
PLUNGE POOL
THE MOUNDS
EL VADO ROAD
PIPELINE
PUEBLO, CO
87.37
88.01
89.01
90.44
CONIACIAN
TURONIAN
U
M
L
U
M
C06
C05
C04
C03
C02
C01
T11
T10
T9
MULATTO SHALE
TOCITO
DILCO/TORRIVIO
GALLUP ESTUARINE
GALLUP "CHANNELS"
GALLUP SHOREFACE
MANCOS
JUANA LOPEZ
MANCOS
GALLUP SHOREFACE
MANCOS
TOCITO "GREENSAND"
MANCOS
MULATTO SHALE
TOCITO
TOCITO
EL VADO SAND STONE
COOPER ARROYO
MULATTO SHALE
TOCITO
MANCOS
JUANA LOPEZ
MANCOS
SMOKY HILL SHALE
FORT HAYS MBR.
?

Figure 4. Chronostratigraphic representation of Turonian-Coniacian strata at selected locations in New Mexico and Colorado. Three seperate unconformities are recorded. See text for details.

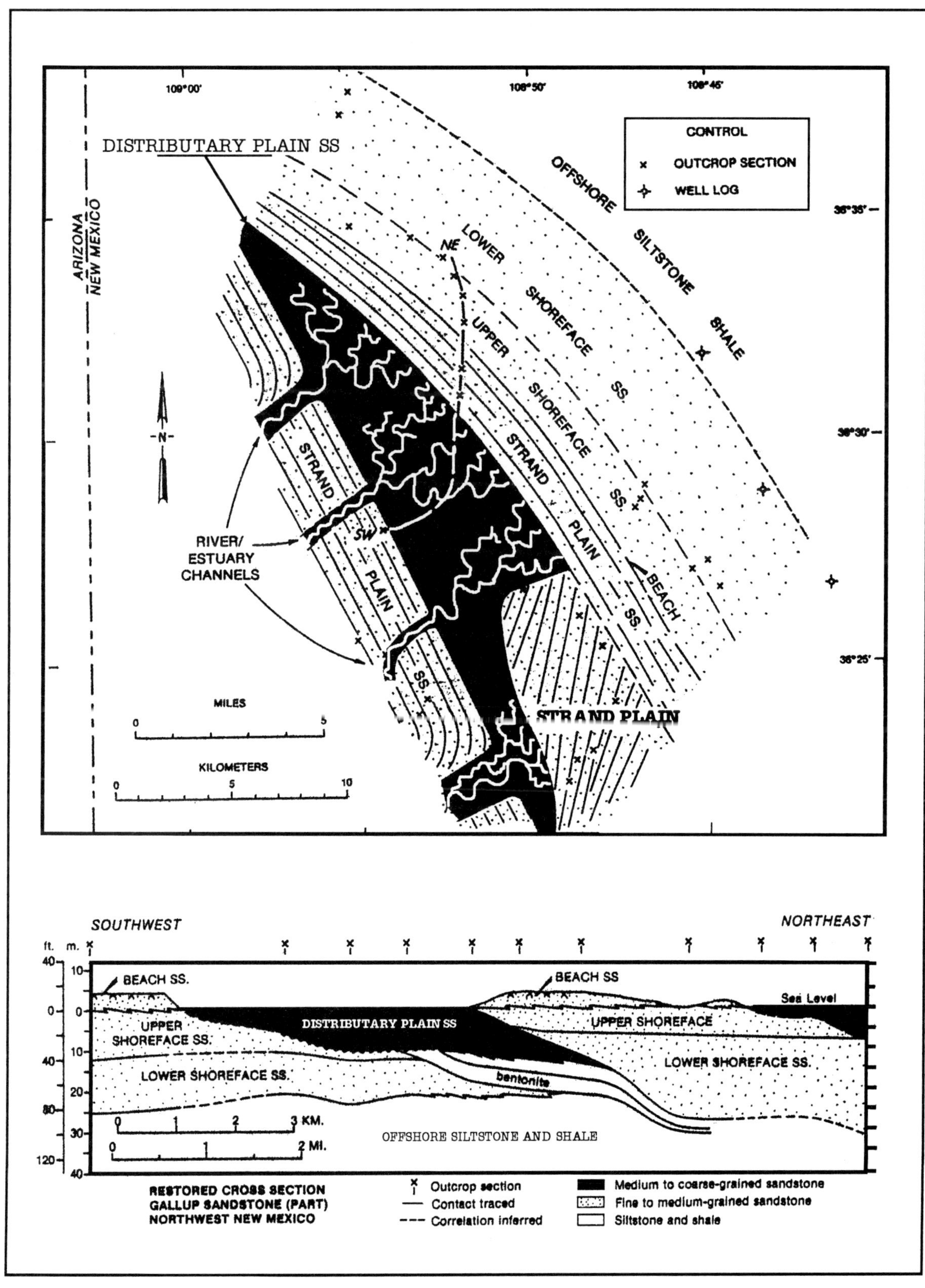

Figure 5. Paleogeographic map (top) and cross section (bottom) showing incision of the previous Gallup strandplain by fluvial channels. We infer that incision of the Gallup strandplain records a Late Turonian eustatic sea level fall. From McCubbin (1982).

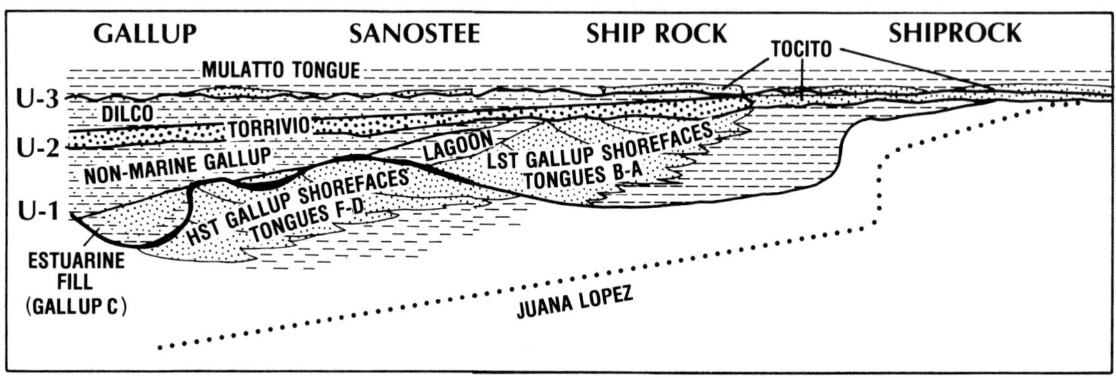

Figure 6. Schematic diagram showing unconformities discussed in text. Unconformity 1 is contained within the Gallup Sandstone and is interpreted as a sequence boundary. Unconformity 2 is at the base of the "main" Torrivio sandbody and is correlated to the base of the Tocito Sandstone in the San Juan Basin. This unconformity is felt to reflect tectonic processes. Unconformity 3 is a transgressive surface, and in updip areas separates the marine Tocito Sandstone from underlying continental facies.